Proceedings of
XXVI Zakopane School on Physics

Zakopane, Poland 13 – 21 April 1991

Proceedings of
XXVI Zakopane School on Physics

Zakopane, Poland 13 – 21 April 1991

condensed matter studies by nuclear methods

JAGIELLONIAN UNIVERSITY AND INSTITUTE OF NUCLEAR PHYSICS KRAKÓW

Edited by:
Jan STANEK
Antoni T. PĘDZIWIATR
Institute of Physics
Jagiellonian University

World Scientific
Singapore • New Jersey • London • Hong Kong

Published by

World Scientific Publishing Co Pte Ltd
5 Toh Tuck Link, Singapore 596224
USA office: 27 Warren Street, Suite 401-402, Hackensack, NJ 07601
UK office: 57 Shelton Street, Covent Garden, London WC2H 9HE

British Library Cataloguing-in-Publication Data
A catalogue record for this book is available from the British Library.

PROCEEDINGS OF XXVI ZAKOPANE SCHOOL ON PHYSICS

ISBN-13 978-981-02-0681-9
ISBN-10 981-02-0681-X

SCHOOL SPONSORS

Institute of Physics, Jagiellonian University, Cracow
Institute of Nuclear Physics, Cracow

SCHOOL ADDRESS:

DW "ZREMB", Zakopane, ul. Piłsudskiego 10A

ORGANIZING COMMITTEE

Chairman: A. Z. Hrynkiewicz

Members: J. Stanek (Scientific Secretary)
H. Bińczycka
J. Grębosz
W. Kwiatek
K. Łątka
M. Marszałek
A. Pędziwiatr

EDITORIAL

This volume contains lectures and seminars given at the XXVI Zakopane School on Physics. The volume includes all manuscripts which were received by organisers before May 20, 1991.

CONTENTS

LECTURES

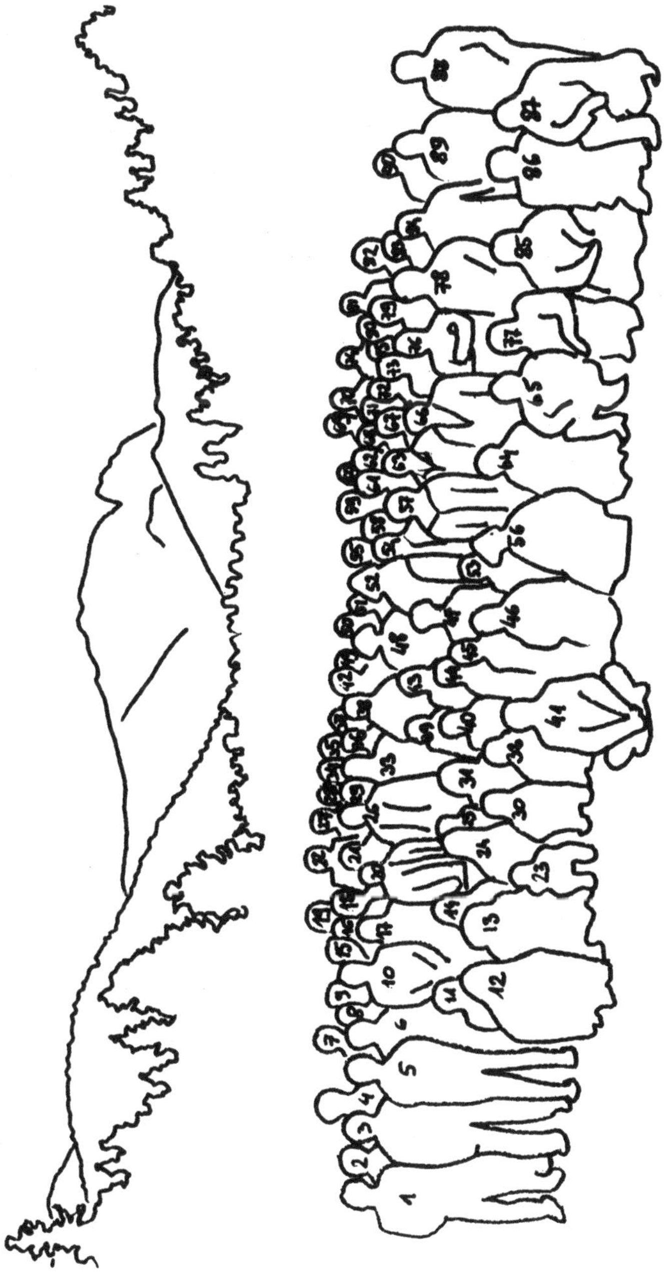

PERSONS IN THE PHOTOGRAPH

1. Zhang Li
2. Tadeusz Panek
3. Józef Bara
4. Jo De Wachter
5. Guido Langouch
6. Marta Marszałek
7. Marcel Greuter
8. Hans Lütgemeier
9. Harald Glückler
10. Günter Schatz
11. Karen Milants
12. Izabela Sosnowska
13. Anna Kwiatek
14. Nathalie Millard
15. Joao G. M. Correia
16. Wojciech Kwiatek
17. Stanisław Szymczyk
18. Marek Sudnik
19. Frits Pleiter
20. Andrzej Hrynkiewicz
21. Antoni Pędziwiatr
22. Friedrich E. Wagner
23. Joasia Kwiatek
24. Barbara Miczko
25. Tomasz Goworek
26. Ulrich Zimmermann
27. Edward Görlich
28. Kazimierz Łątka
29. Józef kajfosz
30. Jan Marszałek
31. Elżbieta Jartych
32. Rafał Marszałek
33. Janusz Oleniacz
34. Bogdan Bogacz
35. Karol Krop
36. Hartmut Spiering
37. Wigcher Spanninga
38. Dariusz Satuła
39. Jan Kaczanowski
40. Małgorzata Sternik
41. Jurek Grębosz
42. Hans J. Lauter
43. Albert Metz
44. Barbara Kopcewicz
45. Ewa Jackiewicz
46. Wiesława Zarek
47. Fernande Grandjean
48. Valeria Pasyuk
49. Paweł Wodniecki
50. Lubos Lestak
51. Denes L. Nagy
52. Renata Wasiewicz
53. Halina Bińczycka
54. Ewa Japa
55. Krzysztof Królas
56. Carolyn Gardner
57. Jagoda Urban
58. Małgorzata Niewiara
59. John Gardner
60. Jacek Marzec
61. Klaus-Peter Lieb
62. Jan Stanek
63. Waldemar Górniak
64. Ewa Jędryka
65. Nika Spiridis
66. Elżbieta Marczewska
67. José C. Soares
68. Michael Frank
69. Janusz Przewoźnik
70. Wacek Musiał
71. Janusz Frąckowiak
72. Mieczysław Subotowicz
73. Johan De Bakker
74. Jan Sarzyński
75. Mirosław Łukasiak
76. Jean Marc Greneche
77. Kwietosława Burda
78. Michał Kopcewicz
79. Dirk Peter van der Werf
80. Eustachy Popiel
81. Wojciech Osoba
82. Wojciech Kwaśny
83. Marek Tuszyński
84. Józef Komraus
85. Lech Nowicki
86. Jakub Cieślak
87. Bogusław Rajchel
88. Jan Żukrowski
89. Roman Hajduk
90. Janusz Kraczka

lectures

POSITRONS IN SOLIDS

U. Zimmermann
Paul Scherrer Institute CH-5232 Villigen PSI

7 April 1991

To Fritz Heinrich

Abstract

The basic elements of positron annihilation in solids are described. The theoretical background is given as well as the usual techniques and some examples of applications in different fields of solid state physics with positrons.

1 INTRODUCTION

The positron is the *antiparticle* of the electron. Both particles have the same rest mass and spin ($\frac{1}{2}$) and the same magnitude of charge and magnetic moment. As isolated particles they are *stable*. If the two particles are in contact, they annihilate into γ-rays, predominantly into two or three, which carry away the rest energy of the two particles. In condensed matter, very often only 2γ-annihilation is important and these photons are emitted in opposite directions. Small deviations in this directions and in other characteristics of the γ-rays are due to the electron motion in the energy bands of the solid. In this way, the annihilation photons can give information about the *electronic structure* of condensed matter.

The motion of the positron is less important and this is due to the fact that positrons in solids are in the lowest energy band in the solid which is close to thermal energies, since in most experiments only one positron at a time is present in the solid and hence the positrons are not restricted by the exclusion principle. This is in contrast to electrons, which occupy higher energy levels according to Fermi-Dirac statistics. Neglection of the positron motion remains a valid approximation even if they are implanted from external β^+-sources with maximum energies of $\sim$1MeV.

While positrons do not interact with the electrons by *exchange* interaction, there is strong *correlation* with the electrons and nuclei in the solid due to the positive charge of the positron. This can affect the experimental observations significantly and considerable attention to positron states in the many-electron system of a solid is often required. Although this may be regarded as a disadvantage of the positron as a probe in condensed matter, the positive charge is the origin of the sensitivity of the positron to different types of *lattice imperfections* such as vacancies, dislocations and surfaces. Indeed, defect physics is one of the most successful fields of application for positron annihilation techniques.

The above mentioned phenomena are outlined in more detail in section 2 below, in order to enable the reader to follow the discussions on the usual experimental techniques (sect. 3) and the interpretations of a limited choice of experimental results given in section 4. Although, this choice is arbitrary to a great extend, it may be convenient to give an idea of the kind of results that can be expected from positron annihilation experiments.

Any representation of positron annihilation techniques in solid state physics is incomplete if the rapid development of *slow-positron beam techniques* is omited. Unfortunately, I feel that it would be too ambitious here to cover this ever growing field with sufficient completeness. This is due to the great variety of different experimental techniques that were developed during the last decade. Therefore, I only briefly touch these applications in the final section and refer the reader to recent reviews and the references stated therein.

2 BASIC THEORY OF POSITRON ANNIHILATION

The fundamental treatment of the annihilation process of e^+/e^- pairs at low relative velocities $v \ll c$ is described in detail in textbooks on Quantum Electrodynamics, e.g. [1,2]. In this chapter I give only a brief summary of the basic principles and results such as the various selection rules and modes of annihilation as well as the different states in which positrons in solids can exist. For further details the reader is referred to previous reviews [3,4,5].

2.1 The Free Electron-Positron Pair

In the annihilation process the total rest mass of $2m_0c^2 = 2\times 511$ keV of the annihilating pair is converted into electromagnetic radiation. The γ-rays carry away this energy and also the kinetic energy of the two particles as well as the potential energy of the Coulomb attraction between the particles and between the particles and an external field, if there is any. In solids the rest mass is the dominating part in the *total γ-ray energy* and in practice the other contributions are negligible, since they are of the order of ~1eV to ~10keV only. However, *kinetic energies* of the order of 1eV can lead to measurable deviations of the γ-ray characteristics in the *laboratory frame* from that in the *center-of-mass frame*. If we consider, e.g., 2γ-annihilation of a positron at rest in the laboratory frame and an electron with a momentum p_- assuming $p_+ \ll p_- \ll m_0c$, which is often the case in solids, then conservation of momentum in that frame requires

$$\mathbf{p}_- = \hbar\mathbf{k}_1 + \hbar\mathbf{k}_2 \qquad (1)$$

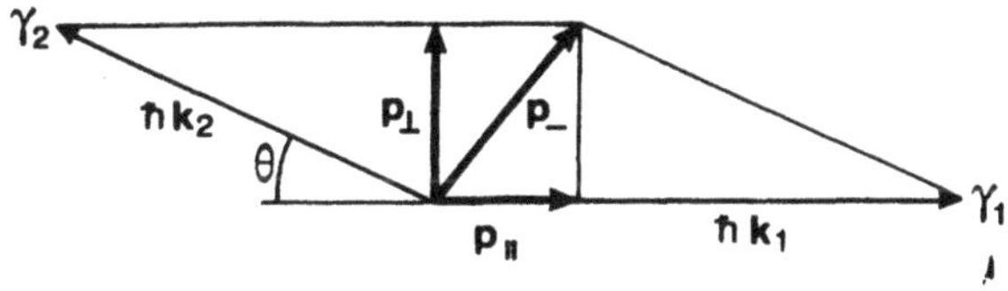

Figure 1: Momentum diagramme.

where $\hbar\mathbf{k}_i$ are the momenta of the two γ-rays and conservation of energy writes as

$$\sqrt{p_-^2c^2 + m_-^2c^4} + m_+c^2 = \hbar c(k_1 + k_2) \tag{2}$$

with $m_- = m_+ = m_0$ the electron and positron rest mass. Neglecting higher order terms in the transversal $\mathbf{p}_\perp$ and parallel $\mathbf{p}_{||}$ component of the electron momentum $\mathbf{p}_-$ (Figure 1), equations 1 and 2 can be solved to give

$$\theta \simeq p_\perp/m_0c \tag{3}$$

$$E_1 \simeq m_0c^2 + \frac{1}{2}\,cp_{||} \tag{4}$$

$$E_2 \simeq m_0c^2 - \frac{1}{2}\,cp_{||}. \tag{5}$$

Equation 3 shows that, for small velocities of the annihilating particles, the momentum component $p_\perp$ of the pair can be measured by means of the angle between the two quanta emitted in 2γ-annihilation. This is the basis for the technique of *angular correlation of annihilation radiation* (ACAR). For a kinetic energy of 10eV, which is a typical value for an electron at the Fermi surface in a metal, the angle θ is about 6mrad (~0.4°).

Equations 4 and 5 show that there is a Doppler shift in the energy E_i of the two emitted quanta which are opposite in sign and proportional to the parallel component $p_{||}$. This is the basis for the technique of *Doppler broadening* (DB) which is also a commonly used method for applications in solids. For 10eV-electrons the maximum Doppler shift is $\Delta E = cp_-/2 \sim 1.6$ keV, hence very large compared to the electron energy which causes the shift but small compared to m_0c^2.

Conservation of energy and momentum can not be fulfilled simultaneously in an annihilation process of a pair that results in emission of a single photon ($k_2 = 0$ in eqs. 1 and 2). Therefore 1γ-annihilation is forbidden for an isolated e^+/e^- pair. In a solid a third particle, electron or atom, could in principle participate in the process and carry away the recoil momentum, but for particle densities in real solids this is very unlikely compared to 2γ-annihilation ($\leq 10^{-8}$ [3]).

The spin averaged cross section for 2γ-annihilation according to Dirac [2] for velocities $v \ll c$ is

$$\sigma_2 = \pi r_0^2 c/v \tag{6}$$

where $r_0 = e^2/m_0c^2 = 2.8\times10^{-15}$ m is the classical electron radius. For 3γ-annihilation the spin averaged cross section for $v \ll c$ is

$$\sigma_3 = \frac{4(\pi^2-9)}{3} \alpha r_0^2 c/v \tag{7}$$

where α is the fine structure constant (1/137). Hence $\sigma_2/\sigma_3 \sim 370$ and one would conclude that 2γ-annihilation is the favoured process in solids. However, there are conditions under which 2γ-annihihilation is forbidden due to violation of charge parity conservation. This quantity is given by $C = P_i P_l P_s$, where P_i is the intrinsic parity which is -1 for a particle-antiparticle pair, $P_l = (-1)^l$ is the spatial parity and l the angular momentum quantum number, $P_s = (-1)^{s+1}$ is the spin parity. The charge parity for the particle pair is therefore $C = (-1)^{l+s}$. For n photons $C = (-1)^n$ and hence for even (odd) $l+s$ we have annihilation under emission of an even (odd) number of photons.

These selection rules are already included in the spin averaged cross sections in equations 6 and 7 for unbound pairs but there are possible states of the e^+/e^- pair for which averaging over spins is not correct. An example is Positronium (Ps), the analogue to the hydrogen atom with the proton replaced by a positron. For ortho-Ps (triplet o-Ps) we have $l = 0$ and $s = 1$ in the ground state, hence $C = -1$ and accordingly, o-Ps can decay under emission of an odd number of photons only. Since each additional photon in the decay scheme adds a factor of the order of the fine structure constant α, o-Ps only very rarely decays under emission of more than three photons ($\sigma_5/\sigma_3 \simeq \alpha^2$).

Since we are interested in the behaviour of the particles on an atomic distance scale in a solid, we need the description of the electrons and the positron in the *wave mechanical* picture for a positron/many-electron system. The *Hamiltonian operator* for 2γ-annihilation and the annihilation rate was first derived by Ferrell [6] in the *second-quantization* formulation.
The interaction of the e^+/e^- pair with the radiation field is a second-order process, i.e. an intermediate state occurs which consists of the two particles and a photon (Figure 2). In this state one of the two particles carries away the recoil momentum m_0c of the photon before annihilation of the two particles takes places and the second photon is created.

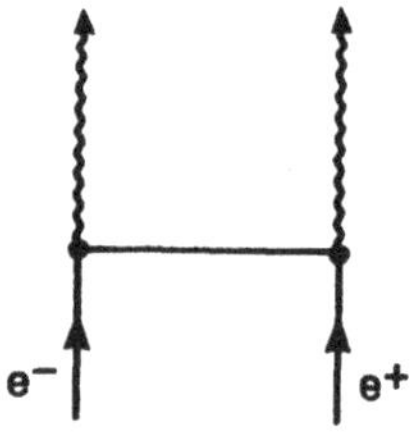

Figure 2: Feynman graph for 2γ–annihilation.

Since in both steps of the process the change in momentum for the respective particles is large compared to their initial momentum, the second-order Hamiltonian which creates the final two-photon state is in a good approximation

$$\sum_{\mathbf{p}_1}\sum_{\mathbf{p}_2} a_{\mathbf{p}_2} b_{\mathbf{p}_2} \delta_{\mathbf{p}_1+\mathbf{p}_2,\mathbf{p}} \tag{8}$$

where a and b are the annihilation operators for electrons and positrons, respectively, and $\mathbf{p}_1$ and $\mathbf{p}_2$ their momenta. The Kronecker δ-function expresses conservation of total momentum. The corresponding point annihilation operators $\hat{\Psi}$ are given by

$$a_{\mathbf{p}_1} = (2\pi)^{-3} \int d\mathbf{x}_1 e^{-i\mathbf{p}_1\mathbf{x}_1} \hat{\Psi}_-(\mathbf{x}_1) \tag{9}$$

and

$$b_{\mathbf{p}_2} = (2\pi)^{-3} \int d\mathbf{x}_2 e^{-i\mathbf{p}_2\mathbf{x}_2} \hat{\Psi}_+(\mathbf{x}_2) \tag{10}$$

Substituting these expressions into eq. 8, one obtains the Hamiltonian

$$H(\mathbf{p}) \propto \int d\mathbf{x} e^{-i\mathbf{p}\mathbf{x}} \hat{\Psi}_+(\mathbf{x}) \hat{\Psi}_-(\mathbf{x}) \tag{11}$$

The transition probability between an initial state $\mid i >$, which consists of one positron and n electrons, and a final state $\mid f >$, which consists of $n-1$ electrons and two photons, is then given by

$$\rho_f^{2\gamma}(\mathbf{p}) \propto \left| < f \mid \int d\mathbf{x} e^{-i\mathbf{p}\mathbf{x}} \hat{\Psi}_+(\mathbf{x}) \hat{\Psi}_-(\mathbf{x}) \mid i > \right|^2 \tag{12}$$

Summing over all final states we obtain the *master equation* , i.e. the 2γ-density in momentum space, in the general form

$$\rho^{2\gamma}(\mathbf{p}) \propto \int\int d\mathbf{x}' d\mathbf{x} e^{-i\mathbf{p}(\mathbf{x}-\mathbf{x}')} < i \mid \hat{\Psi}_+^*(\mathbf{x}') \hat{\Psi}_-^*(\mathbf{x}') \hat{\Psi}_+(\mathbf{x}) \hat{\Psi}_-(\mathbf{x}) \mid i > \tag{13}$$

In the *independent particle model* (IPM) with the positron in state $\psi_+(\mathbf{x})$ and with initially n electrons described by a Slater determinant of single-particle states $\psi_{-j}(\mathbf{x})(j = 1,\ldots,n)$ the *IPM master equation* becomes

$$\rho_{IPM}^{2\gamma}(\mathbf{p}) = const \sum_{j=1}^{n} \left| \int d\mathbf{x} e^{-i\mathbf{p}\mathbf{x}} \psi_+(\mathbf{x}) \psi_{-j}(\mathbf{x}) \right|^2 \tag{14}$$

In the master equations 13 and 14 $\rho^{2\gamma}(\mathbf{p})$ is the probability that the final photon pair has a momentum between $\mathbf{p}$ and $\mathbf{p} + d\mathbf{p}$ (we have used here $\hbar \equiv 1$ in $\mathbf{p} = \hbar\mathbf{k}$ for simplicity). For a single electron, equation 14 shows that this probability is proportional to the square of the *Fourier transform* of the *overlapping wave function* $\psi_+(\mathbf{x})\psi_{-j}(\mathbf{x})$ of the electron and positron wave functions. The total 2γ-annihilation rate is obtained after integration of eq. 14 over p-space:

$$\lambda_{IPM}^{2\gamma} = \int \rho_{IPM}^{2\gamma}(\mathbf{p}) d\mathbf{p} \tag{15}$$

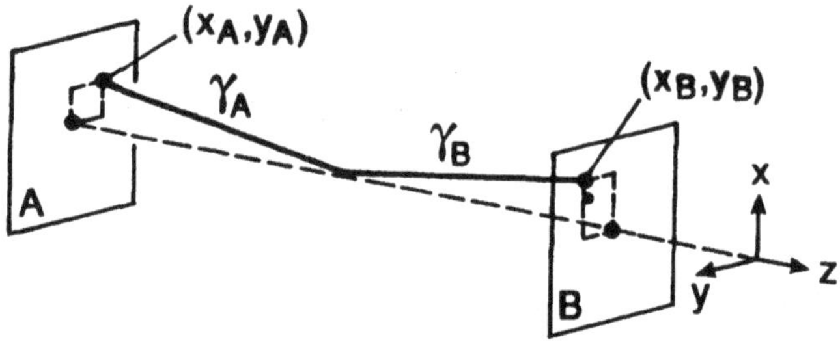

Figure 3: 2D-ACAR.

Equation 15 after integration can be written in the form

$$\lambda_{IPM}^{2\gamma} = const \sum_{j=1}^{n} \int d\mathbf{x}_+ \rho_+(\mathbf{x}_+)\rho_{-j}(\mathbf{x}_+) \tag{16}$$

where we have introduced the positron density $\rho_+(\mathbf{x}) = \psi_+^*(\mathbf{x})\psi_+(\mathbf{x})$ and the single-electron densities $\rho_{-j}(\mathbf{x}) = \psi_{-j}^*(\mathbf{x})\psi_{-j}(\mathbf{x})$. We have denoted real space co-ordinates in the *overlapping density* formula, eq. 16, by $\mathbf{x}_+$ in order to make clear that the electron density at the position of the positron has to be taken.

The factor of proportionality in eq. 14 can be found from Dirac's formula eq. 6 for the spin-averaged two photon cross section. For a non-interacting system of one positron and a free-electron gas of density n we have

$$\lambda^{2\gamma} = \sigma_2 n v = \pi r_0^2 c n \tag{17}$$

Comparing this with eq. 14 by substituting $\psi_+ = 1/\sqrt{V}$ we can find the constant in eq. 14 ($const = \pi r_0^2 c/(2\pi)^3$). Similarly, the constant in eq. 16 is $\pi r_0^2 c$.

We note that the two-photon density $\rho^{2\gamma}$, eqs. 13 and 14, is a measurable quantity and the techniques that can be applied are ACAR or DB. A two-dimensional (2D) ACAR experiment as sketched in Figure 3 consists, in principle, of two 2D position-sensitive detectors A and B and an electronic coincidence circuit such that the locations of detection (x_A, y_A) and (x_B, y_B), which belong to the same 2γ-annihilation event (γ_A and γ_B), define the angles θ_x and θ_y between the two photons. Equation 3 then provides the two components of the transversal component $p_\perp$, $p_x = \theta_x m_0 c$ and $p_y = \theta_y m_0 c$. A 2D-ACAR experiment therefore measures a 2D-projection of the 2γ- momentum density $\rho^{2\gamma}(\mathbf{p})$, hence the experimental 2D-ACAR distribution can be written as

$$N(p_x, p_y) = \int \rho^{2\gamma}(\mathbf{p}) dp_z \tag{18}$$

If we consider a Sommerfeld free-electron gas with Fermi momentum p_F at zero temperature and neglect any interaction between the positron and the electrons, we have $\rho^{2\gamma}(\mathbf{p}) = const$ for $|\,\mathbf{p}\,| \leq p_F$ and zero for $|\,\mathbf{p}\,| > p_F$, i.e. the well-known free-electron Fermi sphere. The measured 2D-ACAR distribution then, according to eq. 18, consists of half an ellipsoid of revolution bounded in the (p_x, p_y) -plane by $p_x^2 + p_y^2 = p_F^2$ and with a principal axis length of $4p_F$, i.e. $N(0,0) = 2p_F$.

Since in a DB experiment we measure only one component of the pair momentum, namely $p_{||}$, DB gives a *one-dimensional* projection of $\rho^{2\gamma}(\mathbf{p})$ and denoting $p_{||}$ by p_z we can write

$$N(p_z) = \int\int \rho^{2\gamma}(\mathbf{p}) dp_x dp_y \tag{19}$$

i.e. a 1D-ACAR. For the Sommerfeld free-electron gas we find for DB a truncated parabola with Fermi cut-off p_F as shown in Figure 4. As we shall see later, there is another ACAR technique called *long-slit* ACAR, which is equivalent to DB as far as the measured quantity is concerned (eq. 19).

The reconstruction of 3D distributions from 1D or 2D projections is a well-known problem in many scientific fields and various numerical procedures have been applied, e.g. computer tomography in medical diagnostics. For the methods used in positron annihilation the reader is referred to the contributions of Berko and of Mijnarends in [5]. Here we consider the full 3D momentum density $\rho^{2\gamma}(\mathbf{p})$ as the measured quantity obtainable in an ACAR experiment and ask us to what extent useful information on the electronic structure of a solid can be extracted from this quantity.

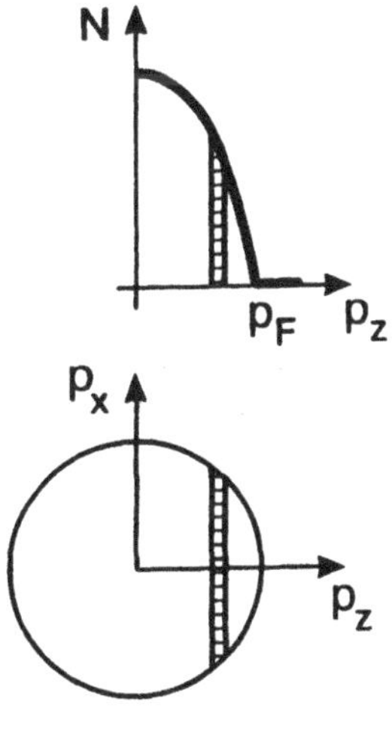

Figure 4: 1D-ACAR for a Fermi sphere.

First we note that according to the master equation 14, the contribution to $\rho^{2\gamma}(\mathbf{p})$ of an electron with wave function $\psi_{-j}(\mathbf{x})$ depends on the wave function $\psi_+(\mathbf{x})$ of the positron. Thus, in addition to $\psi_{-j}(\mathbf{x})$ for each electron, the positron wave function must also be calculated, if one wishes to test a theoretical electronic structure as obtained by an APW construction (Augmented-plane-wave), for example. This may be regarded as a principal disadvantage of the positron method as compared e.g. with Compton profile measurements (see Berko in [5]) which obeys the same type of master equation 14 but with $\psi_+(\mathbf{x})$ replaced by a constant. However, the charge of the positron probe may also be advantageous since positrons often overlap preferentially with those electrons, which are responsible for the electronic porperties of the solid. This follows from the fact that the positron wave function only weakly overlaps with the wave functions of inner electron shells because they are repelled from regions close to the positively charged nuclei. Accordingly, the more interesting outer electron shells, which are responsible e.g. for chemical binding, or the conduction bands in metals, give an *enhanced* contribution to $\rho^{2\gamma}(\mathbf{p})$. The effect is similar to an improved signal-to-noise ratio. However, the charge problem must be treated carefully. For example, if one calculates the annihilation rate for a metal according to eq. 17 neglecting the correlation between the electrons and the positron, the calculated value is 4 to 5 times smaller than the measured one. This effect is taken into account in metals by an *enhancement factor* $h(r_s)$ in eq. 17

$$\lambda^{2\gamma} = \pi r_0^2 cnh(r_s) \tag{20}$$

An approximate expression for the enhancement factor was given by Brandt and Reinheimer [7]

$$h(r_s) = 1 + (r_s^3 + 10)/6 \tag{21}$$

where the dimensionless one-electron radius r_s is defined by $4\pi(r_s a_0)^3/3 = 1/n$ for a conduction-electron density n and $a_0 = \hbar^2/m_0 e^2$ is the Bohr radius.

Similarly, one can introduce a momentum dependent enhancement factor for the conduction electron part in the master equation 14 in order to take account of the correlation effects. However, it turned out that the shapes of measured ACAR distributions in metals show surprisingly weak deviations from the calculations assuming a non-interacting Fermi sea. This is due to the fact that the momentum dependence for e^--e^- correlation and e^+-e^- correlation cancel each other to a great extent. This effect is illustrated in Figure 5 which shows that e^--e^- correlation which leads to the well-known high-momentum tail above the Fermi momentum, is completely canceled out. The net result for $\rho^{2\gamma}$ is the dashed curve, which is usually accounted for by an isotropic enhancement factor of the form

$$\epsilon(p) = a + b\gamma^2 + c\gamma^4 \tag{22}$$

where $\gamma = p/p_F$.

Since a review on these many-body interaction theories is beyond the scope of this article, the reader is referred to the contributions of Carbotte and of Berko in [5] and to [3].

Instead, we proceed by discussing a few more general properties of the above defined 2γ-density in momentum space. If we define an overlapping function in real space for the single-particle wave function of a positron and an electron by

$$\rho_j^{\pm}(\mathbf{x}) = \psi_+(\mathbf{x})\psi_{-j}(\mathbf{x}) \tag{23}$$

the master equation 14 becomes

$$\rho_{IPM}^{2\gamma}(\mathbf{p}) = const \sum_{j=1}^{n} \mid \int d\mathbf{x} e^{-i\mathbf{p}\mathbf{x}} \rho_j^{\pm}(\mathbf{x}) \mid^2 \tag{24}$$

This relation is formally very similar to the well-known scattering formula in X-ray or neutron diffraction, if $\rho_j^{\pm}(\mathbf{x})$ is replaced by the electron density $\rho_{-j}(\mathbf{x})$ and the Fourier

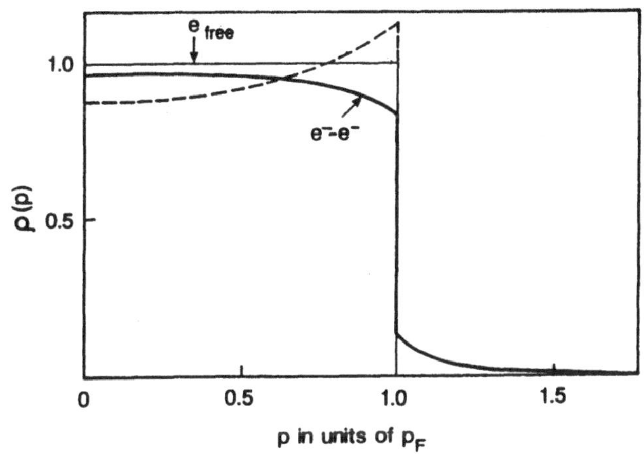

Figure 5: The effect of many-body interactions on $\rho^{2\gamma}(\mathbf{p})$ of the conduction band in Al (from Brandt in [5]).

transform of $\rho_{-j}(\mathbf{x})$ is the structure factor $F(\mathbf{k})$ (see e.g. [8]). $| F |^2$ is the scattering function, which is the measured quantity in usual diffraction experiments. As is well-known in diffraction theory, $\rho_{-j}(\mathbf{x})$ cannot be reconstructed uniquely from $| F |^2$ since the phase of F is usually not measured. Only the autocorrelation function of $\rho_{-j}(\mathbf{x})$, which is called the Patterson function $P(\mathbf{x})$ in X-ray diffraction, can be obtained from $| F |^2$ by Fourier back-transformation. If we define, by analogy, an autocorrelation function of the $\rho_j^{\pm}(\mathbf{x})$ in eq. 23 as

$$B^{2\gamma}(\mathbf{x}) \propto \sum_j \int \rho_j^{\pm}(\mathbf{x}+\mathbf{x}')\rho_j^{\pm}(\mathbf{x}')d\mathbf{x}' \tag{25}$$

we find that $B^{2\gamma}(\mathbf{x})$ can be obtained by Fourier back-transformation of $\rho^{2\gamma}(\mathbf{p})$. The analogue to the structure factor F can also be defined as the Fourier transform of $\rho_j^{\pm}(\mathbf{x})$. If we denote this function by $\Phi_j^{\pm}(\mathbf{p})$ and use the Fourier convolution theorem we have

$$\Phi_j^{\pm}(\mathbf{p}) \propto \int \varphi_{+}(\mathbf{p}')\varphi_{-j}(\mathbf{p}-\mathbf{p}')d\mathbf{p}' \tag{26}$$

where $\varphi_{+}(\mathbf{p})$ and $\varphi_{-j}(\mathbf{p})$ are the Fourier transforms of $\psi_{+}(\mathbf{x})$ and $\psi_{-j}(\mathbf{x})$, respectively. The relations between the different quantities discussed above are summarized in Figure 6 together with their analogues in diffraction theory. The arrows in Figure 6 indicate the directions for which the quantities are associated in a unique way. It is evident that $\rho^{\pm}(\mathbf{x})$ cannot be obtained from the knowledge of $\rho^{2\gamma}(\mathbf{p})$ alone, but $B^{2\gamma}(\mathbf{x})$ can be calculated from $\rho^{2\gamma}(\mathbf{p})$ by Fourier back-transformation.

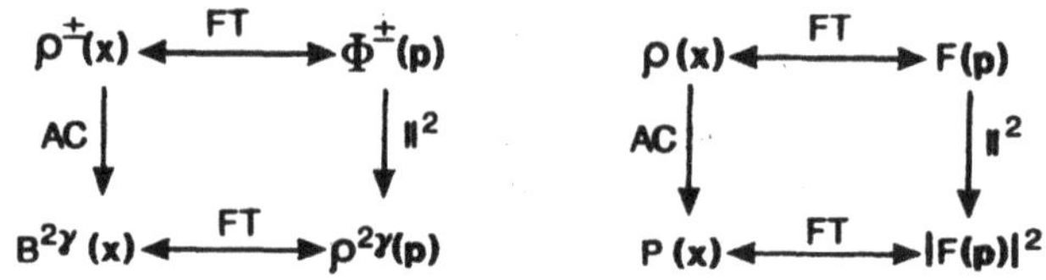

Figure 6: Relations between different quantities in real and reciprocal space. FT≡ Fourier transform, $||^2$≡ magnitude square, AC≡autocorrelation.

One can also show that $B^{2\gamma}$ is useful as a tool to reconstruct full 3D two-photon densities $\rho^{2\gamma}(\mathbf{p})$ from measured 1D or 2D projections. If, for example, a 2D-ACAR $N(p_x, p_y)$ was measured $B^{2\gamma}(x, y, 0)$ is obtained by 2D Fourier back-transformation FT_2

$$B^{2\gamma}(x, y, 0) \propto FT_2\{N(p_x, p_y)\} \tag{27}$$

By means of several measured 2D-ACAR distributions in different crystallographic directions one can therefore use eq. 27 to reconstruct the full 3D Distribution of $B^{2\gamma}(x, y, z)$ and $\rho^{2\gamma}(\mathbf{p})$ can then be obtained by 3D Fourier transformation. However, serious problems in using Fourier series for image reconstruction can arise from the finite number of measured projections and also from the limited number of data points in each of these projections as well as the scatter in the data points. Symmetry relations, which are often also applied in image reconstruction procedures, can also lead to artificial patterns. For a summary of the various reconstruction techniques which are currently used with ACAR, the reader is referred to the contribution of Berko in [5], from which I may recall the statement: "Pure noise can also lead to nice-looking patterns due to the use of symmetry."

2.2 Positronium

Positron annihilation via positronium formation may be an important decay channel for positrons in solids, especially at metallic surfaces and in ionic and molecular substances. A brief description of this analogue to the hydrogen atom may be sufficient here, although this topic is of interest in its own right, since free Ps is a purely leptonic two–body particle–antiparticle system, which makes it ideal for the study of the two–body problem in relativistic quantum electrodynamics [9].

Since the reduced mass in Ps ($m_+ m_- / (m_+ + m_-) = \frac{1}{2} m_0$) is half of the electron mass in hydrogen, the binding energy of the Ps ground state is 0.5 Rydberg (6.8 eV) and the 'Bohr radius' is two times larger ($2a_{Bohr} \simeq 0.106$nm). More generally, taking into account pure Coulomb interaction only, the term scheme of Ps is obtained from that of hydrogen by dividing all energies by 2 and multiplying all distances by this scale factor. Accordingly, the ground state (n = 1) of Ps is split into a triplet spin state

1^3S_1 (total spin $F = 1$, $F_z \equiv M = -1,0,1$, ortho–Ps) and a singlet state 1^1S_0 ($F = 0$, para–Ps).

In contrast to the hydrogen atom, spin–spin hyperfine splitting (hfs) in Ps is of the same order of magnitude as spin–orbital coupling. This is mainly due to the ~660 times larger magnetic moment of the positron as compared with that of a proton. The effect is reduced by the increased distance scale in Ps and an additional increase results from QED corrections and virtual annihilation/re–creation which is of the same order of magnitude as spin–spin hfs. The net result is a hfs of $\Delta\nu$ = 203 GHz (ΔE = 0.84 meV = 10K) between the triplet state (upper level) and the singlet state (lower level). Optical transitions between singlet and triplet states are obviously allowed since the selection rule for total spin is ΔF = 0 or $\pm$ 1.

The eigenstates of ground–state Ps in the notation $| m_s^+, m_s^- >$ are $| 1/2, 1/2 >$ and $| -1/2, -1/2 >$ for the two paired–spin triplet states with $M = +1$ and $M = -1$, respectively. The unpaired–spin states are mixed states of the form $\sqrt{1/2} \mid 1/2, -1/2 >$ $+ \sqrt{1/2} \mid -1/2, 1/2 >$ (triplet $M = 0$) and $\sqrt{1/2} \mid 1/2, -1/2 > - \sqrt{1/2} \mid -1/2, 1/2 >$ (singlet $M = 0$). Since the magnetic moment of a positron with spin +1/2 is μ_{Bohr} and -μ_{Bohr} for an electron (neglecting the g-factor anomaly), it follows that the magnetic moment $\mu = 0$ for all four ground states of Ps. The triplet states annihilate with a rate of $\lambda_t \sim 7 \times 10^6 s^{-1}$ and the singlet state with $\lambda_s \sim 8 \times 10^9 s^{-1}$, i.e. the lifetime of o–Ps is longer than that of p–Ps by a factor of more than 10^3. This is essentially due to the fact that o–Ps decays into three photons and p–Ps into two photons as shown in the previous section.

In solids there are some limits for positrons to form positronium. One limit is given by the electronic density. If the volume per electron in a solid is smaller than or equal in magnitude to the effective volume occupied by Ps in the solid, there is a rapid exchange of the electron bound to the positron and spin–averaging of the annihilation rate occurs, hence it is impossible to distinguish between o–Ps and p–Ps. In metals such a correlated motion between the positron and a single electron cannot last for times longer than $1/\nu_p \sim 10^{-16}$s where ν_p is the plasma frequency. This time is much smaller than observed positron lifetimes in metals (100 – 500 ps). If, however, there is a free volume (i.e. free of electrons) of a size greater than say 0.1nm, as is possible in insulators or at metallic surfaces or internal cavities (voids), Ps formation may be the dominant process.

Another limit is given by energy considerations, namely, Ps formation may be energetically favoured or not. The initial kinetic energy E_+^{kin} of the positron together with the Ps ionization energy E_B (6.8eV) must exceed the ionization energy E_i of the solid in order to form Ps. If on the other hand E_+^{kin} exceeds the lowest excitation energy E_e of the solid, this process will compete with Ps formation. We conclude that Ps formation is most probable if

$$E_i - E_B \leq E_+^{kin} \leq E_e \tag{28}$$

This is the *Ore gap* [10], which was originally formulated for Ps formation in gases. In an insulator with energy gap E_g, a positron work function ϕ_+, an electron work function ϕ_- for an electron at the bottom of the conduction band and a Ps work function ϕ_{Ps}, the Ore gap is (Dupasquier in[5])

$$E_g + \phi_- + \phi_+ - E_B - \phi_{Ps} \leq E_+^{kin} \leq E_g \tag{29}$$

Here we have assumed that the final state of Ps is a *delocalized* ground state in the crystal (*'quasi-Ps'*, see Dupasquier in [5]).

A modified picture is that of Mogensen's *spur model* [11], in which Ps formation of thermalized positrons with electrons in the conduction band, created during slowing–down of energetic positrons, is also taken into account.

Redgardless of the mechanism of Ps formation, if Ps exists in a solid or at the surface of a solid there will be a more or less important overlap of the positron wave function with the electron wave function of the *surrounding medium* and, accordingly, annihilation with these electrons becomes an alternative decay channel which is called *pick–off* annihilation. Since Ps is in thermal equilibrium with the solid, the photons created in pick–off annihilation have similar characteristics as in the case of thermalized free positrons, with the exception of different overlapping densities between the positron and electron wave functions. With respect to the vacuum decay characteristics of Ps, pick–off results in an increased 2γ / 3γ–ratio (> 1/3), an increased annihilation rate ($> \lambda_t = 7 \times 10^6 s^{-1}$) and a broader ACAR distribution, since the photons carry away the additional momentum of the electron in the band structure of the solid, whereas annihilation with the electron of p–Ps creates a photon pair with the low momentum corresponding to the center–of–mass energy of the thermalized Ps.

If pick–off annihilation is efficient enough, the annihilation characteristics of Ps can become indistinguishable from that of free positrons. A useful technique to decide whether there is Ps formation or not, is then to apply an external magnetic field. This so–called *magnetic quenching* has the effect of mixing the two $M = 0$ states of ortho and para–Ps, i.e. the spin character of $M = 0$ $^3S_1$ state changes partially to $|\frac{1}{2}, -\frac{1}{2}>$ and for $M = 0$ $^1S_0$ to $|-\frac{1}{2}, \frac{1}{2}>$, depending on the field strength and the $M = \pm 1$ triplet states remain unaffected. The mixing of $M = 0$ states also has a strong effect on the triplet and singlet annihilation rates and branching ratio, since the 2γ decay channel is now open also for the $M = 0$ triplet state (for a summary of the theory see [9] and Dupasquier in[5]). This is especially true for the triplet rate, which, for small magnetic interaction energy $2g'\mu_B B$ compared with the hfs energy Δ, is given by

$$\lambda_t' = \lambda_t + \frac{1}{4}x^2\lambda_s \qquad (x \ll 1) \tag{30}$$

with

$$x = 2g'\mu_B B/\Delta$$

As is evident from eq. 30, the sensitivity of the triplet annihilation rate λ_t' to magnetic quenching is due to the fact that $\lambda_t \ll \lambda_s$.

2.3 Positron Annihilation in Perfect Lattices

In a perfect crystal at zero temperature, the potential for *both* particles, electrons and positrons, is strictly periodic, i.e. $V(\mathbf{x}+\mathbf{R}_i) = V(\mathbf{x})$ where $\mathbf{R}_i$ is any lattice vector in real space. Accordingly, $\psi_{-j}(\mathbf{x})$ and $\psi_+(\mathbf{x})$ in eq. 14 become *Bloch waves*:

$$\psi_{\mathbf{k},j} = u_{\mathbf{k},j}(\mathbf{x})e^{i\mathbf{kx}} \tag{31}$$

and

$$\psi_+ = u_+(\mathbf{x}) \tag{32}$$

where j here denotes the band index and $\mathbf{k}$ the wave vector of the electron. $u_{\mathbf{k},j}(\mathbf{x})$ and $u_+(\mathbf{x})$ are periodic in $\mathbf{x}$ as $V(\mathbf{x})$. We assume $\mathbf{k}_+ = 0$ in eq. 32 at zero temperature. Substituting eqs. 31 and 32 into the IPM master equation 14 we find

$$\rho^{2\gamma}(\mathbf{p}) = const \sum_{\mathbf{k},n,j} \mid A_j^{\pm}(\mathbf{k},\mathbf{G}) \mid^2 \delta_{\mathbf{p}-\mathbf{k}-\mathbf{G}_n} \tag{33}$$

where $A_j^{\pm}(\mathbf{k},\mathbf{G})$ are the Fourier coefficients of the product of the two Bloch waves eq. 31 and 32, i.e.

$$A_j^{\pm}(\mathbf{k},\mathbf{G}) = \int d\mathbf{x} u_+(\mathbf{x})u_{\mathbf{k},j}(\mathbf{x})e^{-i\mathbf{Gx}} \tag{34}$$

which can also be expressed in terms of the FT of the single-particle Fourier coefficients $A^+(\mathbf{G}) = FT_{-3}\{u_+(\mathbf{x})\}$ and $A_j^-(\mathbf{k},\mathbf{G}) = FT_{-3}\{u_{\mathbf{k},j}(\mathbf{x})\}$:

$$A_j^{\pm}(\mathbf{k},\mathbf{G}) = \int A^+(\mathbf{G}')A_j^-(\mathbf{k},\mathbf{G}-\mathbf{G}')d\mathbf{G}' \tag{35}$$

The $\mathbf{G}'s$ are wave vectors and $\mathbf{G}_n$ is any reciprocal lattice vector. The Kronecker δ-function in eq. 33 displays the important fact that the annihilation photons not only carry away the momentum $\mathbf{k}$ of the electrons but there are possible contributions with a photon momentum $\mathbf{p} = \mathbf{k}+\mathbf{G}_n$, i.e. these photons carry away an additional momentum $\mathbf{G}_n$, which is due to the interaction with the crystal lattice as a whole. Since the coefficients $A^{\pm}$ decrease with increasing $\mathbf{G}_n$ the corresponding so-called high-momentum components or "Umklapp"-annihilations in the ACAR distribution rapidly decrease with increasing $\mathbf{G}_n$.

It is clear that the sum in eq. 33 must be taken over occupied electron states $(\mathbf{k},j)$ only. This can be accomplished by adding a factor n_j to each term of the sum, where $n_j = 1$ if that state is occupied by one electron (or 2 for two electrons with opposite spin) and zero for unoccupied states. n_j is the *occupation number*. It is easy to see that a step in $\rho^{2\gamma}(\mathbf{p})$ occurs when a band is cut by a Fermi surface (n_j is a step function), while occupied bands give flat contributions.

In principle, the wave functions and eigenvalues for positrons in ideal crystals can be treated in a similar way as in electron band structure calculations, i.e. by expanding the wave functions into suitable sets of plane waves, such as OPW or APW.

Without going into any detail in these calculational methods (see e.g. the contribution of Mijnarends in [5]), we mention here only some important differences between electron and positron states in solids.

The Coulomb potential in the stationary Schrödinger equation for single electrons in a crystal is usually written as a sum of different terms

$$V_-(\mathbf{r}) = V_n(\mathbf{r}) + V_e(\mathbf{r}) + V_{xc}(\mathbf{r}) \tag{36}$$

where $V_n(\mathbf{r})$ is the attractive Coulomb potential of the nuclei, $V_e(\mathbf{r})$ the averaged repulsive Coulomb potential of the electrons and $V_{xc}(\mathbf{r})$ denotes electron exchange and correlation interaction. In the corresponding picture for positrons, this potential becomes

$$V_+(\mathbf{r}) = -V_n(\mathbf{r}) - V_e(\mathbf{r}) - V_c(\mathbf{r}) \tag{37}$$

$V_c(\mathbf{r})$ now contains the electron-positron correlation interaction alone, since positrons and electrons are distinguishable particles and hence there is no exchange interaction. The opposite sign for all terms is due to the different charge, which has the effect to repel the positrons from the positively charged nuclei. Its spatial distribution has a kind of "Swiss cheese" structure with holes around each nucleus and maxima in the interstitial regions. This effect makes a representation of the positron wave function by plane waves more suitable than for electrons. As is known, the rapid oscillation of the electron wave functions near the nuclei leads to slow convergence if plane-wave expansions are applied. The exclusion of the positron wave function from the ion cores and the resulting character closer to plane waves is also important in other positron applications in solids. For example, low energy diffraction of positrons at solid surfaces seems to be easier to describe theoretically than in the case of electrons (LEED) [12].

In eqs. 31 - 35 we have considered electrons and positrons at zero temperature, which explicitly entered into the description as zero positron wave vector $k_+ = 0$ and in a step shaped occupation number $n(\mathbf{k})$. Both assumptions lose their validity at $T \neq 0$. For $n(\mathbf{k})$ this results in a T-dependence which is given by a *Fermi-Dirac* distribution $f_-(E_\mathbf{k}, T)$ for the *electrons* and a *Boltzmann* distribution $f_+(E^+, T)$ for the *positrons*:

$$n(\mathbf{k}, \mathbf{k}_+, T) = f_-(E_\mathbf{k}, T) f_+(E^+, T) \tag{38}$$

$$f_-(E_\mathbf{k}, T) = 1/[1 + e^{(E_k - \mu)/k_B T}] \tag{39}$$

$$f_+(E^+, T) = (\pi m^* k_B T)^{-3/2} e^{-E^+/k_B T} \tag{40}$$

Boltzmann statistics is valid for positrons since there is only one positron at a time in the solid and, therefore, no exchange occurs as for electrons. Since $k_+ \neq 0$ at $T \neq 0$ we have to replace ψ_+ in eq. 32 by $\psi_{\mathbf{k}_+} = u_{\mathbf{k}_+}(\mathbf{x}) e^{i\mathbf{k}_+\mathbf{x}}$ and the overlapping density according to eq. 35 will also depend on k_+ (see Mijnarends in [5]). Similarly, the Kronecker δ-function in eq. 33 becomes k_+-dependent, $\delta_{\mathbf{p}-\mathbf{k}-\mathbf{k}_+-\mathbf{G}_n}$. This results in a smeared Fermi cut-off in measured ACARs due to positron motion.

2.4 Positrons in Defects and at Surfaces

Since positrons are repelled from the atomic cores, any region in a solid with lower than average atomic density has the effect of attracting positrons and hence locally increases the positron density. Screening of these sites by conduction electrons may be important but is not complete. The attraction may eventually be so strong that the positron wave function loses its Bloch state character and the positron becomes localized or *trapped* in or close to such sites. Due to the different electronic structure in the vicinity of a trapping site as compared to the bulk, this can lead to measurable changes in the annihilation parameters, e.g. the annihilation rate. For a crystal containing a certain concentration c of identical traps, e.g. vacancies in a metal, the positrons then have a certain probability ϕ_t for annihilation from a trapped state and a probability $(1-\phi_t)$ for annihilation from the free state, where ϕ_t depends on the concentration of the traps and on the rates for annihilation from the free and trapped states. In the simplest *two-state trapping model* one assumes that the free state is depopulated with a rate $(\lambda_f+\sigma c)$, where λ_f is the free annihilation rate and $\sigma c \equiv \kappa$ is the trapping rate which is assumed to be proportional to c.

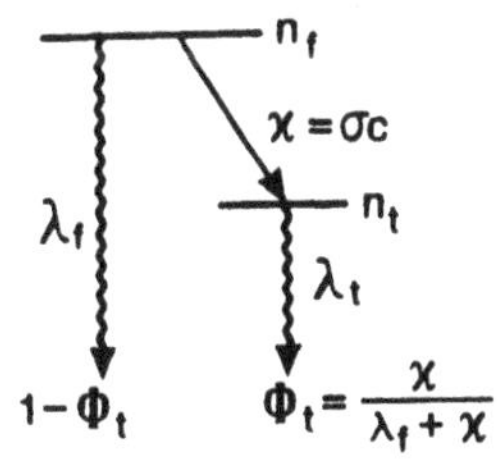

Figure 7: Two-state trapping model.

The trapped states are then populated with a rate σc and depopulated with λ_t, the annihilation rate from the trapped state. The situation is sketched in Figure 7. One then considers a number n_0 of thermalized non-interacting positrons which were implanted in the solid all at the same time $t=0$. The number of free positrons at time t, $n_f(t)$, and the number of trapped positrons, $n_t(t)$, are then given by a system of differential equations

$$dn_f/dt = -(\lambda_f+\kappa)n_f \tag{41}$$

$$dn_t/dt = \kappa n_f - \lambda_t n_t \tag{42}$$

and the condition that at $t=0$ $n_f(0)=n_0$ and $n_t(0)=0$. With the definition of the annihilation probability P at a time t $P(t) = (\lambda_f n_f + \lambda_t n_t)/n_0$, which is called the *lifetime spectrum*, the solution of eqs. 41 and 42 leads to

$$P(t) = \lambda_1 I_1 e^{-\lambda_1 t} + \lambda_t I_2 e^{-\lambda_t t} \tag{43}$$

where

$$I_2 = \frac{\kappa}{\lambda_f+\kappa-\lambda_t} = 1 - I_1 \tag{44}$$

and

$$\lambda_1 = \lambda_f + \kappa \tag{45}$$

Equation 43 is a two-component lifetime spectrum with lifetimes $\tau_1 = 1/\lambda_1$ and $\tau_t = 1/\lambda_t$ and $P(t)$ is the measured quantity in positron lifetime spectrometry (LT). The parameters τ_1, τ_t and I_2 are obtained by means of suitable fitting procedures [13]. The model can easily be extended to more than one type of trapping rates, see e.g. [14]. The trapping rate $\kappa = \sigma c$ can be calculated from the fit parameters by combining eqs. 44 and 45

$$\kappa = \sigma c = I_2(\lambda_1 - \lambda_t) \tag{46}$$

Equations 43 to 46 hold for $\lambda_t \leq \lambda_1$ and, since this must be true for any c, $\lambda_t \leq \lambda_f$. The reason is that the model is meaningful only for positive values of n_t. As is evident, the knowledge of the specific trapping rate σ is a prerequisite, if one wishes to know the concentration of trapping sites c.

The specific trapping rate is determined by different possible processes depending on various parameters such as the temperature, the size of the traps and their spatial distribution. If we consider the positron in its initial state ψ_i extended over distances much larger than the effective trapping size of the defect, the trapping rate is given by the quantum mechanical transition probability per unit time between ψ_i and the localized final state ψ_f:

$$\sigma \propto | M_{if} |^2 \, \delta_{E_i - E_f} \tag{47}$$

The matrix element M_{if} depends on ψ_i and ψ_f and on the mechanism by which the binding energy $\Delta E = E_i - E_f$ is dissipated. If ΔE is of the order of 1eV, as for vacancies in many metals, electron-hole creation is the dominant mechanism (see Nieminen in [5]). In this case, σ is essentially temperature-independent. If, however, ΔE is low, optical or acoustic phonons may be important and a T-dependence of σ results from the T-dependence of the phonon occupation number. Trapping, limited by this type of mechanisms, is called *transition-limited*. For iron λ_f=9ns^{-1} and $\sigma = 1.1 \times 10^{15}$ s^{-1} [15], and with the trapping probability $\phi_t = \sigma c/(\lambda_f + \sigma c)$ we find, that for a trap concentration of $c = 10^{-5}$, $\phi_t \sim 92\%$ and for $c = 5 \times 10^{-7}$ still $\sim 6\%$ of the positrons are trapped, i.e. that positrons are very sensitive to trapping in vacancies.

If, on the other hand, this transition rate is fast enough the mobility of the free positrons may eventually set a limit to the total trapping rate. This may be the case in larger voids. Then, on a length scale larger than the mean free path l^+ of the positrons, the process can be described by diffusion of classical particles in a similar way as for electron-holes in semiconductors (see e.g. [16]). This is the case of *diffusion-limited* trapping. The mobility μ^+ due to elastic scattering of the positrons by *acoustic phonons* has a T-dependence given by

$$\mu^+ = const\ T^{-3/2} \tag{48}$$

and with the *Nernst-Einstein* equation

$$D^+ = \mu^+ k_B T \tag{49}$$

the T-dependence of the *diffusivity* D^+ becomes

$$D^+ = const\ T^{-1/2} \tag{50}$$

i.e., diffusion is faster at lower temperature. The validity of the $T^{-1/2}$ law for the diffusivity over a wide temperature range in metals was confirmed by means of positron beam techniques [17]. Possible deviations at low temperature and close to the melting point are discussed in [18].

The above treatment of the positron also implies that their behaviour can be described by a diffusion equation similar to Fick's equations, i.e. by a density $n(\mathbf{x},t)$ (e^+/cm^3) which obeys

$$\frac{\partial n}{\partial t} = div(D^+ grad(n)) - \lambda_f n \tag{51}$$

The term $\lambda_f n$, in addition to Fick's 2nd law, takes account for annihilations in the bulk. The positron flux $\mathbf{j}$ is given by Fick's 1st law

$$\mathbf{j} = -D^+ grad(n) \tag{52}$$

As a simple example, we apply Fick's equations 51 and 52 to the problem of diffusion-limited trapping by a spherical trap of radius R_0 for constant D^+ and λ_f with the initial condition that at t=0 the positron density is constant for $x \geq R_0$. Transition trapping is assumed to be fast enough to ensure that each positron is trapped immediately after reaching the trap, i.e. it is removed from the ensemble. Equation 51 then reads as

$$\frac{\partial n}{\partial t} = D^+ \Delta n - \lambda_f n \tag{53}$$

The boundary condition is

$$n(\mathbf{x},t) = 0 \qquad |\mathbf{x}| \leq R_0 \tag{54}$$

and the initial condition

$$n(\mathbf{x},0) = n_0 \qquad |\mathbf{x}| > R_0 \tag{55}$$

The solution of eq. 53 under the conditions 54 and 55 in spherical co-ordinates is

$$n(r,t) = n_0 \frac{r - R_0}{r}\left[1 - 2\Phi\left(\frac{r - R_0}{\sqrt{2D^+ t}}\right)\right] e^{-\lambda_f t}, \quad r \geq R_0 \tag{56}$$

where Φ is the error function defined by

$$\Phi(z) = (2\pi)^{-1/2} \int_0^z e^{-\frac{1}{2}y^2} dy \tag{57}$$

The annihilation rate at a distance r is $N(r,t) = \lambda_f n(r,t)$ and with the definition $p(r) = \int_0^\infty N(r,t)dt/n_0$ for the annihilation probability at distance r we find

$$p(r) = 1 - \frac{R_0}{r} e^{-\sqrt{\lambda_f/D^+}(r - R_0)}, \quad r \geq R_0 \tag{58}$$

We can also calculate a time dependent trapping rate $\sigma_t = -\oint \mathbf{j} ds/n_0$ with the integral taken over the surface S of the trap and $\mathbf{j}$ is given by Fick's 1st law (eq. 52), i.e. by

$grad(n)$. The total trapping rate is then obtained after integrating σ_t over the time t. We can, however, also integrate first over t and then calculate the gradient. The trapping rate σ is then

$$\sigma = -D^+ \oint_S grad(p) ds = 4\pi R_0 D^+ \tag{59}$$

This is the commonly used formula for diffusion limited trapping [19]. With respect to atomic concentration of traps c, the trapping rate becomes

$$\sigma = \frac{4\pi R_0 D^+}{V_A} \tag{60}$$

where V_A is the atomic volume.

We note that for a metal typically $D^+ = 1$ cm^2/s and $\lambda_f = 10$ ns^{-1} and hence the *diffusion length* $L^+ = \sqrt{D^+/\lambda_f} = 100$ nm. This means that the positron samples $\simeq 5 \times 10^8$ atoms before annihilation and for traps separated by distances $\gg$ 100nm trapping can be neglected.

For vacancies in iron we find the diffusion limited trapping rate (eq. 60) $\sigma \sim 1.6 \times 10^{16}$s^{-1} ($V_A = 1.2 \times 10^{-23}$cm^3, $R_0 \sim 0.15$nm), which is large compared to the transition rate of 10^{15}s^{-1}. Hence, trapping of positrons in vacancies in metals is transition-limited and essentially temperature independent, as stated above. However, in vacancy clusters the transition rate may be faster than the diffusion rate, since the two rates increase with R_0^3 and R_0, respectively. A temperature dependence proportional to $T^{-\frac{1}{2}}$ is then expected according to the T-dependence of the diffusivity D^+ (eqs. 50 and 60). We note that the transition and diffusion rates become equal for a cluster of 64 vacancies in iron (cluster radius $4R_0$).

Positron trapping in semiconducters has been much less studied than in metals. There are theoretical arguments [20] and experimental evidence [21] that the trapping rates are considerably different for different defect types in semiconductors. For example, an increase by a factor of 10 in the trapping rate for vacancies in Si was observed when the temperature decreased from 120K to 20K [21].

For a capture radius $\ll L^+$ eq. 58 becomes $p \simeq 1 - R_0/r$ and, therefore, the extension of the zone around the defect, depleted in positrons, has a FWHM of $\sim 4R_0$.

If we take the limit $R_0 \rightarrow \infty$ in eq. 58 we have

$$p(r) = 1 - e^{-r/L^+} \tag{61}$$

This means that at a surface the depleted zone has a depth equal to the diffusion length L^+. Positrons implanted in a solid to a depth not extending substantially beyond L^+ therefore have a certain chance to move back to the surface. Here, the positron has the possibility to escape from the solid, either as a bare positron or as Ps. In both cases the emitted particle can exist in a free vacuum state or it may form a bound surface state. Emission of slow positrons or Ps from solid surfaces is a frequently observed process. This is due to the fact that the work function for positrons and Ps are often negative and of the order of 1eV. The positron work function ϕ_+ is defined as the minimum energy required to remove a positron from its lowest energy level in the bulk of the solid to the vacuum level outside the solid. The situation is sketched in Figure 8 for a positron in the vicinity of a metal surface.

The positron work function ϕ_+ in a metal is

$$\phi_+ = -D - \mu_+ \tag{62}$$

where D is the surface barrier and μ_+ the chemical potential. The surface dipole barrier D is the same as for electrons but repulsive for positrons and attractive for electrons

$$\phi_- = +D - \mu_- \tag{63}$$

The dipole is due to the tailing of the electron distribution into the vacuum and includes effects of surface reconstruction and relaxation of the surface layer. The negative sign of the dipole term for positrons is the main reason for the observed negative work function in many metals and hence for positron emission from metallic surfaces. The chemical potential μ_+ includes repulsion from the ion cores (zero-point energy $\varepsilon_0 > 0$) and attraction to the electrons (correlation energy, $\varepsilon_{corr} < 0$)

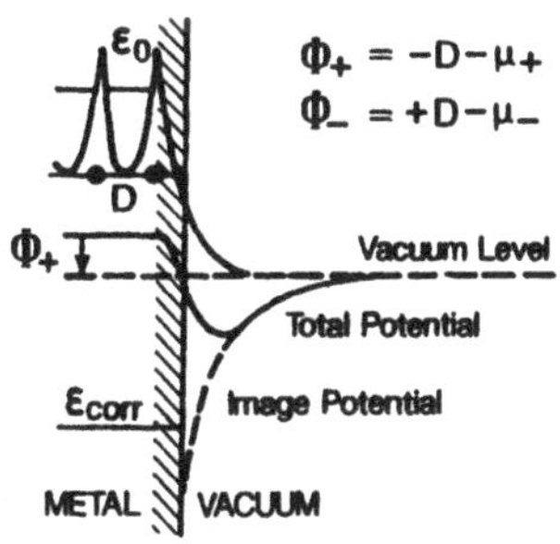

Figure 8: Positron work funtion ϕ_+.

$$\mu_+ = \varepsilon_0 + \varepsilon_{corr} \tag{64}$$

and with eq. 62

$$\phi_+ = -D - \varepsilon_0 - \varepsilon_{corr} \tag{65}$$

Formation of Ps requires escape of a positron *and* an electron from the solid and subsequent combination to form Ps. At this step the Ps binding energy E_B=6.8eV is released. Thus the Ps work function is given by a Born-Haber cycle as

$$\phi_{Ps} = \phi_+ + \phi_- - E_B \tag{66}$$

Values of ϕ_+ and ϕ_{Ps} were listed by Nieminen in [5]. For metals and semiconductors the electron work function ϕ_- is generally less than E_B=6.8eV and therefore $\phi_{Ps} < \phi_+$. Hence Ps formation is energetically always more favoured than positron emission. However, measured branching ratios indicate that both channels are usually open and comparable fractions of Ps and e^+ are emitted. The reason is that Ps formation at a surface is a dynamic process and non-adiabatic effects have to be considered (see Nieminen in [5]).

Convincing evidence for positron surface states was obtained from the discovery of temperature dependent Ps formation on metallic surfaces, which could be explained by thermally activated Ps formation of positrons trapped at the surface [22]. With the binding energy of the positron surface state, E_s, which is due to the image potential (Figure 8), the activation energy E_a for this process is $E_a = E_s + \phi_- - 6.8eV$. Since generally $\phi_- < 6.8eV$, as already mentioned, thermal desorption of positrons with surface binding energies E_s of several eV becomes energetically possible by virtue of Ps formation.

The surface state may, however, be considered as an ideal situation. In practice, the purely two-dimensional delocalized surface state is probably disturbed if surface defects or impurities are present, which have to be considered as possible traps. Nevertheless, this shows that the positron can also be an interesting probe for surface studies.

2.5 Thermalization of Positrons in Solids

In the preceding sections we have assumed that the positrons at any time are in the state of lowest possible energy in the solid, i.e. at thermal equilibrium with the solid. However, in practice the positrons are implanted in the solid at much higher energy, in many cases even at energies up to $\sim$ 1 MeV as in experiments using radio–active β^+–emitters. The validity of the ideas developed so far therefore strongly depends on the assumption that positrons of this high energy are slowed down sufficiently fast, i.e. on a time scale comparable to the lifetime of the positron in the solid ($\sim$ 100 ps).

A detailed discussion of the different slowing–down processes is beyond the scope of this lecture (see e.g. Paulin and Brandt in [5]). I just summarize here a few results and some differences to the slowing–down of β^-–particles. At very high energies, $E \sim$ 10 MeV, the dominant process for energy loss is bremsstrahlung, which can be ignored below $\sim$ 1 MeV. In the range of $\sim$ 20 eV to $\sim$ 1 MeV energy loss is mainly due to ionization and excitation, i.e. by electron scattering. Elastic nuclear scattering (Mott), however, leads to directional randomization for positrons in this energy range. Electron–electron scattering (Møller) and positron–electron scattering (Bhabha) have different cross–sections (virtual annihilation and recreation). Since electrons are not distinguishable from each other, the maximum energy transfer in a collision of two electrons is only 50 % whereas this is 100 % for a collision between a positron and an electron.

The detailed treatment of positron–electron scattering shows that energy loss from 1 MeV to $\sim$ 100 eV is very fast ($\sim$ 1 ps) compared to the positron lifetime (100 ps) and can be neglected. Thus for most practical β^+–sources all positrons can be assumed to start at $\sim$ 100 eV in the solid irrespective of their origin. Below $\sim$ 0.5 eV phonon excitation becomes the dominant stopping process. The combined action of phonon and electronic excitations leads to thermalization times of the order of $\sim$ 10 ps in metals.

In spite of the different slowing–down mechanism for positrons and electrons, the *penetration* of energetic positrons is usually treated in a similar semi–empirical way as for electrons. For spectra of radio–active β^+–emitters the implantation profile is assumed as

$$P(x) = \frac{1}{x_0} e^{-x/x_0} \tag{67}$$

where x is the distance from the surface and x_0 is given by an empirical relation of the form

$$x_0 = \frac{A}{\rho} E_{max}^n \tag{68}$$

where E_{max} is the maximum energy of the β–spectrum in MeV, A a constant in g/cm² and n = 1.43. For positrons the exponent n ranges between 1.14 [23] and 1.6 (Brandt in [5]). A detailed discussion of measured and calculated implantation profiles shows a weak dependence on the shape of the spectra and on Z of the sample material (see Paulin in [5]). For *monoenergetic positron beams* a modified implantation profile of the form

$$P(z) = -\frac{d}{dx}\left(e^{-(x/x_0)^m}\right) \tag{69}$$

was found to be appropriate (see [12] and references therein). From Monte Carlo calculations [24] and experimental data [25] $m \approx 2$ in eq. 69 and $A = 45$ μg/cm², $n \approx$ 1.4 ... 1.6 in eq. 68 was suggested in [26] (E in keV).

One finds for a ^{64}Cu source (E_{max} = 0.65 MeV) $x_0 \simeq 120$ μm for Al and $x_0 \simeq 35$ μm for Fe using $x_0 = E_{max}^{1.43} / 17\rho$ as for β^- (E_{max} in MeV, ρ in g/cm³), which gives in these cases also reasonable values for β^+.

3 EXPERIMENTAL TECHNIQUES

We only briefly describe in this section the basic principles of the different 'conventional' techniques without going into technical details, because in some cases the necessary experimental equipment is now commercially available (LT and DB), together with the required data aquisition and analysis programmes. For more details the reader is referred to the literature, e.g. the contributions in [5] by Vehanen and Rytsölä for LT, MacKencie for LT, DB, long-slit ACAR and compound techniques, and Manuel for 2D-ACAR.

3.1 Angular Correlation

The basic principles of 2D-ACAR experiments have been described in previous sections (Figure 3). Different types of 2D-sensitive detectors have been applied.

A *multi-counter* system was developed at Brandeis University [27] consisting of two detector arrays mounted on modified milling machine bases, which allowes for precise positioning by step motors. The distance between the detector planes is 20m. Each detector consists of a 2D-array of 32 NaI scintillation counters (crystal size: 5cm long, 3.8cm diameter) behind a collimating system, which defines a square grid of 3.6×3.6 mrad^2 and an angular resolution of 0.5×1.5 mrad^2. Typical coincidence counting rates are 0.2c/s per counter pair in the center of the 2D-ACAR distribution and the total coincidence rate 50c/s, obtained with a 500mCi ^{58}Co source. A full 2D-ACAR distribution takes 10-14 days of acquisition time for a peak accuracy of ±0.4%.

At Arlington, Texas, a system using position sensitive Anger cameras was first developed [28]. Each of the two commercially available Anger cameras consists of a large NaI crystal (400×400mm^2, 13mm thick) and an array of 37 phototubes. The system was used with a mono-energetic beam at the Livermore 120MeV LINAC (2×10^7 e^+/s at variable e^+-energies 740eV to 18keV) to give a total coincidence counting rate of ~5c/s [29].

Another type of 2D-sensitive detectors used are the *high-density proportional chambers* (HDPC) applied first for 2D-ACAR at the University of Geneva, see Manuel in [5]. These detectors are equipped with a multi-wire chamber sandwiched between two 1cm thick gamma-to-electron converters [30]. A multi-wire chamber containes crossed cathode and anode planes of gold-plated tungsten wires spaced at 1mm distance. The converters are stacks of thin lead sheets interleaved with Teflon foils. Drilled holes, 0.6mm diameter and spaced at 1mm, act as electron multiplier channels with a drift voltage across the lead-Teflon stack. The electron showers are detected in the crossed-wire chambers, which are filled with iso-butane gas. The advantage of this type of detectors is their high spatial resolution and small size. At present, HDPCs with a sensitve area of 300×300mm^2 and a sensitivity of 15% are used in Geneva. With a sample-to-detector distance of 7.75m and a 50mCi ^{22}Na source the total coincidence count rate is 20c/s and the angular resolution is 0.18×0.22mrad^2.

A crucial point in these experiments is always the coincidence resolution time τ, which may set a limit to the achievable counting rate due to accidental coincidences, given by $N_a = 2\tau N_1 N_2$, where N_1 and N_2 are the single-channel count rates of the two detectors. This leads to a signal-to-noise ratio $S \equiv N/N_a \propto 1/\tau I_0 \Omega$ in an ACAR experiment, where N is the true coincidence rate, Ω the solid angle sustended by the detectors and I_0 the source intensity [31].

Since HDPCs have large resolution times (200ns), scintillation counters in principal can be used at higher count rates than HDPCs in ACAR experiments. However, since in most experiments the count rate is limited by the available source strength (10 to 100mCi), this limitation for HDPCs is of minor importance and HDPCs are then

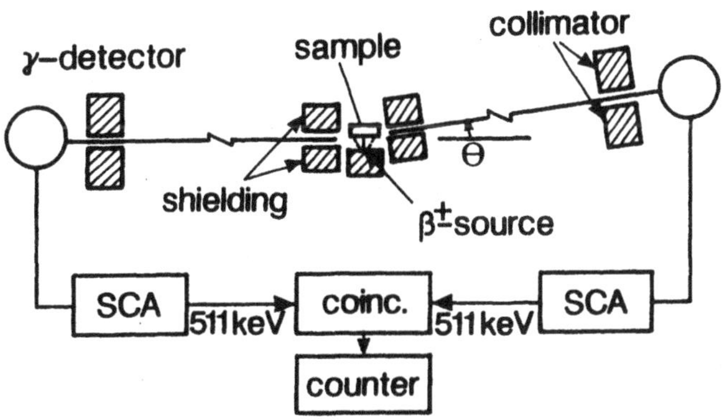

Figure 9: 1D-ACAR, long-slit geometry (SCA: single-channel analyzer).

advantageous because of their high spatial resolution.

An advantage of scintillation counters is also their good energy resolution, which is not possible with HDPCs. This has the effect of further decreasing their signal-to-noise ratio by a factor of ~8 [31]. For a more detailed comparison of the above 2D-ACAR detectors the reader is referred to [32]. An interesting new development of position sensitive photo multiplier tubes in combination with arrays of small BGO scintillators, mainly for application in positron emission tomography (PET) [33], has recently opened new possibilties also for applications in 2D-ACAR. These detectors combine high resolution in space and time ($\leq$ 3.2mm and $\leq$6ns, resp. [33]). A total coincidence counting rate of 5200c/s at a signal-to-noise ratio of 100 was predicted [31] for such a detector system (7×7cm^2 at 7.75m), provided that a high intensity positron beam would be available ($10^{10}e^+$/s).

Generally, 2D-ACAR experiments are taking up much time at least for laboratory scale experiments with limited source strength. Therefore the more conventional 1D-ACAR technique with 'long-slit' geometry is still frequently used. The basic principles of this technique are sketched in Figure 9. Typical data for such an equipment are: Pb-collimators 1mm$\times$250mm, sample-detector distance 2500mm (angular resolution 0.4mrad, ideally), NaI crystals 50mm dia. and 250mm long, resolving time 2τ=90ns, ~8c/s obtained with 1Ci ^{64}Cu at θ=0 for a Fe sample.

3.2 Doppler Broadening

As mentioned earlier, the Doppler shift for annihilation of an electron with the Fermi momentum is ~1.5keV. In a typical Doppler broadening experiment using Ge(Li) or high-purity Ge detectors (see MacKencie in [5]), the resolving power is only 1.15keV,

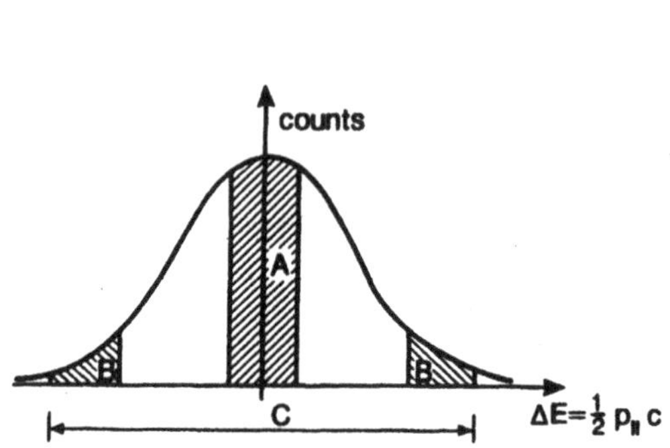

Figure 10: Doppler paramters S=A/C and W=B/C.

hence the momentum resolution in DB does not compete with ACAR. Nevertheless, a DB-experiment is often the right choice, since this method is fast, it needs only low source activities ($\sim 10\mu$Ci) and the equipment is simple and commercially available. These qualities of DB especially meet the requirements of defect spectroscopy, where the evolution of the defects with temperature and time is of great interest. Usually, an 'S-parameter' or a 'W-parameter', is defined and its change with temperature, time or deformation is then taken as the measured quantity to indicate the evolution of the defects. The parameters are defined, as shown in figure 10, by $S = A/C$ or $W = B/C$, where A, B and C are the areas under the measured Doppler line as indicated in the figure. The limits of integration for these areas are somewhat arbitrary but values $\sim$0.5 for the 'lineshape-parameter' S and $\sim$0.2 for the 'wing- parameter' W are recommended.

Although, there is no direct physical meaning of these parameters, they have been used successfully in defect spectroscopy, where they are considered as mean values with respect to the trapping probability ϕ_t of Figure 7. The S-parameter, for example, is then

$$S = (1 - \phi_t)\hat{S}_f + \phi_t\hat{S}_t \tag{70}$$

where $\hat{S}_f$ and $\hat{S}_t$ are the S-parameters for the free state ($\phi_t = 0$) and for saturation trapping ($\phi_t = 1$), respectively.

3.3 Lifetime

For this method ^{22}Na sources are commonly used. The decay scheme (Figure 11) shows that each β-decay is followed by the emission of a photon of $\sim$1.3MeV. Since the intermediate state of ^{22}Ne has a lifetime of 3ps, which is short compared to mean positron lifetimes in solids ($\geq$100ps), the 1.3MeV photon can be used as the start signal of a timing spectrometer, the stop signal is one of the 511keV annihilation photons.

In a lifetime spectrometer (Figure 12) the source ($\leq 10\mu$Ci^{22}Na) is sealed between thin Mylar or Ni foils (1μm) and sandwiched between two pieces of the sample material, which are thick enough to stop all positrons. The source-sample sandwich is

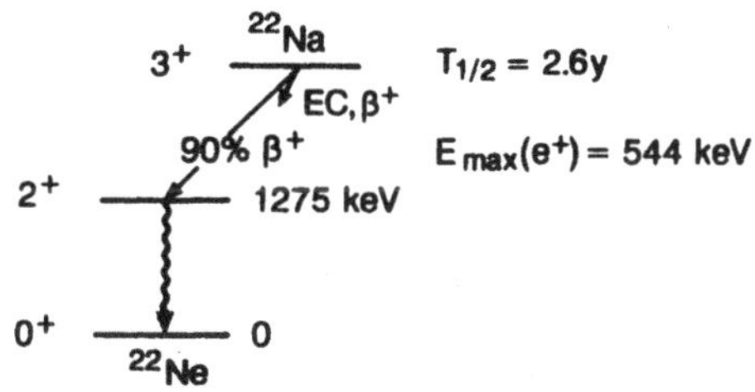

Figure 11: Decay scheme of ^{22}Na.

placed between a pair of fast scintillation counters (Vehanen and Rytsölä in [5]). The electronic circuit essentially consists of two fast timing descriminators (CF) and a time-to-amplitude converter (TAC), which is usually also gated by an additional coincidence signal, obtained from a separate slow energy discrimination circuit (not shown in the figure). The time resolution of a typical lifetime spectrometer is ~200ps. This means that deconvolution of the measured spectra is necessary and the resolution function must be known accurately. The γ-cascade of ^{60}Co is frequently used to determine this function, but known lifetime spectra of well defined sample materials have also been applied [34].

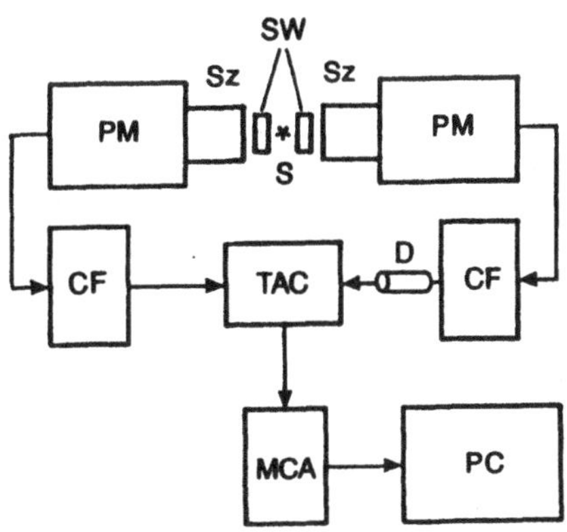

Figure 12: Lifetime spectrometer (fast circuit only), S source, SW sample sandwich, Sz scintillator, PM photo-multiplier tube, CF constant fraction discriminator, TAC time-to-amplitude converter, MCA multichannel ananlyzer, PC personal computer.

The time resolution is mainly determined by the scintillator material and size. Fast detectors are usually made of plastic scintillators (NE 111, Pilot U, KL236). To obtain reasonable efficiency, dimensions of ~1 inch are required. The usual shape is cylindrical, however truncated cone geometry was reported to be the optimum [35].

More efficient BaF_2 scintillators have also been applied [36] and a five times higher coincidence count rate was achieved. These high-density fast scintillators also show improved energy resolution as compared to plastics. Making use of the high efficiency of BaF_2, a time resolution of 140ps FWHM was obtained with scintillator dimensions 30mm diameter and 8mm thickness and with a Hamamatsu R2083Q phototube [37]. The available data analysis programs [38] are based on procedures that decompose the spectra into a finite number of lifetime components [13]. Other techniques based on inverse Laplace transforms [39] and on Fourier transforms [40] were also proposed.

4 EXPERIMENTAL RESULTS

In this chapter we give only few examples of results which were obtained mainly in three different Swiss laboratories, namely ETH/Zurich, University of Geneva and PSI/Villigen. The examples should not be considered as a review. The reader is referred to the literature for further information.

4.1 Electron Structure

In Figure 13 a 2D-ACAR distribution of potassium is shown [41], as calculated (right) from band structure based on the Korringa-Kohn-Rostocker (KKR) method (for KKR see Mijnarends in [5]) and as measured (left) by means of the 2D-ACAR technique with HDPCs described in sect. 3.1. The central part of the 2D-ACAR was cut at 8%

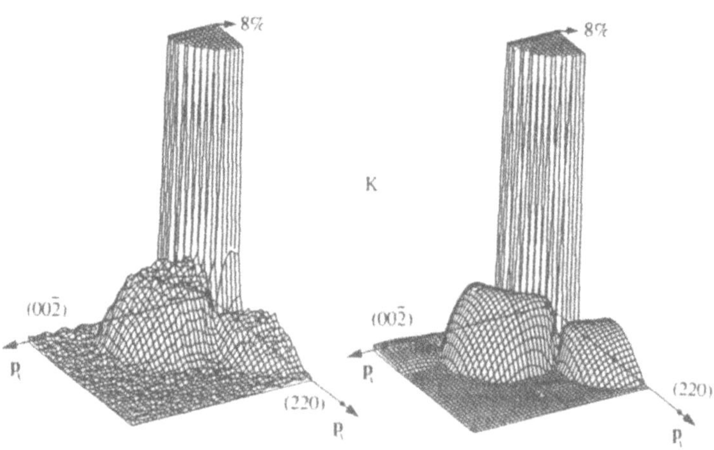

Figure 13: 2D-ACAR of mono-crystalline potassium showing high-momentum components, measured (left) and calculated (right) [41].

intensity in order to show the weak but distinct high-momentum components (Umklapp-Annihilation) which are due to modified momentum conservation in a crystal lattice, as discussed in sect. 2.3.

An impressive example for such side-peaks is given in Figure 14, which displays positronium formation in quartz [42]. This example shows that Ps in ionic solids also interacts with the lattice and a delocalized 'quasi-Ps' state is formed, which is described by a generalized Bloch wave (see Dupasquier in [5]). The Figure also indicates the excellent angular resolution of the HDPC-apparatus.

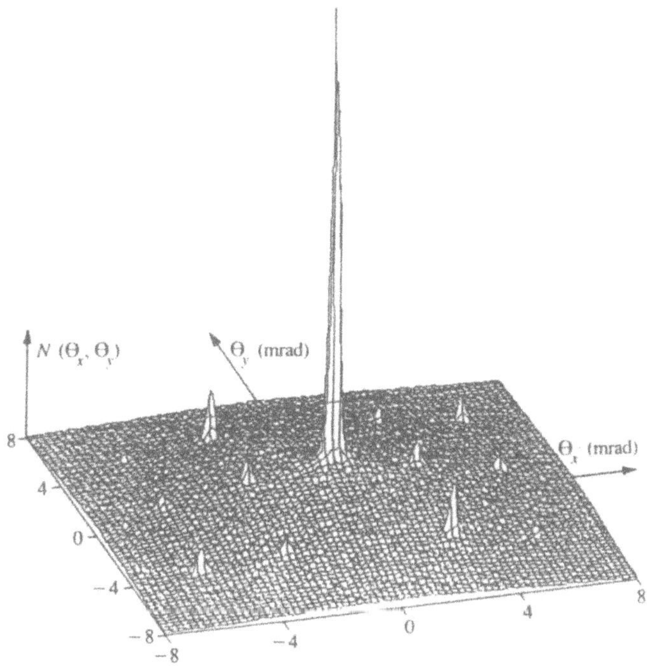

Figure 14: 2D-ACAR of mono-crystalline quartz at 77K showing high-momentum components [42].

In Figure 15, the Fermi surface of vanadium as reconstructed from 2D-ACAR and as calculated are compared [42], [43]. The agreement is remarkable, although band calculations in transition metals are more difficult than for simple metals.

An example for observation of Ps in a 1D-ACAR experiment using long-slit geometry is shown in Figure 16 [44]. The appearance of the strong para-Ps peak in

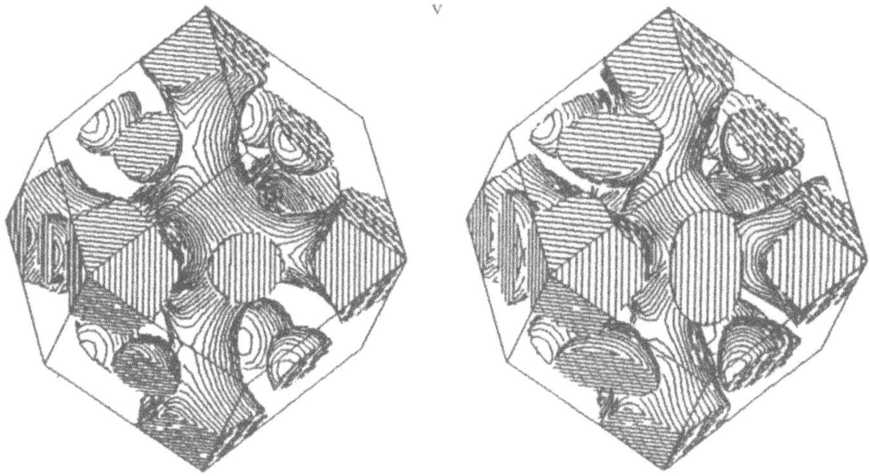

Figure 15: Fermi surface of vanadium as reconstructed from 2D-ACAR (left) and calculated (right) [42], [43].

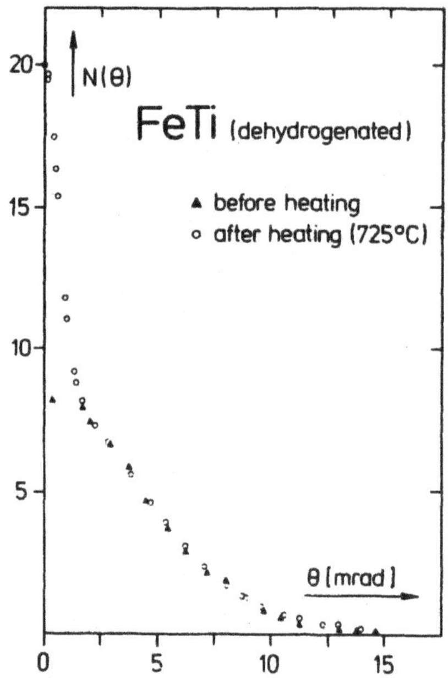

Figure 16: 1D-ACAR (long-slit) of dehydrogenated FeTi before and after annealing, normalized to equal intensities for $2 \leq \theta \leq 9$mrad [44].

dehydrogenated FeTi was explained by Ps formation on clean iron surfaces of micro-cracks which were introduced during hydrogenation. The peak appeared only after heat treatment because of trapping of the positrons in smaller defects also produced during cyclic loading with hydrogen. After annealing of these defects the positrons are free to diffuse to the surfaces of the cracks and form positronium. The appearance of the para-Ps peak on clean oxide-free iron surfaces confirmed the picture that oxygen is

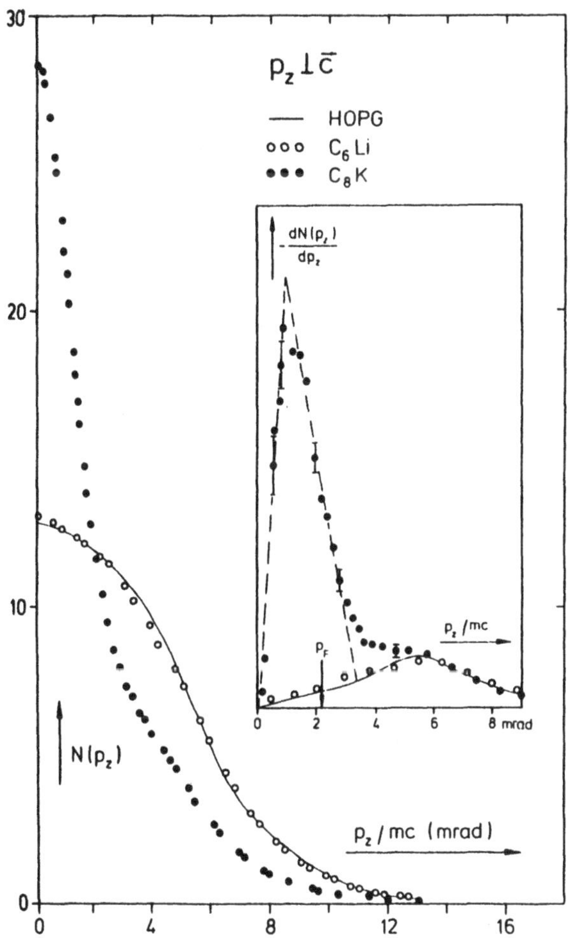

Figure 17: 1D-ACAR (long-slit) of highly oriented pyrolytic graphite HOPG and intercalated with Li and K to stage 1 (normalized to equal area) [45]. Inset: Derivatives normalized to equal values at $p_z/mc > 5$mrad.

bound by Ti and thus protects the iron surfaces from oxydation. The clean metallic Fe surfaces are also responsible for the H-absorption mechanism and the protective action of Ti in this compound, which remains effective after many cycles of hydrogenation-dehydrogenation, which is the reason that FeTi can be used as a hydrogen storage material on an industrial scale.

1D-ACAR using long-slit technique was applied for studies of graphite intercalation compounds (GIC), prepared from highly ordered pyrolytic graphite (HOPG) [45]. In GIC's ordered sequences of graphite and intercalated layers (I-layers) are formed, which may consist of alkali metals, metal halides a.o. Ordering occurs within an I-layer and between the layers as well, i.e. the I-layers are separated by a certain number n of graphite layers, which is called the stage number of the GIC. In Figure 17, the results are shown for the stage 1 GIC's C_6Li and C_8K, which differ in the number of C-atoms, 6 or 8, in the planar unit cell of the alkali atoms in the I-layer.

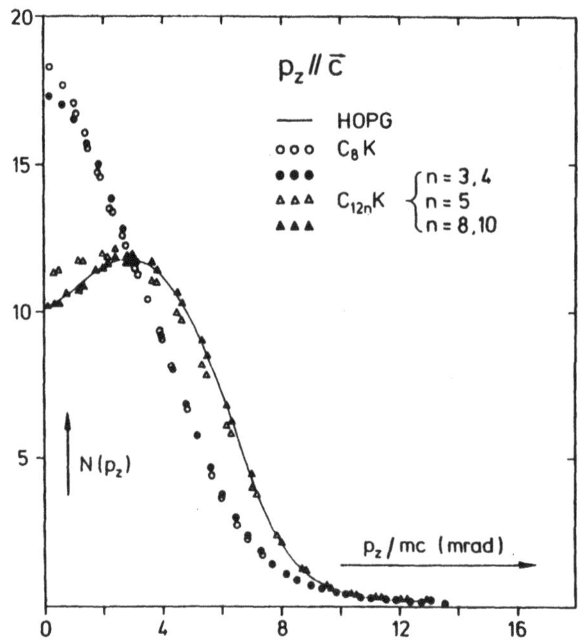

Figure 18: 1D-ACAR (long-slit) of HOPG and K-intercalated at different stages n [45]. Curves are normalized to equal area.

The ACAR for stage C_6Li is very similar to that of pure graphite and there is now evidence for a free-electron contribution. Pure graphite has been measured with ACAR as early as 1957 by Berko et al [46] and the observed anisotroy and the dominant contributions of the σ- and π-electrons are well understood (see also [47]). In sharp contrast to this, stage 1 C_8K shows an intense narrow component, which is due to a free-electron band occupied by as much as 0.7 e^-/K-atom. This was in agreement with calculated band structures which predicted that the Fermi level is below the Li 2s-band while the K 4s-band is cut by the Fermi level. Accordingly, the *donor* GIC's are classified into s(+π) band metals (C_8K) and π-band metals (C_6Li).

The free-electron contribution in C_8K is, however, very anisotropic, which follows from the broader ACAR for $p \parallel c$ shown in Figure 18. Also shown in this figure are the ACAR's for higher stages n=3-10 of the K-intercalated HOPG's. Stage 5 is still dominated by π- and σ-electrons very similar to graphite. The remarkable change in shape on going from stage 5 to stage 4 is due to Coulomb interaction which repels the positrons from the charged K-layers. Hence, for $n \geq 5$ the positrons predominantly annihilate in the carbon layers giving ACAR's very similar to pure graphite.

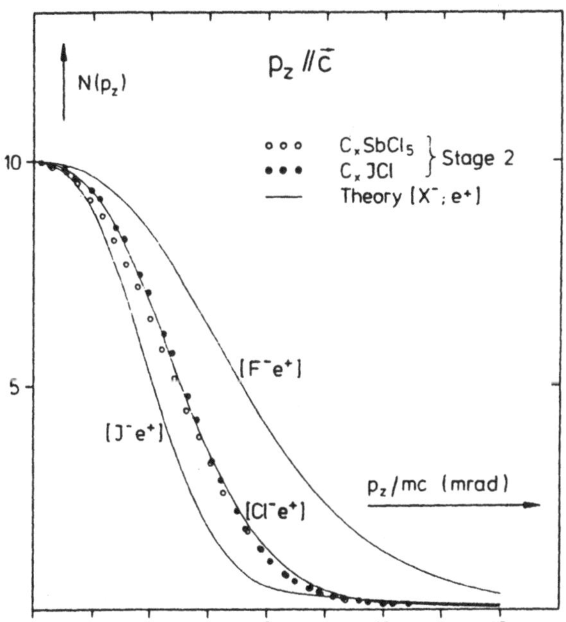

Figure 19: 1D-ACAR (long-slit) of HOPG intercalated with $SbCl_5$ and ICl at stage 2 [45], compared with theoretical distributions for the Ps halides $[I^- e^+]$, $[Cl^- e^+]$ and $[F^- e^+]$ from [48].

in a standard annihilation experiment (ACAR or lifetime), if pick-off annihilation of quasi-Ps is an important decay channel. In this case, a narrow para-Ps peak does not appear and the longest lifetime component may be as short as the low-density limit of free e^+/e^- pairs (500ps). Magnetic quenching is then an efficient tool to decide wether Ps was formed or not. An example of magnetic quenching is given in Figure 20 for Ps formation in color centers in KCl [49], [50]. A distinct narrowing in the 1D-ACAR distribution, obtained with the long-slit arrangement, was observed with a static magnetic field of 20kG. This is due to mixing of the M=0 triplet and the singlet state, and the corresponding reduction in pick-off annihilations, as dicussed in section 2.2.

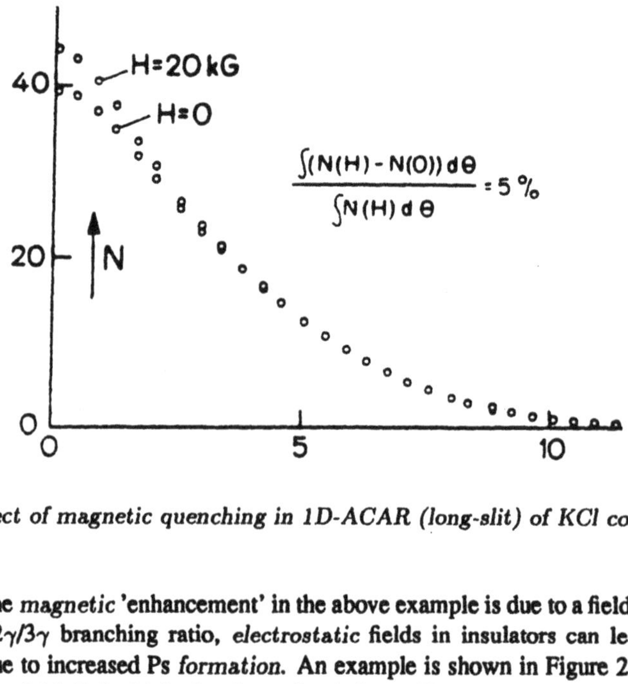

Figure 20: Effect of magnetic quenching in 1D-ACAR (long-slit) of KCl containing F-centers [50].

While the *magnetic* 'enhancement' in the above example is due to a field induced change in the $2\gamma/3\gamma$ branching ratio, *electrostatic* fields in insulators can lead to an enhancement due to increased Ps *formation*. An example is shown in Figure 21, which displays the Ps yield Φ in high density polyethylene (HDPE) as a function of the electric field E [51]. The Ps yield was obtained from decomposition of both lifetime spectra and 1D-ACAR distributions [52]. The sharp encrease in Ps formation Φ at ~0.8MV/cm is due to *heating* of the positrons in the electrostatic field E to an energy large enough to lift them into the *Ore gap* (see eq. 29 in section 2.2). This energy is ~5eV in HDPE. As is seen in the figure, nearly all positrons form Ps at fields above 1MV/cm. The behaviour of the Ps yield Φ at lower fields, showing a minimum at ~0.3MV/cm, is difficult to explain by the Ore gap model, since measurements below the glas transition temperature (-30°C) showed no such variations at low fields and in

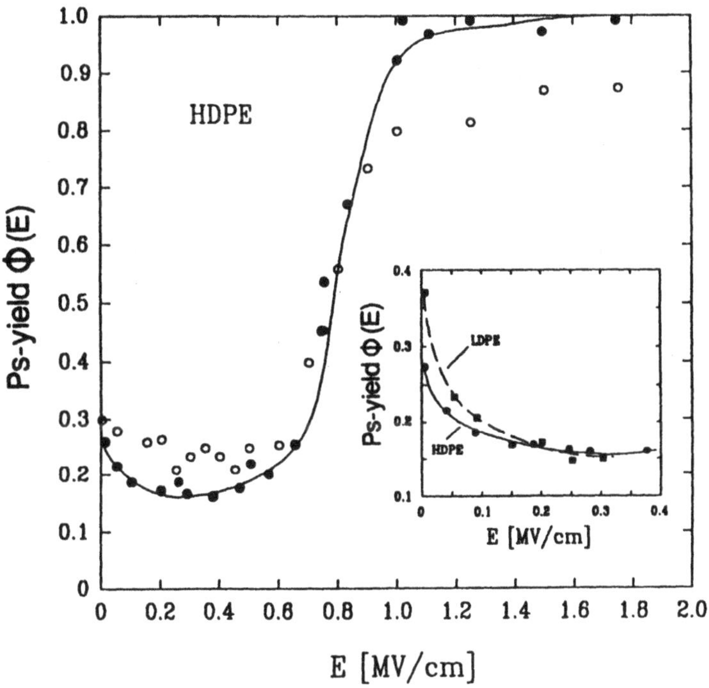

Figure 21: Effect of a static electric field E on Ps yield Φ in high density polyethylene (HDPE). Inset: low field behaviour [52].

low density PE (LDPE) the decrease of Φ is much larger than in HDPE (see inset in Figure 21). The authors therefore explain the decrease with E by Mogensen's *spur model* (see section 2.2), which predicts such a decrease with increasing E because the positrons are separated from the spur electrons under the action of the field.

4.2 Defect Studies

One of the most successful applications for positrons is certainly in defect physics. In fact, a great part of available data on the formation and diffusion of defects in metals are due to the positron method. As an example, we show in Figure 22 the effect of positron trapping in thermal vacancies in pure lead on the 1D-ACAR at zero correlation angle θ [53]. The curve was obtained using a long-slit equipment similar to that described in section 3.1 but with fixed angle θ=0 and with a broad angular resolution (3.4mrad FWHM). This allowed higher counting statistics and *in situ* variation of the temperature from ambient temperature to the melting point of Pb during 25 hours (constant heating rate 12°C/h), i.e. within two halflives of a ^{64}Cu source (12.7h). The measured quantity N is similar to the S-parameter in Figure 10 but without normalization to total area under the 1D-ACAR curve. We call N the *peak height*.

In analogy to eq. 70 we can set the peak height proportional to the trapping

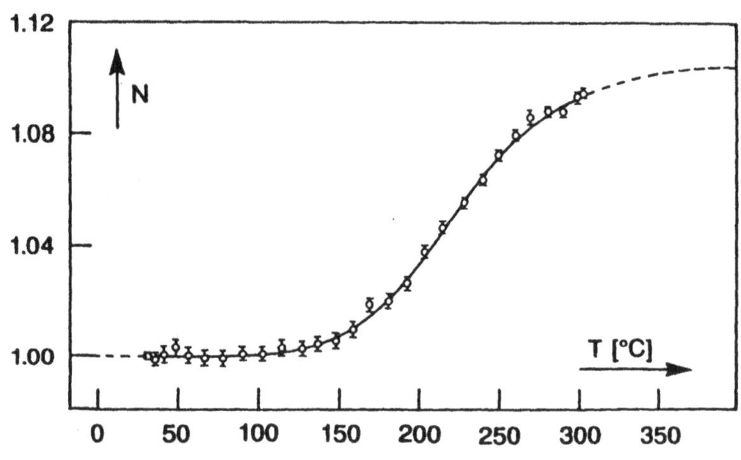

Figure 22: Peak height for pure Pb with trapping in thermal vacancies [53].

probability ϕ_t

$$N = (1 - \phi_t)\hat{N}_f + \phi_t\hat{N}_t \tag{71}$$

With the trapping probability given by $\phi_t = \sigma_v c_v/(\lambda_f + \sigma_v c_v)$ as in Figure 7 and the temperature dependence for vacancies in metals with formation enthalpy H_v given by (see Maier in [5])

$$c_v(T) = c_{\infty}e^{-H_v/k_BT} \tag{72}$$

in *thermal equilibrium*, we find the T-dependence of the peak height

$$N(T) = \frac{\hat{N}_f + \hat{N}_v A e^{-H_v/k_B T}}{1 + A e^{-H_v/k_B T}} \tag{73}$$

where $A = \sigma_v c_\infty / \lambda_f$. The parameters A, H_v, $\hat{N}_f$ and $\hat{N}_v$ in eq. 73 can be determined from measured points by least-square fit procedures. The curve plotted in Figure 22 was a result of such a fit and a vacancy formation energy H_v=0.631±0.012eV was obtained. One can, of course, also plot the result as an Arrhenius plot by forming

$$\phi \equiv \frac{N - \hat{N}_f}{\hat{N}_v - N} = A e^{-H_v/k_b T} \tag{74}$$

This was done in Figure 23 for the data of Figure 22.

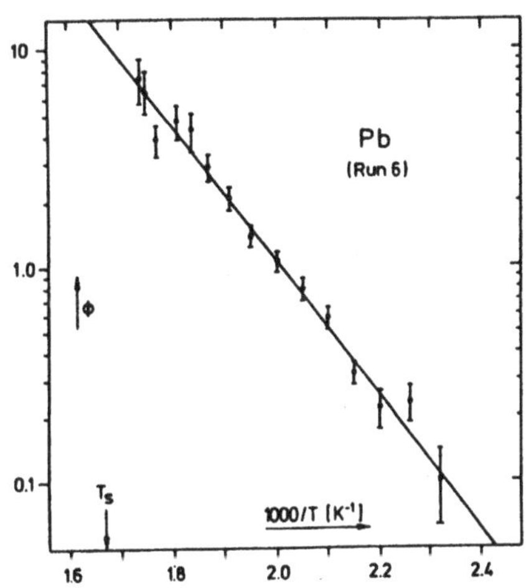

Figure 23: Arrhenius plot for Pb [53].

The study of non-equilibrium defects in metals is also a field of application of positron trapping. It has been applied to quenched-in vacancies, plastic deformation and radiation damage (see Petersen in [5]). An example of radiation induced defects is given in Figure 24 for neutron irradiated iron (neutron spectrum of a light water reactor, flux 5×10^{10}n/cm^{2}s, fluence 5×10^{18}n/cm^2, T=290^{0}C) [54]. The annealing stage observed during heating with constant heating rate 1^{0}C/minute, is due to the decay of small vacancy clusters, which was indicated in lifetime experiments by a lifetime component of τ_2=237±2ps and I_2=39±2% [55].

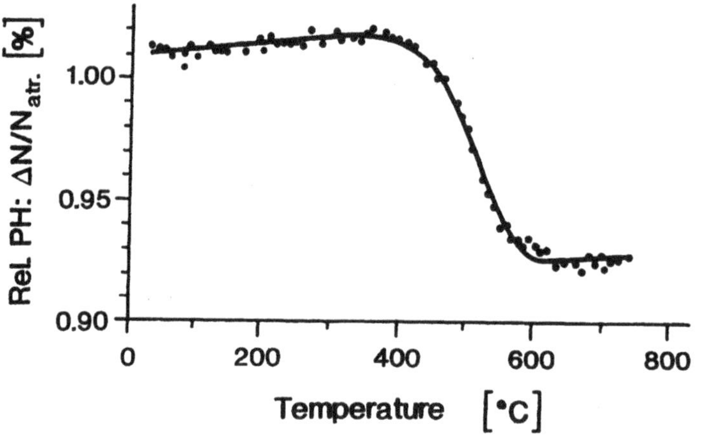

Figure 24: Peak height of neutron irradiated Fe.

As is usual in metal physics, an activation energy H is ascribed to the annealing stage at ~500°C. H can be extracted from the measured points by least-square fitting of a suitable model function as was done for Pb above. The simplest model for the defect concentration c is found by assuming first order kinetics for the annealing mechanism

$$\frac{dc}{dt} = -Kc \tag{75}$$

where K is a reaction rate with a T-dependence given by the actvation energy H

$$K = K_0 e^{-H/k_B T} \tag{76}$$

and $K_0 = \nu_D c_S$ a constant given by the Debye frequency ν_D ($\sim 10^{13} s^{-1}$) and a fixed concentration c_S of sinks, evenly distrubuted in the sample, which absorb the defects. For an isothermal annealing experiment K is constant in t and the solution of eq. 75 is simply an exponential $c = c_0 exp(-Kt)$. Since we measure with a constant heating rate $T = \alpha t$, K in our experiment is time-dependent (eq. 76) and the solution of eq. 75 is not as easy. One finds with the above assumptions

$$c(T) = c_0 e^{-K_0 T E_2(H/k_B T)/\alpha} \tag{77}$$

where

$$E_2(x) = e^{-x} - xE_1(x) \tag{78}$$

and

$$E_1(x) = \int_x^{\infty} \frac{e^{-y}}{y} dy \tag{79}$$

E_1 and E_2 are tabulated exponential integrals. If we use again the trapping probability $\phi_t = \sigma c/(\lambda_f + \sigma c)$ and eq. 71 as for thermal vacancies in Pb, we finally calculate the model function for the peak height as

$$N(T) = \frac{\hat{N}_f + \hat{N}_t A e^{-K_0 T E_2(H/k_B T)/\alpha}}{1 + A e^{-K_0 T E_2(H/k_B T)/\alpha}} \tag{80}$$

The least-square fit of this model function, eq. 80, to the measured points is plotted as a solid line in Figure 24. The value obtained for the activation energy is H=0.94±0.02eV.

5 FUTURE DEVELOPMENTS

In recent years, new ideas have been developed for applications of the positron as a probe in solids. Many of these new techniques make use of *slow* or *mono-energetic positron beams*. Although a review of the present status of these rapidly expanding activities can not be given here, a few remarks may be desirable. For more details the reader is referred to the recent reviews [12], [56], [57].

The possible applications in solids of these new beam techniques may be grouped into four categories: electronic structure, surfaces and interfaces, defect spectroscopy and micro-probe analysis. In any application using positron beams, it is certainly necessary to make use of the unique properties of the positron in order to obtain information that cannot be achieved by means of electron beams, which are easier to produce.

In *electronic structure* studies this unique information is due to the annihilation radiation which can lead to a more detailed knowledge of the electronic structure than secondary electron emission or X-ray emission induced by an electron beam, for example. The advantage of positron beams over ACAR studies using β^+-sources is due to the need of large crystals for the conventional method, and the growth of large homogeneous untwinned crystals is a delicate procedure for many compounds. The use of focussed mono-energetic positron beams with variable energy reduces the required sample volume to less than 1mm^3. A positron beam with a cross-section of ~1mm^2 and an intensity of $10^{10}e^+$/s combined with fast 2D-detectors would make possible, for example, 2D-ACAR experiments on many high-T_c superconductors which are difficult to obtain in larger crystal sizes, as well as the study of Guinier-Preston zones and epitaxial layers by means of 2D-ACAR [31].

Improved lifetime techniques based on relativistic beams (for a summary see [18]), or based on pulsed beams (see [58]) have also been proposed and even simultaneous lifetime and momentum measurements of single annihilation events (age-momentum correlation) were already carried out [59].

Various new developments for applications of positron beams in solid state physics are due to the negative work functions for positrons and positronium in many solids. This means that positrons implanted in a solid at a depth not exceeding substantially the mean diffusion length L^+ (sect. 2.4), have a certain probability to diffuse back to the surface and escape from the solid. This process of *re-emission* of positrons depends on the surface conditions as well as on the defect structure within the depth $\sim L^+$. A variable-energy slow positron beam, therefore, offers a new and independent means of defect spectroscopy with positrons in addition to the 'conventional' positron techniques described in sections 2.4 and 3. This is accomplished by measuring the re-emitted positrons or Ps by a suitable detection system or, more easily, by measuring the $2\gamma/3\gamma$ branching ratio. Since this can be done at different energies of the incident positron beam, depth profiling of defects becomes possible as well as the study of defect structures at interfaces between overlayers and a substrate. For reviews of this topic and examples see [12], [17], [60], [61], [62]. For a comparative study of Doppler S-parameter and Ps-fraction measurements see [26]. This technique has already reached some technological relevance since systems as complicated as defect-impurity interactions in compound semiconductors have been studied successfully [63].

In addition to depth profiling of defects, re-emission of positrons also offers the possibility of positron microscopy by imaging of the re-emitted positrons [64], [65], [66]. *Positron re-emission microscopy* (PRM) may be regarded as the analogue of electron field emission microscopy or photoelectron emission microscopy but with smaller transverse energy spread of the emitted particles. This is mainly due to the lower energy of re-emitted positrons ($\sim$1eV) as compared to electron emission ($\sim$10eV). A transverse energy of 0.2eV in the geometry of a field emission microscope with magnification M=10^6 will result in a resolution limit of 0.2nm.The de Broglie wavelength is $\sim$1.2nm for 1eV positrons. Hence a principal resolution limit of 1.2nm in PRM can be expected, and this is due to the de Broglie wavelength.

The *contrast mechanism* may, however, be very different for electrons and positrons. This is especially true for trapping of positrons in defects close to the surface. As was shown in sect. 2.4, a defect of trapping radius R_0 will give an image of radius $\sim 2R_0$ if diffusion limited trapping is valid. Thus a vacancy cluster in a metal with R_0=0.6nm will give an image size of 2.4nm (FWHM) due to a zone depleted in positrons by diffusion. The contrast is lower for single vacancies, since trapping is limited by transition. For defects below the surface, further smearing of the contrast is to be expected. Other contrast mechanisms with re-emitted positrons giving better contrast at higher resolution are also possible, e.g. adsorbed molecules. Nevertheless, imaging of defect clusters with modest resolution may be a valuable tool in different fields such as in ion-implanted semiconductor devices or surface segregation in alloys.

The above principal resolution limit was never reached so far, mainly because of limitations in the primary positron beam current, which was only $5\times10^5 e^+/s$ in [64] and typically 30h of accumulation time was required to achieve a resolution of $2.3\mu m$ with M=56× in a back reflection PRM arrangement.

The limited beam current in available positron beams for applications in positron microscopy and for positron microbeam experiments (e.g. with ACAR as discussed above) as well as for other applications (see [12]), is the motivation for the development of high-intensity positron beams, which are now studied in different countries [67]. It seems that currents of $10^{10} e^+/s$ can be expected in the future.

There are different methods of production and formation of such intense slow positron sources [68]. Primary sources may be based on intense radioactive isotopes or on pair production. These primary positrons are then slowed down by solid moderators in which most of the positrons are lost by annihilation, only a small amount is re-emitted from the moderator at energies of 1-2eV. The moderation efficiencies are usually less than 10^{-2} to 10^{-7} depending on the energy of the primary positrons and the moderator used.

For the formation of high-brightness beams remoderation stages are necessary which are similar to the primary moderation stages but work at lower energies of the incoming positrons. Brightness enhancement is achieved by acceleration of the slow positrons to several keV and focusing this beam to an additional moderator to a spot size as small as possible. The re-emitted positrons then have a higher brightness and may be focused to the desired size onto the sample or onto the next brightness enhancement stage. Due to the lower energy, re-emission efficiencies are of the order of 10-20%. Intense sources are now at the stage of feasibility studies at different institutes:

- JAERI-Takasaki, Japan: A LINAC based *positron factory* is under study (bremsstrahlung/pair-production) [69], [70]. With a 100kW/100MeV LINAC 10^{10} slow e^+/s seems to be the limit for this technique.

- ILL-Grenoble, France: The high-flux nuclear reactor at ILL could be used with the $^{113}Cd(n,\gamma)^{114}Cd$ nuclear reaction to produce a high flux of energetic γ-rays which will create $\sim10^{13}$ fast e^+/s by pair production. Using W moderators a slow positron flux of $10^{10}/s$ can be expected [58], [71].

- ORNL-Oak Ridge, USA: The ORNL high-flux reactor could be used for the reaction $^{78}Kr(n,\gamma)^{79}Kr$ (6.8% β^+, 35h). Krypton is led to a cold trap and with a solid Ne moderator 10^{10} slow e^+/s are expected. With the planned Advanced Neutron Source even 10^{12} slow e^+/s will be possible [72].

- PSI-Villigen, Switzerland: A new concept consisting of a thin pre-moderator and magnetic and electrostatic confinement of the positrons emitted from accelerator produced ^{18}F is expected to give a moderation efficiency of 25% and a slow positron intensity exceeding $2\times10^{10}/s$ [73].

The development of these positron 'factories' would certainly change the 'way of live' in the positron community, due to an expected transition from laboratory scale experiments to inter-laboratory collaboration, based on large and expensive facilities, as is now already usual in many other fields of physics.

Acknowledgements: I am very grateful to G. Solt for stimulating discussions and to A. Cornaz, J. Brunner and H. Grimmer for helpful suggestions. I also like to thank W. B. Waeber for continuous support.

References

[1] Akhiezer, A. J. and Berestetskii, V. B., "Quantum Electrodynamics", Interscience Publishers, New York 1965.

[2] Berestetskii, V. B., Lifshitz E. M. and Pitaevkii, L. P., "Relativistic Quantum Theory", Pergamon, Oxford 1971.

[3] West, R. N., Adv. Phys. 22, 263-383 (1973).

[4] Hautojärvi, P. (edt.), "Positrons in Solids", Topics in Current Physics 12, Springer, New York 1979.

[5] Brandt, W. and Dupasquier, A. (edts.), "Positron Solid-State Physics", North-Holland, Amsterdam 1983.

[6] Ferrell, R. A., Rev. Mod. Phys. 28, 308 (1956).

[7] Brandt, W. and Reinheimer, J., Phys. Rev. B 2, 3104 (1970).

[8] Champeney, D. C., "Fourier Transforms and their Physical Applications", Academic Press, London 1973.

[9] Rich, A., Rev. Mod. Phys. 53, 127-165 (1981).

[10] Ore, A., Univ. Bergen, Arbok, Naturvitsensk. Rekke 9 (1949).

[11] Mogensen, O., J. Chem. Phys. 60, 998 (1974).

[12] Schultz, P. J. and Lynn, K. G., Rev. Mod. Phys. 60, 701-779 (1988).

[13] Kirkegaard, P. and Eldrup, M., Comp. Phys. Comm. 3, 240 (1972).

[14] Sharma, S. C., Johnson, R. M. and Diana, L. M. in "Novel NDE Methods for Materials", Rath, B. (edt.), the Metallurgical Society of AIME 1983.

[15] Vehanen, A., Hautojärvi, P., Johansson, J., Yli-Kauppila, J. and Moser, P., Phys. Rev. B 25, 762 (1982).

[16] Busch, G. und Schade, H., "Vorlesung über Festkörperphysik", Birkhäuser, Basel 1973.

[17] Hautojärvi, P. and Huomo, H., Helvetica Physica Acta 63, 429 (1990).

[18] Seeger, A. and Banhart, F., Helvetica Physica Acta 63, 403 (1990).

[19] Frank, W. and Seeger, A., Appl. Phys. 3, 61 (1974).

[20] Puska, M. J., Corbel, C. and Nieminen, R. M., Phys. Rev. B 41, 9980 (1990).

[21] Mäkinen, J., Corbel, C., Hautojärvi, P., Moser, M. and Pierre, F., Phys. Rev. B 39, 10162 (1989).

[22] Mills jr., A. P. and Pfeiffer, L., Phys. Rev. Letters 43, 1961 (1979).

[23] Evans, R.D., "The Atomic Nucleus", McGraw-Hill, New York 1955.

[24] Valkealahti, S. and Nieminen, R. M., Appl. Phys. A 35, 51 (1984).

[25] Vehanen, A., Saarinen, K., Hautojärvi, P. and Huomo, H., Phys. Rev. B 35, 4606 (1987).

[26] Huomo, H., Soininen, E. and Vehanen, A., Appl. Phys. A 49, 647 (1989).

[27] Berko, S., Haghooie, M. and Mader, J. J., Phys. Lett. A 63, 355 (1977).

[28] West, R. N., Mayers, J. and Walters, P. A., J. Phys. E 14, 478 (1981).

[29] Howell, R. H., Fluss, M. J., Rosenberg, I. J. and Meyer, P., "Positron Annihilation", Jain, P. C., Singru, R. M. and Gopinathan, K. P. (edts.), World Scientific, Singapore, p. 801 (1985).

[30] Jeavons, A. P., Townsend, D. W., Ford, N. L., Kull, K., Manuel, A., Fischer, Ø. and Peter, M., IEEE Trans., NS- 25, 164 (1978).

[31] Manuel, A. A. and Peter, M.,Helvetica Physica Acta 63, 397 (1990).

[32] West, R. N., "Positron Annihilation", Jain, P. C., Singru, R. M. and Gopinathan, K. P. (edts.), World Scientific, Singapore, p. 11 (1985).

[33] Uchida, H., Yamashita, T., Iida, M. and Muramatsu, S., IEEE Trans. Nucl. Science 33, 464 (1986).

[34] Kirkegaard, P. and Eldrup, M., 2nd Int. Conf. Pos. Ann. (1971).

[35] Kögel, G., "Positron annihilation", Hasiguti, R. R. and Fujiwara, K. (edts.), 1979.

[36] Bauer, W., Major,J., Weiler, W., Maier, K. and Schaefer, H. E., "Positron Annihilation", Jain, P. C., Singru, R. M. and Gopinathan, K. P. (edts.), World Scientific, Singapore, p. 805 (1985).

[37] de Vries, J., Zecca, A., Brusa, R. S., Grisenti, R. G., Oss, S. and van Veen, A., "Positron Annihilation", Dorikens-Vanpreat, L., Dorikens, M. and Seegers, D. (edts.), World Scientific, Singapore, p. 654 (1989).

[38] Kirkegard, P., Pedersen, N. J. and Eldrup, M., Ibid, p. 642 (1989).

[39] Schrader, D. M., "Positron Annihilation", Coleman, P. G., Sharma, S. C. and Diana, L. M. (edts.), North-Holland, Amsterdam, p. 912 (1982).

[40] Yuan-Jin He and Bi-Song Cao, "Positron Annihilation", Jain, P. C., Singru, R. M. and Gopinathan, K. P. (edts.), World Scientific, Singapore, p. 813 (1985).

[41] Oberli, L., Manuel, A. A., Sachot, R., Descouts, P., Peter, M., Rabou, L. P. L. M., Mijnarends, P. E., Hyodo, T. and Stewart, A. T., Phys. Rev. B 30, 1147 (1985).

[42] Peter., M., IBM J. Res. Develop. 33, 333 (1989).

[43] Jarlborg, T. and Arbman, G., J. Phys. F 6, 189 (1976).

[44] Zimmermann, U., Stucki, F. and Heinrich, F., Physics Letters 74A, 346 (1979).

[45] Cartier, E., Heinrich, F., Pfluger, P. and Güntherodt, H.-J., "Positron Annihilation", Coleman, P. G., Sharma, S. C. and Diana, L. M. (edts.), North-Holland, Amsterdam, p. 607 (1982).

[46] Berko, S., Kelly, R. E. and Plaskett, J. S., Phys. Rev. 106, 824 (1957).

[47] Colombino, P., Fiscella, B. and Trossi, L., Nuovo Cimento 27, 589 (1963).

[48] Farazdel, A. and Cade, P. E., J. Chem. Phys. 66, 2612 (1977).

[49] Herlach, D. and Heinrich, F., Helvetica Physica Acta 45, 10 (1972).

[50] Herlach, D., Helvetica Physica Acta 45, 894 (1972).

[51] Cornaz, A., Brunner, J. and Cartier, E., "Positron Annihilation", Dorikens-Vanpreat, L., Dorikens, M. and Seegers, D. (edts.), World Scientific, Singapore, p. 793 (1989).

[52] Cornaz, A., PhD Thesis work, unpublished.

[53] Heinrich, F. and Zimmermann, U., Helvetica Physica Acta 52, 200 (1979).

[54] Ghazi-Wakili, K., Tipping, Ph., Zimmermann, U. and Waeber, W. B., Z. Phys. B, Condensed Matter 79, 39 (1990).

[55] Zimmermann, U., Saarinen, K. and Hautojärvi, P., unpublished.

[56] Waeber, W. B. (edt.), Proc. PSI Workshop on Intense Slow Positron Beams and Applications in Condensed Matter Physics, Helvetica Physica Acta 63, 377-470 (1990).

[57] Schultz, P. (edt.), 4th International Workshop on Positron Beam Techniques for Solids and Surfaces, July 3-6, 1990, London/Ontario (Canada), to be published.

[58] Triftshäuser, W. and Kögel, G., Helvetica Physica Acta 63, 378 (1990).

[59] Kishimoto, Y., Tanigawa, S., Morinaga, H. and Matsuoka, Y., "Positron Annihilation", Coleman, P. G., Sharma, S. C. and Diana, L. M. (edts.), North-Holland, Amsterdam, p. 168 (1982).

[60] Lynn, K. G., Helvetica Physica Acta 63, 389 (1990).

[61] Vehanen, A., Hyperfine Interactions 45, 179 (1989).

[62] Saarinen, K., Hautojärvi, P., Vehanen, A., Krause, R. and Dlubek, G., Phys. Rev. 39, 5287 (1989).

[63] Saarinen, K., Hautojärvi, P., Keinonen, J., Rauhala, E., Räisänen, J. and Corbel, C., Phys. Rev. B 43, 4249 (1991).

[64] Van House, J. and Rich, A., Phys. Rev. Letters 61, 488 (1988).

[65] Brandes, G. R., Canter, K. F. and Mills Jr., A. P., Phys. Rev. Letters 61, 492 (1988).

[66] B. Goss Levi, Physics Today 12, 17 (1988).

[67] Peter, M. and Manuel, A. A., Helvetica Physica Acta 63, 458 (1990).

[68] Zimmermann, U., Helvetica Physica Acta 63, 435 (1990).

[69] Tanigawa, S., Helvetica Physica Acta 63, 385 (1990).

[70] Tabata, Y. and Okada, S., "Positron Annihilation", Dorikens-Vanpreat, L., Dorikens, M. and Seegers, D. (edts.), World Scientific, Singapore, p. 34 (1989).

[71] Krusche, B. and Schreckenbach, K., Nucl. Instr. Meth. Phys. Res. A 295, 155 (1990).

[72] Okada, S., Proc. of the 2nd Workshop on Researchers Using Positrons, Tokyo 28/2-1/3 1991, to be published.

[73] Taqqu, D., Helvetica Physica Acta 63, 442 (1990).

PERTURBED ANGULAR CORRELATIONS
A TOOL FOR STUDYING POINT DEFECTS IN METALS

F. Pleiter
Laboratorium voor Algemene Natuurkunde, Materials Science Centre
University of Groningen, The Netherlands

ABSTRACT. In this paper we give a brief description of the Perturbed Angular Correlation (PAC) technique, and its application to the study of elementary or point defects in metals. The theoretical part is followed by the presentation of some recent experimental results on (i) cubic defects in nickel, (ii) hydrogen-vacancy interaction in tungsten, and (iii) trap mutation in ion-implanted helium-doped tungsten.

1. INTRODUCTION

Metals contain elementary or point defects such as vacancies and interstitials. Under suitable conditions these combine to form larger defects such as dislocations, loops, and clusters consisting of numerous elementary defects. The study of elementary defects, their production, migration, agglomeration, and trapping at sinks has been the object of extensive experimental and theoretical work. The study of larger defects has been of interest wherever technologists have been concerned with work-hardened, temperature-shocked, or irradiated metals.

To a considerable extent measurements on point defects have been made by observing thermal annealing of residual resistivity. That method works because it is sensitive to the aggregate concentration of vacancies and interstitials. It is attractive because it is widely applicable and relatively simple to carry out. Through a large number of systematic experiments it has produced a model of resistivity recovery that defines a common language for all point-defect research.

Many of the obtained results, however, depend as much on the interpretation of the experimental facts as on the facts themselves. A particular problem arises out of the fact that in resistivity-recovery experiments and other similar techniques, the behaviour of defects is defined solely in terms of the recovery phenomenon, *i.e.*, there is no specific information which characterizes the structure of the defect responsible for a given annealing stage. To overcome this problem, a number of approaches have been developed which 'flag' a particular defect structure via a characteristic signal or 'fingerprint'. With some notable exceptions these fingerprints do not solve the problem of structure, but permit discrimination between structures. In analogy to the view from Plato's cave, they give the events outside the cave a distinctive shape, but are generally unable to lead to a full description without the intervention of extensive theorizing.

One of these approaches is based on the observation of defect-induced hyperfine interactions by means of the Perturbed Angular Correlation (PAC) technique. This paper describes the PAC technique and its application to the study of point defects in metals, and is organized as follows. A semiclassical description of the technique in sect. 2 is followed by a more fundamental quantum-mechanical treatment in sect. 3. The discussion is restricted to the case of static hyperfine interactions. For completeness, however, we give in the Appendix a brief introduction to stochastic interactions, and especially to Blume's theory of randomly reorienting fields. The application of the PAC technique is illustrated in sect. 4 by presenting some recent experimental results on (i) cubic defects in nickel, (ii) hydrogen-vacancy interaction in tungsten, and (iii) trap mutation in ion-implanted helium-doped tungsten.

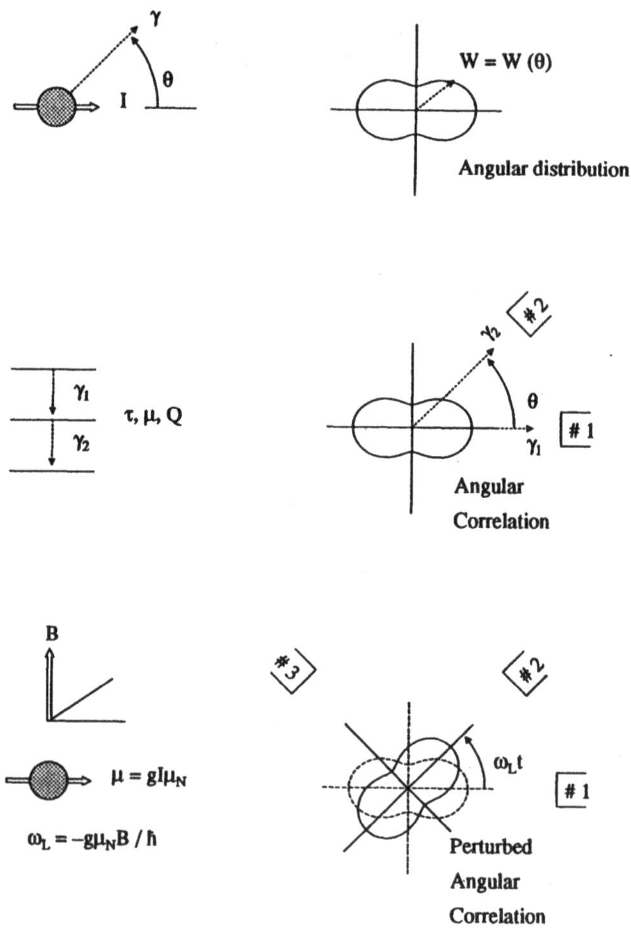

Fig. 1. See text for explanation.

2. SEMICLASSICAL DESCRIPTION OF PAC

When an excited atomic nucleus decays by γ-ray emission, the emission probability will have an anisotropic distribution with respect to the direction of the nuclear spin. The directional distribution can be expressed as a series of Legendre polynomals

$$W(\theta) = 1 + A_2P_2(\cos\theta) + ... \qquad (1)$$

where the ellipsis indicates higher-order terms, θ is the angle between the nuclear spin direction and the direction of γ-ray emission, and A_2 is the anisotropy coefficient. In the example shown at the top of fig. 1, the γ-quanta are most likely emitted into a direction parallel to the nuclear spin. We call this a positive anisotropy.

When measuring on a radioactive sample, however, one normally does not observe this anisotropy because the spins of the nuclei in the sample will be randomly oriented in space. To overcome this, one may align the nuclear spins at a low temperature in a strong magnetic field (nuclear orientation) or, alternately, one may select a subensemble of nuclei with a preferential spin direction, in the following way. Suppose that our nucleus emits two successive γ-quanta, γ_1 and γ_2 (see fig. 1, middle). If the directional distribution for γ_1 is similar to the pattern shown in the upper part of fig. 1, the detector will selectively register γ_1-quanta that were emitted by nuclei with spins pointing towards the detector. Since the emission probability of γ_2 is anisotropic with respect to the nuclear spin direction, it will also be anisotropic with respect to the direction of observation of γ_1. The directional distribution of γ_2 detected in coincidence with γ_1 will, therefore, have the angular dependence given by eq. (1). We call this an angular correlation.

It is easily recognized that the angular correlation is altered in the presence of a magnetic field. We will assume that the nucleus possesses a magnetic moment $\mu = gI\mu_N$, and that the magnetic field is perpendicular to the selected spin direction and, thus, to the magnetic moment (see fig. 1, bottom). Under influence of the magnetic interaction the latter will perform a Larmor precession with frequency $\omega_L = -gB\mu_N/\hbar$. Since the angular correlation pattern is fixed to the selected spin direction, the pattern will rotate with the same frequency. Unless the lifetime τ of the intermediate nuclear level is such that $\omega_L\tau << 1$, this rotation is observable. We say that the angular correlation is perturbed. Because the angular correlation pattern consists of two symmetric lobes, the count rate of γ_2 detected in coincidence with γ_1 will be modulated with the double Larmor frequency. Note that the modulations observed with the detector pairs #1–#2 and #1–#3 have opposite phases.

At the top of fig. 2 we show the result obtained in the simple experiment that we have described in the preceding paragraph. For this measurement we used a nickel foil in which we had implanted radioactive ^{111}In decaying by electron capture to ^{111}Cd, with a half-life of 2.8 d. The coincidence count rates measured with detector pairs positioned at relative angles of 180° and 90°, respectively, are plotted *vs.* the time lag between detection of γ_1 (171 keV) and detection of γ_2 (245 keV). The slopes of the curves correspond with the lifetime, $\tau = 123$ ns, of the intermediate nuclear level. The modulation of the count rate is caused by the precession of the nuclear magnetic moment, $\mu = -0.765\mu_N$, in the local magnetic hyperfine field of nickel, $B_{hf} = -7.1$ T. The modulation of the curves have opposite phases, indeed. The coincidence count rates, N(180) and N(90), are commonly combined to form the following ratio

$$R(t) = \frac{N(180) - N(90)}{N(180) + N(90)} \qquad (2)$$

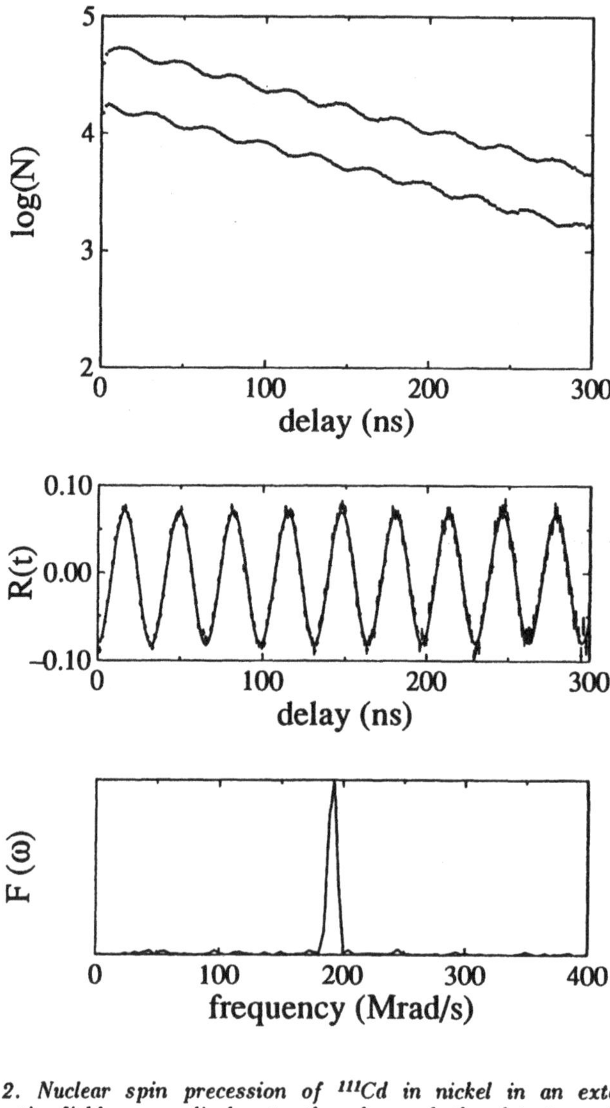

Fig. 2. Nuclear spin precession of ^{111}Cd in nickel in an external magnetic field perpendicular to the plane of the detectors. Top: coincidence count rates measured with detectors positioned at relative angles of 180° and 90°, respectively. The magnetic perturbation is visible as a modulation superimposed on the nuclear decay curves. Middle: count rate ratio $R(t)$ defined by eq. (2). Bottom: Fourier transform of $R(t)$.

By this procedure the nuclear decay is factored out and the modulation is enhanced (see fig. 2, middle). The Larmor frequency ω_L may be obtained either by fitting a cosine function to the experimental data, or by calculating the Fourier transform of R(t) (see fig. 2, bottom).

More information about data reduction can be found in Refs. [1,2].

3. QUANTUM MECHANICAL DESCRIPTION OF PAC

3.1. The Quantum Toolbox

In this section we recall a number of concepts that ought to be well known from the quantum mechanics text books, and which will prove to be extremely useful in the following discussion of the perturbation factor. In our presentation we will freely draw on the review article published by Fano [3].

We shall introduce a quantum-mechanical operator O and a complete set of eigenvectors $|i\rangle$. Any pure-state wave function $|\psi\rangle$ may be expanded into this set of eigenvectors

$$|\psi\rangle = \sum_i c_i|i\rangle = \sum_i \langle i|\psi\rangle|i\rangle \tag{3}$$

The expectation value of O for this state is

$$\langle O\rangle = \langle\psi|O|\psi\rangle = \sum_{ij} c_i^* c_j \langle i|O|j\rangle \tag{4}$$

where the coefficients $c_i^* c_j$ depend only on the pure state $|\psi\rangle$, and the matrix elements $\langle i|O|j\rangle$ depend solely on the operator O and the chosen set of eigenvectors. Defining the projection operator $P \equiv |\psi\rangle\langle\psi|$ with matrix elements $\langle i|P|j\rangle = c_i c_j^*$, we may rewrite eq. (4) as

$$\langle O\rangle = \sum_{ij} \langle j|P|i\rangle\langle i|O|j\rangle = \mathrm{Tr}(PO) = \mathrm{Tr}(OP) \tag{5}$$

Often we have to deal with an ensemble of particles in different states. While the coefficients $c_i^* c_j$ will vary from particle to particle, the matrix elements $\langle i|O|j\rangle$ are common. If we use a bar to denote the ensemble average, we may write

$$\overline{\langle O\rangle} = \sum_{ij} \overline{c_i^* c_j}\langle i|O|j\rangle = \mathrm{Tr}(\rho O) = \mathrm{Tr}(O\rho) \tag{6}$$

where we have introduced the density operator $\rho \equiv \overline{P}$. It is easy to show that ρ is a Hermitian operator with non-negative diagonal elements ρ_{ij} which represent the probability of finding the eigenstate $|i\rangle$. Since we assumed the eigenvectors to form a complete set, we have $\mathrm{Tr}(\rho) = 1$.

We shall now introduce a Hamiltonian $\mathcal{H}$ that changes the state $|\psi\rangle$ into $\mathcal{H}|\psi\rangle$. It follows from the above definitions that the density operator ρ transforms according to

$$\rho' \equiv \overline{|\psi'\rangle\langle\psi'|} \equiv \mathcal{H}\overline{|\psi\rangle\langle\psi|}\mathcal{H}^\dagger = \mathcal{H}\rho\mathcal{H}^\dagger \tag{7}$$

The state vector $|\psi\rangle$ will develop in time under influence of the Hamiltonian $\mathcal{H}$, and so will the density operator ρ. Since the eigenvectors $|i\rangle$ are independent of time, the coefficients c_i must carry the time dependence. Starting from the time-dependent Schrödinger equation, one finds the following equation of motion

$$\frac{\partial}{\partial t}\rho = \frac{i}{\hbar}[\rho,\mathcal{H}] \tag{8}$$

with the formal solution

$$\rho(t) = \Lambda(t)\rho(o)\Lambda^{\dagger}(t) \tag{9}$$

$$\Lambda(t) = \exp(-i\mathcal{H}t/\hbar) \tag{10}$$

provided that the Hamiltonian $\mathcal{H}$ does not explicitely depend on time. We call $\Lambda(t)$ the evolution operator.

Fano [3] has shown that each operator can be expanded into a properly chosen orthogonal set of base operators U_i, with the following properties

$$Tr(U_iU_j^{\dagger}) = \delta_{ij} \tag{11}$$

In analogy with eq. (3) one may write

$$O = \sum_i Tr(U_i^{\dagger}O)U_i \tag{12}$$

This leads immediately to the important relation

$$Tr(OO') = \sum_i Tr\left\{ Tr(U_i^{\dagger}O)\, U_iO' \right\} = \sum_i Tr(U_i^{\dagger}O)Tr(U_iO') \tag{14}$$

In words: the trace of the product of two operators can be expressed as a sum of products of two separate traces, each containing one of the operators.

The base that we will choose in the following discussion of the perturbation factor is set up by spherical tensors, which transform under space rotations like spherical harmonics and are defined through their matrix elements

$$\langle I'm'|U_\lambda^\mu|Im\rangle = (-)^{I'+m'}\sqrt{2\lambda+1}\begin{pmatrix} I' & I & \lambda \\ -m' & m & \mu \end{pmatrix} \tag{15}$$

We may expand the density operator itself into this base, yielding

$$\rho = \sum_{\lambda\mu} \rho_\lambda^{\mu\dagger}U_\lambda^\mu \tag{16}$$

In the literature the expansion coefficients ρ_λ^μ are often called "statistical tensors"

3.2 The Perturbation Factor

We now turn back to the perturbation factor. The interested reader is referred to the rather complete overviews on this subject presented by Frauenfelder and Steffen [4] and by Steffen and Alder [5].

We shall consider an ensemble of nuclei, all of them being in the initial state with spin I_i, which is described by the density operator ρ_i (see fig. 3). For a radioactive sample, ρ will be proportional to the unity operator. The Hamiltonians $\mathcal{H}_1$ and $\mathcal{H}_2$ shall induce transitions from the initial state via an intermediate level with spin I and lifetime τ to the ground state with spin I_f. These transitions may be any of γ-ray emission, internal conversion, β-decay, particle reaction, *etc.* The density operator in the intermediate state, $\rho(t)$, shall change with time due to a static hyperfine interaction $\mathcal{K}$ until, at time t, the transition to the ground state takes place. Using our quantum toolbox, we may write the final-state density operator as

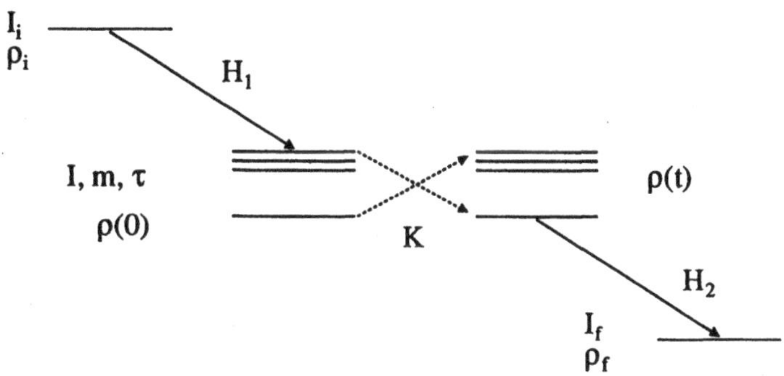

Fig. 3. Schematic nuclear transition scheme. The transitions from the initial level I_i to the intermediate level I and from the intermediate level to the ground state level I_f are described by the Hamiltonians $\mathcal{H}_1$ and $\mathcal{H}_2$, respectively. In the intermediate state the m-sublevels are continuously repopulated due to a hyperfine interaction $\mathcal{K}$.

$$\rho_f = \mathcal{H}_2\rho(t)\mathcal{H}_2^\dagger = \mathcal{H}_2\Lambda(t)\rho(0)\Lambda^\dagger(t)\mathcal{H}_2^\dagger = \mathcal{H}_2\Lambda(t)\mathcal{H}_1\rho_1\mathcal{H}_1^\dagger\Lambda^\dagger(t)\mathcal{H}_2^\dagger$$

The angular correlation function W(1,2;t) gives us the probability that after the first transition $\mathcal{H}_1$, after the hyperfine interaction $\mathcal{K}$ during a period of time t, and after the second transition $\mathcal{H}_2$, we will find the system in any of the possible final states. Using our quantum toolbox once more, we may completely separate the Hamiltonians $\mathcal{H}_1$, $\mathcal{H}_2$ and $\mathcal{K}$

$$\begin{aligned}
W(1,2;t) &= \mathrm{Tr}(\rho_f) \\
&= \mathrm{Tr}\left(\mathcal{H}_2\Lambda(t)\mathcal{H}_1\rho_1\mathcal{H}_1^\dagger\Lambda^\dagger(t)\mathcal{H}_2^\dagger \right) = \mathrm{Tr}\left(\mathcal{H}_1\rho_1\mathcal{H}_1^\dagger\ \Lambda^\dagger(t)\mathcal{H}_2^\dagger\mathcal{H}_2\Lambda(t) \right) \\
&= \sum_{\lambda\mu} \mathrm{Tr}(U_\lambda^{\mu\dagger}\mathcal{H}_1\rho_1\mathcal{H}_1^\dagger)\ \mathrm{Tr}\left(U_\lambda^\mu\Lambda^\dagger(t)\mathcal{H}_2^\dagger\mathcal{H}_2\Lambda(t) \right) \\
&= \sum_{\lambda\lambda'\mu\mu'} \mathrm{Tr}(U_\lambda^{\mu\dagger}\mathcal{H}_1\rho_1\mathcal{H}_1^\dagger)\ \mathrm{Tr}(U_{\lambda'}^{\mu'}\mathcal{H}_2^\dagger\mathcal{H}_2)\ \mathrm{Tr}(U_{\lambda'}^{\mu'\dagger}\Lambda(t)U_\lambda^\mu\Lambda^\dagger(t)) \\
&= \sum_{\lambda\lambda'\mu\mu'} B_\lambda^{\mu\dagger}(1)\ A_{\lambda'}^{\mu'}(2)\ G_{\lambda\lambda'}^{\mu\mu'}(t)
\end{aligned}$$

The factors $B_\lambda^{\mu\dagger}$, $A_{\lambda'}^{\mu'}$ and $G_{\lambda\lambda'}^{\mu\mu'}$ are calculated in different coordinate systems. For $G_{\lambda\lambda'}^{\mu\mu'}$ this is usually the system in which $\mathcal{K}$ is diagonal or can easily be diagonalized. A_λ^μ and B_λ^μ are conveniently calculated if the z-axis of the reference system coincides with the direction of the observed radiation, after which a proper rotation of the coordinate system must be carried out. The final result is

$$W(1,2;t) = \sum_{\lambda\lambda'\mu\mu'} B_\lambda A_{\lambda'} G^{\mu\mu'}_{\lambda\lambda'}(t) Y^{\mu *}_{\lambda}(1) Y^{\mu'}_{\lambda'}(2) [(2\lambda+1)(2\lambda'+1)]^{-\frac{1}{2}} \tag{17}$$

$$G^{\mu\mu'}_{\lambda\lambda'}(t) = \mathrm{Tr}(U^{\mu'\dagger}_{\lambda'} \Lambda(t) U^{\mu}_{\lambda} \Lambda^\dagger(t)) \tag{18}$$

The anisotropy coefficients B_λ and A_λ are real and depend on λ but not on μ. They are tabulated in Ref. [6]. The perturbation factor $G^{\mu\mu'}_{\lambda\lambda'}$ contains all information about the hyperfine interaction. The spherical harmonics Y^{μ}_{λ} describe the orientation of the detectors with respect to the principal axes system of the Hamiltonian $\mathcal{K}$. For a $\gamma\gamma$ angular correlation the indices must fulfil the selection rules λ = even, $|\mu| \le \lambda$ and $0 \le \lambda \le \min(2I, 2L_1, 2L_2)$, where I is the spin value of the intermediate nuclear state and L_1 and L_2 are the multipole orders of the γ–ray transitions.

In the special case that the perturbing hyperfine interactions are invariant under a three–dimensional space rotation (*e.g.* powder sources in the absence of an externally applied field), the expression for the perturbation factor reduces to (see Refs. [4,5])

$$G_{\lambda\lambda}(t) = \frac{1}{2\lambda+1} \sum_{\mu} G^{\mu\mu}_{\lambda\lambda}(t) \tag{19}$$

$$W(1,2;t) = \sum_{\lambda} B_\lambda A_\lambda G_{\lambda\lambda}(t) \left(\sum_{\mu} Y^{\mu *}_{\lambda}(1) Y^{\mu}_{\lambda}(2) \right) (2\lambda+1)^{-1}$$

$$= (4\pi)^{-1} \sum_{\lambda} B_\lambda A_\lambda G_{\lambda\lambda}(t) P_\lambda(\cos \Theta) \tag{20}$$

where Θ is the angle between the directions (1) and (2). In the last step we have used the well-known addition theorem for spherical harmonics.

The more general case of an angular correlation in the presence of a texturized hyperfine interaction has been formulated by Rots *et al.* [7,8].

3.3. Examples

Zero hyperfine field.

We have $\mathcal{K} = 0$ and $\Lambda = \mathbf{1}$. Substitution in eq. (18) yields

$$G^{\mu\mu'}_{\lambda\lambda'}(t) = \mathrm{Tr}(U^{\mu'\dagger}_{\lambda'} U^{\mu}_{\lambda}) = \delta_{\lambda\lambda'} \delta_{\mu\mu'}$$

With this, eqs. (17) and (20) reduce to the unperturbed angular correlation function.

Decoupling field.

In this case we have an axial field that is aligned parallel to one of the detectors. If the field points towards detector #1, then $Y^{\mu}_{\lambda}(1) = 0$ for $\mu \neq 0$, and the sum in eq. (17) is restricted to terms with $\mu = 0$. Using eq. (15), we obtain

$$G^{0\mu'}_{\lambda\lambda'}(t) = \sum_{mm'm''} (-)^{I+m'} (2\lambda'+1)^{\frac{1}{2}} \begin{pmatrix} I & I & \lambda' \\ -m' & m & \mu' \end{pmatrix} \Lambda_{m'm''} (-)^{I+m''} (2\lambda+1)^{\frac{1}{2}} \begin{pmatrix} I & I & \lambda \\ -m'' & m'' & 0 \end{pmatrix} \Lambda^\dagger_{m'm}$$

Remember that in the case of an axial field $\mathcal{K}$ as well as Λ is diagonal. Putting $m' = m'' = m$ and $\Lambda^\dagger_{mm} = \Lambda^{-1}_{mm}$, and using the orthogonality properties of the 3j–symbols, one obtains

$$G^{0\mu'}_{\lambda\lambda'}(t) = [(2\lambda'+1)(2\lambda+1)]^{\frac{1}{2}} \sum_{m} \begin{pmatrix} I & I & \lambda' \\ -m & m & \mu' \end{pmatrix} \begin{pmatrix} I & I & \lambda \\ -m & m & 0 \end{pmatrix} = \delta_{\lambda\lambda'} \delta_{0\mu'}$$

$$W(1,2;t) = \sum_{\lambda} B_\lambda A_\lambda Y_\lambda^{0*}(1) Y_\lambda^0(2)(2\lambda+1)^{-1}$$

$$= \sum_{\lambda} B_\lambda A_\lambda \left(\sum_{\mu} Y_\lambda^{\mu*}(1) Y_\lambda^{\mu}(2) \right) (2\lambda+1)^{-1}$$

$$= (4\pi)^{-1} \sum_{\lambda} B_\lambda A_\lambda P_\lambda(\cos\Theta)$$

Again, one observes the unperturbed angular correlation.

Perpendicular magnetic field.

We have $\mathcal{K} = -\mu B$, $\Lambda_{\mu\mu'} = \delta_{\mu\mu'}\exp(-i\mu\omega_L t)$ with $\omega_L = -gB\mu_N/\hbar$. If the radiations are emitted at an azimuthal angle of 90°, only spherical harmonics with even values of μ and μ' contribute to the sum in eq. (17)

$$G_{\lambda\lambda'}^{\mu\mu'}(t) = \sum_{mm'} (-)^{I+m'}(2\lambda'+1)^{\frac{1}{2}} \begin{pmatrix} I & I & \lambda' \\ -m' & m & \mu' \end{pmatrix} e^{-im'\omega_L t} \times$$

$$\times (-)^{I+m'}(2\lambda+1)^{\frac{1}{2}} \begin{pmatrix} I & I & \lambda \\ -m' & m & \mu \end{pmatrix} e^{+im\omega_L t}$$

$$= \delta_{\mu\mu'} e^{-i\mu\omega_L t}(2\lambda+1) \sum_{mm'} \begin{pmatrix} I & I & \lambda' \\ -m' & m & \mu' \end{pmatrix} \begin{pmatrix} I & I & \lambda \\ -m' & m & \mu \end{pmatrix}$$

$$= \delta_{\lambda\lambda'}\delta_{\mu\mu'} e^{-i\mu\omega_L t} \qquad (21)$$

In the last step we used again the orthogonality properties of the 3j–symbols. Substituting this result and the spherical harmonics

$$4\pi Y_0^{0*}(1)Y_0^0(2) = 1$$

$$4\pi Y_2^{0*}(1)Y_2^0(2) = \frac{5}{4}$$

$$4\pi Y_2^{\pm2*}(1)Y_2^{\pm2}(2) = \frac{15}{8} e^{\pm 2i(\phi_2-\phi_1)}$$

in eq. (17), we finally obtain for $\lambda \le 2$

$$4\pi W(1,2;t) = \sum_{\lambda} B_\lambda A_\lambda (2\lambda+1)^{-1} \left(\sum_{\mu} e^{-i\mu\omega_L t} Y_\lambda^{\mu*}(1) Y_\lambda^{\mu}(2) \right)$$

$$= B_0A_0 + B_2A_2 \left[\frac{1}{4} + \frac{3}{4}\cos\{2(\omega_L t - \phi_2 + \phi_1)\} \right] \qquad (22)$$

The same result was derived in sect. 2 from semi-classical arguments. It is this function that describes the perturbation shown in fig. 2.

Randomly oriented magnetic fields.

From eqs. (19) and (21) we find

$$G_{\lambda\lambda}(t) = (2\lambda+1)^{-1} \sum_{\mu} e^{-i\mu\omega_L t}$$

For $\lambda = 2$ this evaluates to the well known expression

$$G_{22}(t) = \frac{1}{5}[1 + 2\cos(\omega_L t) + 2\cos(2\omega_L t)] \qquad (23)$$

Electric quadrupole interaction.

The interaction Hamiltonian for this case is (see Refs. [4,5])

$$\mathcal{K} = \frac{4}{5}\pi \sum_q (-)^q Q_q^{(2)} V_{-q}^{(2)}$$

The rank-2 tensors $Q^{(2)}$ and $V^{(2)}$ represent the nuclear quadrupole moment and the extra-nuclear electric field gradient, respectively. We restrict the discussion to the case that the asymmetry parameter of the electric field gradient is zero. In the principal axes system the following relations are valid

$$\Lambda_{\mu\mu'} = \delta_{\mu\mu'} e^{-i[3\mu^2 - I(I+1)]\omega_Q t}$$

$$\omega_Q \equiv \frac{eQV_{zz}}{4I(2I-1)\hbar}$$

$$eQ \equiv 4\sqrt{\pi/5}\, \langle II|Q_0^{(2)}|II\rangle$$

Inserting these expressions into eq. (18), one obtains

$$G_{\lambda\lambda'}^{\mu\mu'}(t) = \delta_{\mu\mu'}[(2\lambda'+1)(2\lambda+1)]^{\frac{1}{2}} \times$$

$$\times \sum_{mm'} \begin{pmatrix} I & I & \lambda' \\ -m' & m & \mu \end{pmatrix} \begin{pmatrix} I & I & \lambda \\ -m' & m & \mu \end{pmatrix} e^{-3i(m'^2 - m^2)\omega_Q t} \qquad (24)$$

The perturbation factor for the random case is obtained by substituting this result into eq. (19). It should be remarked that the perturbation factor for the case of an electric quadrupole interaction depends on the spin of the intermediate nuclear level. For a spin $I = 5/2$, the perturbation factor contains in general three different modulation frequencies. For higher spin values, however, the number of frequencies may be considerably larger. It should further be noted that the amplitudes of the various frequency components have different angular dependencies, which property makes PAC to a very sensitive tool for determining the symmetry and the orientation of the electric field gradient tensor.

In fig. 4 we show an experimental result that was obtained for ^{111}Cd in polycrystalline Cd metal.

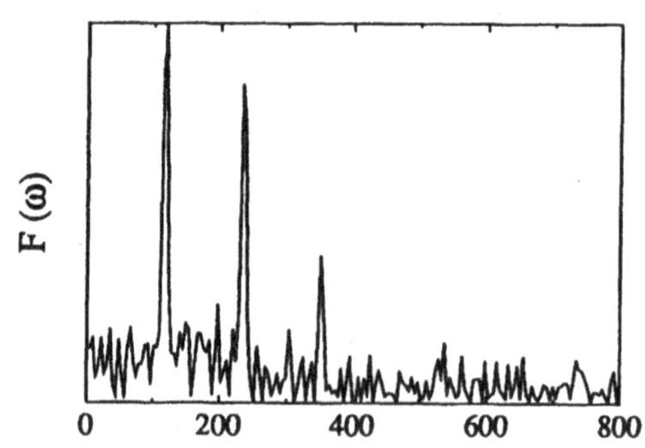

Fig. 4. Fourier transform of the perturbation factor measured for ^{111}Cd in polycrystalline hcp cadmium.

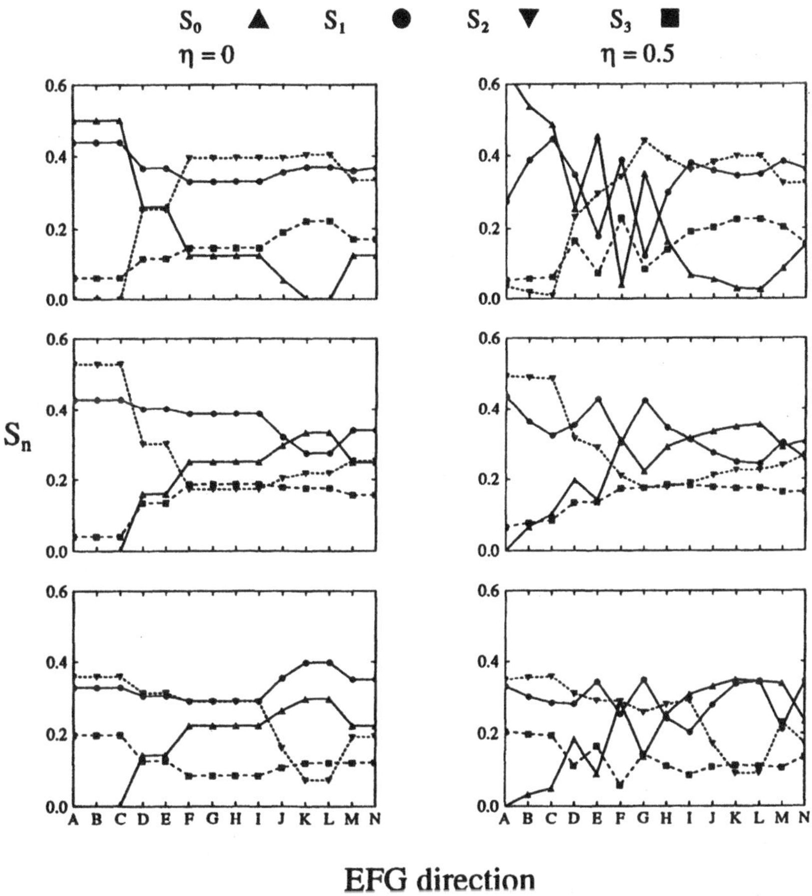

Fig. 5. Relative amplitudes S_n of the different frequency components in the perturbation factor for electric quadrupole interaction and nuclear spin $I = 5/2$. The asymmetry parameter is $\eta = 0$ (left) or $\eta = 0.5$ (right); the detectors are oriented along <100> (top), <110> (middle) or <111> (bottom) crystallographic directions. The labels A till N denote the following directions of the principal z- and y-axes: A [100][010], B [100][012], C [100] [011], D [210][$\bar{1}$20], E [210][001], F [110][$\bar{1}$10], G [110][001], H [110][$\bar{1}$12], I [110][$\bar{1}$11], J [221][1$\bar{2}$2], K [111][$\bar{2}$11], L [111] [$\bar{1}$10], M [211][$\bar{1}$11], N [211][0$\bar{1}$1].

Point defects in metals.

If the electric field gradient is due to a lattice imperfection located at a certain distance from the radioactive probe nucleus, one must average the angular correlation function W(1,2;t) over all possible equivalent lattice positions of the imperfection. In a cubic lattice, this procedure will in general involve 24 different orientations of the electric field gradient tensor and the result will not markedly differ from eq. (19), which is valid for the polytropic case. However, if the principal z-axis is aligned along one of the low-index crystallographic directions, one may be able to infer the tensor orientation by conducting PAC experiments for a proper set of detector geometries (see, *e.g.*, Wegner [9]). In fig. 5 we have plotted the relative amplitudes of the various frequency components that occur in the ratio R(t) for the case of a nuclear spin $I = 5/2$, for a number of field gradient orientations and detector geometries.

The usefulness of the effect is demonstrated in fig. 8. The strong suppression of the higher harmonics in the perturbation factor in the topmost box proves that the principal z-axis of the electric field gradient tensor is oriented along the <100> crystallographic direction, as the reader easily may check with the help of the data in fig. 5.

4. APPLICATIONS TO THE STUDY OF POINT DEFECTS IN METALS

4.1. The Early Period

The first PAC experiments were intended to measure hyperfine fields, either magnetic or electric. In these experiments a suitable nuclear probe was introduced into the sample material, usually being a chemical compound or a dilute metallic alloy. The perturbation factors shown in fig. 2 and fig. 4, though recently measured, could have been the results of such experiments. One of the aims of the experiments in that early period was to collect in a systematic way data on magnetic hyperfine fields in ferromagnetic materials like iron, cobalt, nickel and gadolinium, and on electric field gradients in chemical compounds and noncubic metals.

The samples for those experiments were prepared by means of chemical and metallurgical methods. In the mid-sixties, ion implantation was added to the arsenal of preparative techniques [10]. Although people were aware of the fact that ion implantation can change the local environment of the implanted probe atom, the results of the first implantation experiments did not seem to yield evidence that lattice damage was an important effect. Indeed, in the first PAC study of lattice defects, carried out as early as 1963, Hinman *et al.* [11] bombarded polycrystalline silver foils with 40 MeV α-particles. A few years later, point defects in metals had become a fashionable field of application of the PAC technique.

4.2. First-Generation Defect Studies

During the next period, point defects – and in particular vacancies – were observed in nearly all the cubic and noncubic metals. Ion implantation was the most frequently applied method to introduce the radioactive probe atoms and, at the same time, to create the point defects. But one used also other damaging techniques such as quenching, electron irradiation, cold-working, *etc.* Typical issues that were addressed concerned the nature of the observed elementary defect (vacancy *vs.* interstitial), the number of vacancies in a particular defect cluster, and the geometrical arrangement of these vacancies. Single crystal experiments yielded valuable information about the symmetry of the electric field gradient tensor, which is assumed to reflect the symmetry of the defect cluster. The obtained results have been reviewed by several authors; see, *e.g.*, Refs. [12-14]. We will now discuss two cases.

Vacancies in fcc nickel [15]

Immediately following implantation, the perturbation factor consists of a single precessional signal due to a magnetic hyperfine field of −7.1 T (*cfr.* fig. 2). This has been identified as due to ^{111}Cd in an undisturbed lattice because it has the same magnitude as the field determined in diffused, defect-free samples. After annealing at 400 K, two different magnetic frequencies are present, as illustrated in fig. 6. The larger of the two is the field of the undisturbed ^{111}Cd observed in the as-implanted sample and, consequently, the smaller has been assigned to a defect state. Since no detectable quadrupole perturbation affects the latter, one must conclude that we are dealing with a bound state of cubic symmetry, *i.e.*, one that would not have been seen in nonmagnetic material.

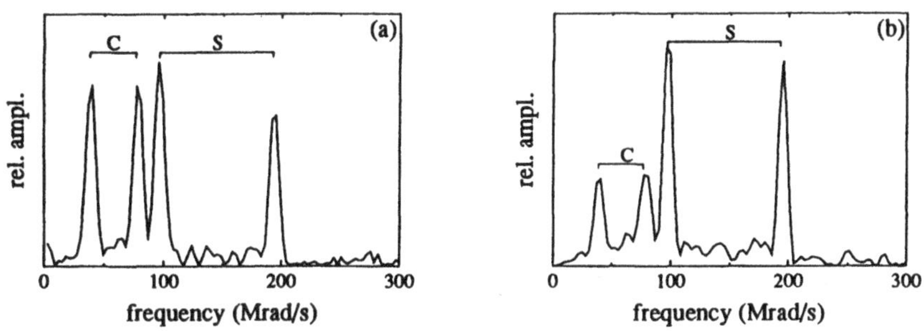

Fig. 6. Fourier transform of the perturbation factor for ^{111}Cd implanted in nickel as measured before (A) and after (B) bombardment with 1×10^{17} Xe/cm^2. The sample had been annealed at 400 K immediately after the implantation. Since the sample was not magnetized, substitutional atoms (S) and cubic vacancy clusters (C) each contribute with two peaks that correspond with single and double Larmor frequencies (cfr. eq. (23)).

Vacancies in bcc tungsten [16,17]

Since tungsten is a nonmagnetic cubic metal, probe atoms on regular undisturbed lattice sites as well as cubic defect configurations do not give rise to hyperfine interactions. Therefore, any perturbation of the angular correlation function must be due to defect trapping. After implantation one precessional signal is present in the perturbation factor, as shown in fig. 7a. The corresponding defect has been identified as a single vacancy. The amplitude of this signal increases with increasing annealing temperature and at the same time a new precessional signal appears (fig. 7b). The new signal has been interpreted as being due to the trapping of a divacancy. On the basis of the annealing behaviour one may conclude that the first vacancy is more strongly bound than the second one (see fig. 7c). A third kind of defect, probably a large vacancy cluster, is observed after annealing at higher temperatures (fig. 7d).

The assignment of the observed quadrupole frequencies to particular defect configurations is partly based on the measured orientations of the electric field gradient tensors. An example of such a measurement is shown in fig. 8; the results unambiguously prove that the principal z-axis of the electric field gradient tensor is oriented along the <100> crystallographic direction.

4.3. Second-Generation Defect Studies

In modern PAC experiments not the elementary defect itself is the object of investigation, but one uses the vacancy clusters, which are now routinely prepared by a variety of methods, to trap other mobile species such as interstitial atoms. In these second-generation experiments it is the trapping ability of the vacancy that keeps the interstitial atom so close to the nuclear probe that it can be studied through the hyperfine interaction, just like the study of vacancy clusters had been possible thanks to the trapping ability of the ^{111}In probe atoms.

In this last section we present some very recent results. But before we do so, we will briefly discuss the process of trapping mobile atoms at immobile traps. The mobile atoms are supposed to be introduced into the lattice by implantation without additional damage production.

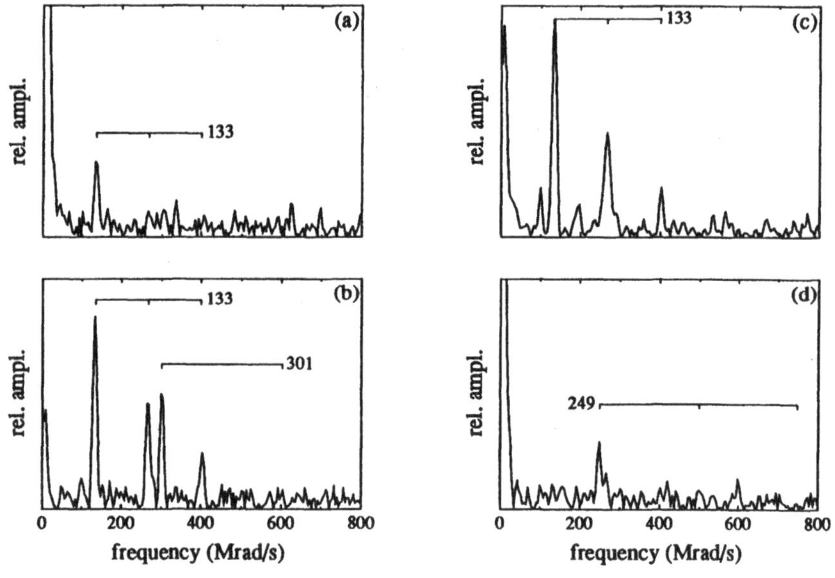

Fig. 7. Fourier transform of the perturbation factor for ^{111}Cd implanted in tungsten as measured after the implantation (a) and after subsequent annealing at 620 K (b), 750 K (c) and 820 K (d). The various peaks are labeled with the value of the quadrupole frequency $\omega_0 = 6\omega_Q$ expressed in Mrad/s.

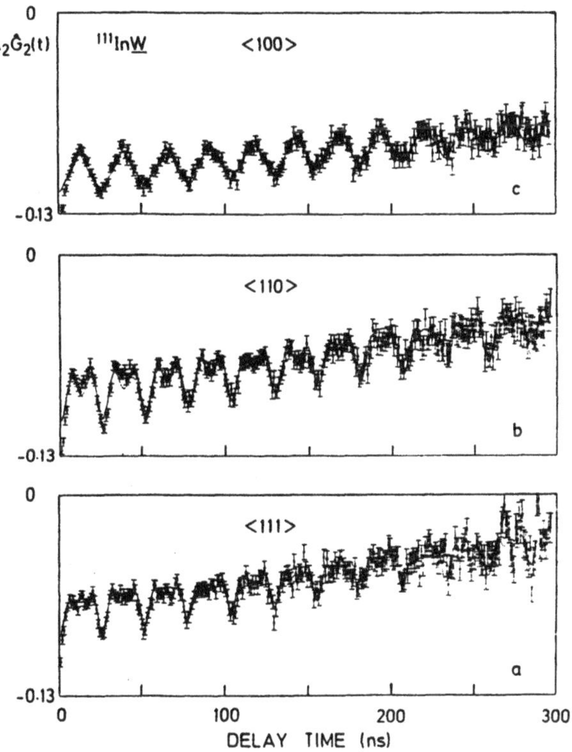

Fig. 8. Perturbation factors for ^{111}Cd implanted in tungsten after annealing at a temperature of about 950 K, measured with detectors orientated along (a) <111>, (b) <110> and (c) <100> crystallographic directions (from Ref. [17]).

Trapping of mobile species at immobile traps [18,10]

The number of particle-trap encounters during a time interval dt, within a layer with an area dS and a thickness dx, is equal to the product of (i) the average number of jumps performed by the mobile particle, (ii) the number of mobile particles in the layer, and (iii) the relative number of traps

$$\nu dt \ c(x)dSdx \ zn(x)/N_0 \tag{25}$$

where ν is the jump frequency, c(x) is the steady-state concentration of mobile particles, n(x) is the trap concentration, and N_0 is the atomic density. The factor z in eq. (25) is related to the effective trapping radius for the trapping process.

The number of injected mobile particles within the same layer is equal to the product of (i) the efficiency η of the injection process and (ii) the number of the impinging particles

$$\eta \ \mathrm{JdSdt} = \eta \left(\int_0^\infty \mathrm{I(x)dx} \right) \mathrm{dSdt} \tag{26}$$

where J is the flux density and I(x) describes the depth distribution of the mobile particles.

The trapping probability in the thin layer is equal to the ratio of the expressions (25) and (26)

$$\frac{z\nu c(x)}{N_0\eta J} n(x)dx \equiv \mu(x)n(x)dx$$

The steady-state concentration of mobile particles, c(x), follows from the balance between diffusion, production and trapping

$$D \frac{\partial^2}{\partial x^2} c(x) + \eta I(x) - \mu(x)n(x)J = 0 \tag{27}$$

If we can neglect the effect of trapping, the solution of eq. (27) is given by

$$\frac{D}{\eta} c(x) = \int_0^x \xi I(\xi)d\xi + x \int_x^\infty I(\xi)d\xi \tag{28}$$

The second term on the right-hand side of eq. (28) vanishes at large depths, provided that $I(\xi) \simeq \xi^{-k}$ with $k > 2$. For x beyond the average deposition depth $\langle x \rangle$ we have, therefore

$$\frac{D}{\eta} c(x) \cong \langle x \rangle J \tag{29}$$

$$\mu(x) \cong \frac{z\langle x \rangle}{\Gamma \lambda^2 N_0} \tag{30}$$

We have used the relation $D = \Gamma\lambda^2\nu$, where Γ is a geometrical factor and λ is the length of an elementary jump. Under the above mentioned restrictions, $\mu(x)$ is a constant and may be considered to be the cross-section for trapping.

Eq. (30) is only valid if the effect of trapping is negligeable. However, for an implantation dose of 2×10^{12} $^{111}In/cm^2$ and an implantation energy of 50 keV, one may expect about 10^{14} vacancies/cm^2. The trapping cross-section is about $10^{-14}\langle x \rangle$ cm^2 if $\langle x \rangle$ is expressed in lattice units. Hence, the ratio of trapping and production is approximately equal to $\langle x \rangle \geq 1$, and the trapping term in eq. (27) may not be neglected. A more detailed calculation shows that in this regime the effective trapping cross-section is roughly inversely proportional to the total number of traps in the sample.

Cubic defects in nickel [20]

There exists a long-standing controversy on the interpretation of recovery stages in fcc metals. Two models have been proposed in the past: the one-interstitial model (OIM) and the two-interstitial model (TIM). This is not the right place to discuss the arguments in favour of or against one or the other model; the interested reader is referred to Refs. [21,22]. Here, we just mention that the crucial difference between the OIM and the TIM is the assignment of stage-III recovery to vacancy migration in the first case, and to migration of the second type of interstitial in the second case.

In a recent paper, Frank and Seeger [23] addressed the configuration of the cubic defect in nickel that we introduced in sect. 4.2. They argued that there are two possibilities for its configuration: (i) an In atom occupying a tetrahedral interstice, and (ii) an In atom located at the centre of a tetrahedron formed by four neighbouring vacant lattice sites (fig. 9). We designed the following experiment, the outcome of which should enable us to decide which of the two proposed models is correct.

A single crystal of nickel was implanted at room temperature with 50 keV ^{111}In ions to an estimated dose of $2\times10^{12}\,cm^{-2}$. The sample was subsequently annealed at 475 K at which temperature cubic defects are known to reach a maximum population [15]. After being annealed, the crystal was bombarded at room temperature with 100 eV Xe ions to stepwise increasing doses ranging from $1\times10^{15}\,cm^{-2}$ to $1\times10^{17}\,cm^{-2}$. The impinging Xe ions transfer at most 3.7 times the minimum displacement energy to the host lattice atoms and eventually initiate a single displacement collision sequence, leaving a monovacancy at the surface. Because recovery stage III in nickel is above room temperature, the OIM and the TIM agree upon the immobility of the vacancies that are formed during the Xe bombardment. However, the self-interstitials injected into the lattice are mobile and will perform a random walk until they get trapped at the surface, lattice vacancies, impurity atoms or other sinks.

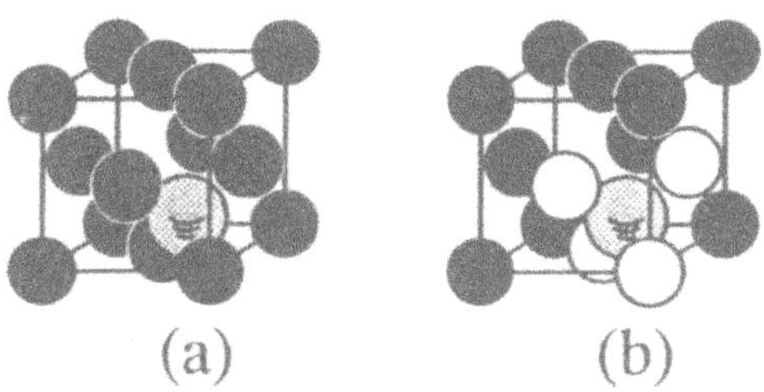

Fig. 9. Proposed configurations of the cubic defect in nickel: (A) an In atom occupying a tetrahedral interstice, and (B) an In atom located at the centre of a tetrahedron formed by four neighbouring vacancies.

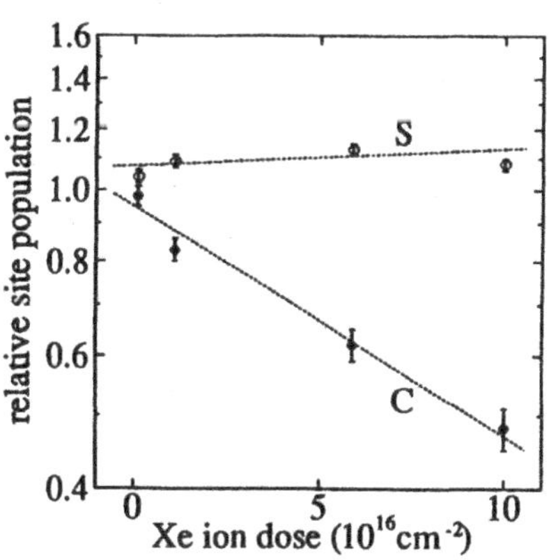

Fig. 10. Relative number of substitutional ^{111}In atoms (S) and cubic defects (C) as a function of the Xe dose.

Some of the PAC results are shown in fig. 6. The integrated peak areas for substitutional In atoms and In atoms in cubic defects are plotted in fig. 10 as a function of the Xe dose. The effective trapping cross-section turned out to be three orders of magnitude smaller than the value calculated from eq. (30). This illustrates the importance of the trap density for the trapping process.

According to model (i), cubic defects are formed by trapping of self-interstitials at substitutional In atoms during stage-III annealing, and the reduction of the cubic-defect fraction is caused by additional trapping of one or more self-interstitials. This will destroy the cubic symmetry of the original defect configuration and create an electric field gradient, which combined with the magnetic hyperfine field gives rise to complex hyperfine interactions and a fragmentation of the Fourier power over a large number of frequencies. However, if model (i) were correct, also the remaining substitutional In atoms should be able to trap self-interstitials and, consequently, the substitutional fraction should decrease at approximately the same rate as the cubic defect fraction. This is not observed.

According to model (ii), the In atom in a cubic defect is surrounded by vacancies. Trapping of self-interstitials will lead to annihilation of one or more vacancies in a defect-antidefect reaction. Thus, also in this model the cubic-defect population is expected to decrease with increasing Xe dose. However, the substitutional fraction should increase, in contrast to case (i). Since the cubic defect is converted already by trapping of a single interstitial whereas trapping of three interstitials is required for complete recovery of the cubic defect, the increase of the substitutional fraction due to this process amounts to at most a few percent of the change of the cubic-defect population.

The results of the experiment disagree with model (i), but they agree with model (ii). Therefore, the cubic defect in nickel must be a vacancy cluster. Since the defect is formed at about 350 K in recovery stage III by thermally activated trapping of migrating lattice defects [15], the migrating species at that temperature can only be of vacancy type. This outcome directly disproves the TIM because the model denies vacancy migration in recovery stage III.

Hydrogen – vacancy interaction in tungsten [24]

There exists theoretical as well as experimental evidence that a vacant lattice site in a metal may trap up to several hydrogen atoms [25]. However, till now nobody has been able to measure the exact number of these atoms. The following PAC experiment was intended to provide an answer to this question.

Polycrystalline tungsten foils were implanted with 50 keV ^{111}In to a dose of less than 1×10^{12} cm^{-2}, and subsequently annealed at 750 K. This treatment creates a situation where the only distinct precessional signal is due to a substitutional In atom with a single vacancy attached to it (see fig. 7c). Next, the samples were implanted with 1 keV H_2 molecules. The maximum energy transferred to the lattice is 11 eV, which is well below the minimum displacement energy.

In a series of experiments the hydrogen dose was varied from 1×10^{13} cm^{-2} up to 1×10^{17} cm^{-2}, resulting in a shift of the quadrupole frequency from the initial value of 133 Mrad/s via 127 Mrad/s to 121 Mrad/s, while the asymmetry parameter remained zero within limits of error (fig. 11). This result suggests that the vacancy in tungsten may trap up to two hydrogen atoms. The observed trapping cross-section is about two orders of magnitude smaller than the value calculated from eq. (30), in agreement with the estimated trap density. However, the cross-section decreased by another two orders of magnitude at higher hydrogen doses because the hydrogen itself acts as a trapping centre. The new centres manifest themselves in fig. 11 as a frequency cluster at about 220 Mrad/s.

The existence of two different hydrogen occupancies was corroborated by the result of an isochronal annealing experiment. As illustrated in fig. 12, hydrogen is released from the vacancy at two well defined temperatures, from which we derive dissociation energies of 1.6 eV and 1.4 eV for the first and the second hydrogen atom, respectively. The fact that we did not observe trapping of more than two hydrogen atoms implies that the dissociation energies for the next hydrogen atoms are smaller than 1.1 eV. These values compare well with theoretical estimates [26,27].

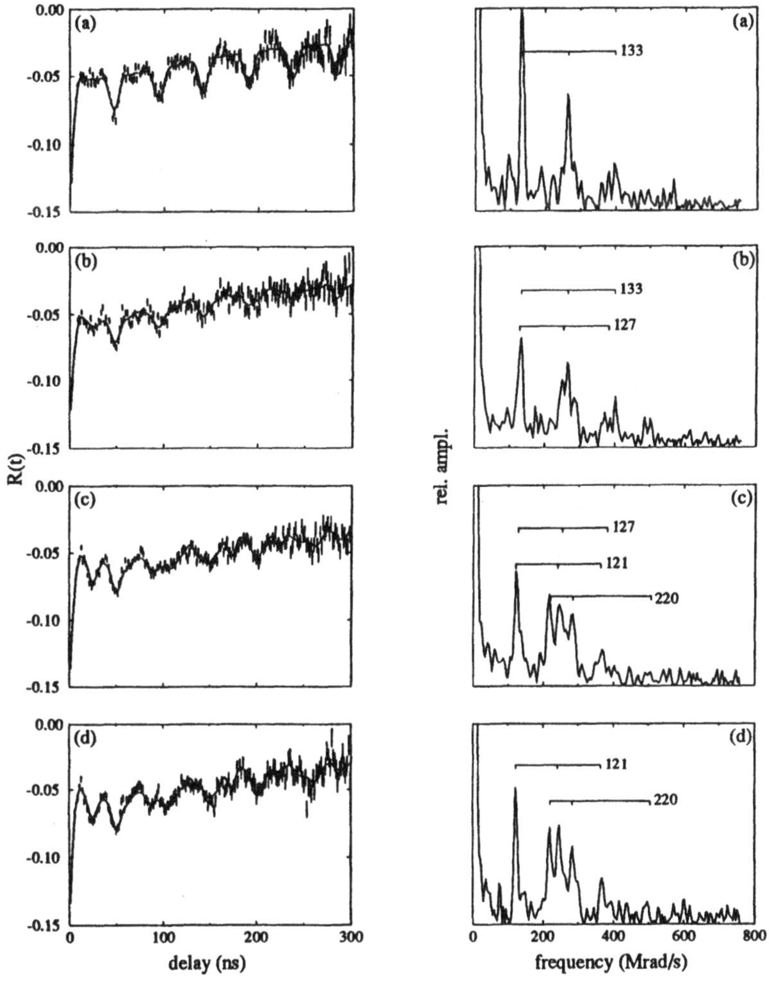

Fig. 11. Perturbation factor (left) and its Fourier transform (right) for ^{111}In in tungsten after annealing at 750 K, before hydrogen implantation (a) and after hydrogen implantation to a dose of 1×10^{14} cm^{-2} (b), 3×10^{15} cm^{-2} (c) and 9×10^{16} cm^{-2} (d).

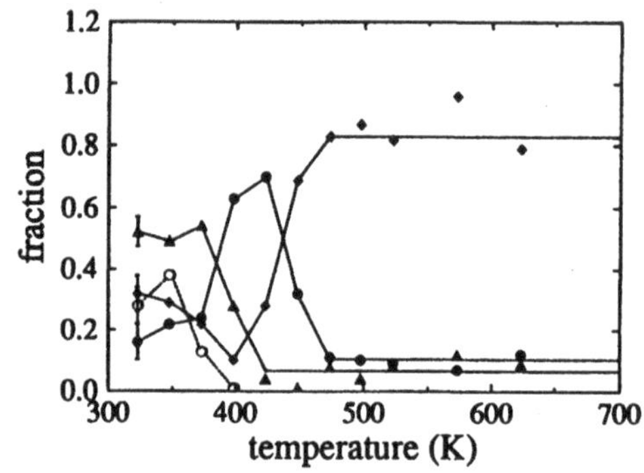

Fig. 12. Defect fractions as a function of the annealing temperature: ■ 133 Mrad/s, ● 127 Mrad/s, ▲ 121 Mrad/s, and ○ 220 Mrad/s. The fractions are normalized to the sum of the first three defect fractions.

The point-charge model with an effective hydrogen charge of +0.33 e (see Ref. [28]) predicts a decrease of the quadrupole frequency by 3–9 Mrad/s per hydrogen atom, for most locations within the unit cell occupied by the vacancy. In fig. 13 we have marked those octahedral and tetrahedral positions for which the calculated hyperfine parameters agree with the PAC data. While this outcome supports the above interpretation, it also shows that the hyperfine interaction is not sensitive enough to allow a more precise site location.

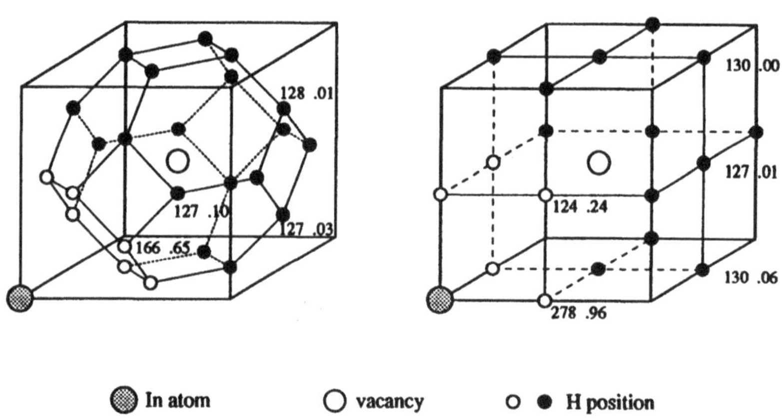

Fig. 13. Calculated values of ω and η for an In–vacancy cluster decorated with a single hydrogen atom on a tetrahedral (left) or an octahedral (right) postion. The results for the filled circles are consistent with the experimental results.

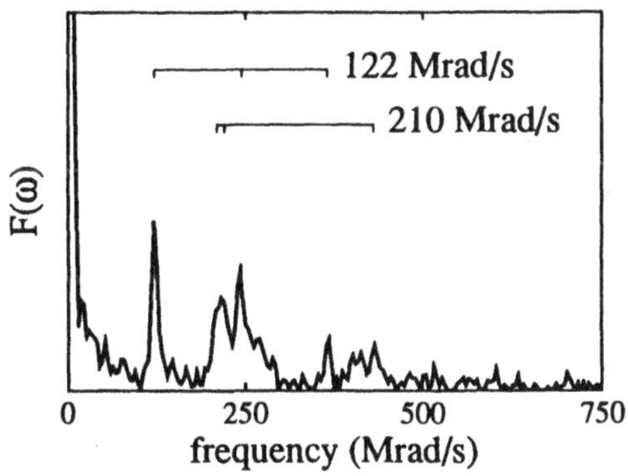

Fig. 14. Fourier transform of the perturbation factor for ^{111}In in tungsten, after injection of 3×10^{15} He/cm^2 and subsequent annealing at 1250 K.

Trap mutation in ion-implanted helium-doped tungsten [29]

The high mobility of helium in metals and its strong binding at defects is utilized in Thermal Helium Desorption Spectrometry (THDS) to probe lattice defects. This technique is based on the following idea: after low-energy helium injection some of the helium atoms will be trapped at defects and, subsequently, released during thermal annealing. The discriminative power of the method is due to the fact that different bound states show up in the desorption spectrum at different and well defined temperatures. THDS has been very successfully applied to study vacancies and vacancy clusters [30].

It has been found that vacancies may be multiply filled

$$He_nV + He \rightarrow He_{n+1}V \qquad n = 0, 1, ...$$

The next helium atom is usually less strongly bound than the previous ones. However, at a certain degree of filling the trend of smaller binding energies for higher helium filling is reversed. To explain this phenomenon, trap mutation has been postulated to occur

$$He_nV \rightarrow He_nV_2I \rightarrow He_nV_2 + I$$

Apparently, the volume has expanded in order to accomodate more helium atoms. The expelled self-interstitial I will be released during the annealing process. Unfortunately, HeV and He_2V_2 have approximately the same dissociation energies [31]. PAC, however, can easily distinguish the two configurations because they possess different symmetries and will give rise to electric field gradient tensors with asymmetry parameters $\eta = 0$ and $\eta = 1$, respectively. To demonstrate this, we currently perform the following PAC experiment.

Polycrystalline tungsten foils were implanted with 50 keV In ions to a dose of 7×10^{12} cm^{-2}. After being annealed at 750 K, the foils were post-implanted with 300 eV helium to a dose of 3×10^{15} cm^{-2}, thus decorating approximately 60% of the vacancies in the samples. No new precessional signals were observed: the Fourier power is apparently distributed over many frequencies because of a large range of helium occupancies. However, an increasing number of helium atoms will be released upon thermal annealing and the Fourier power is expected to concentrate on one or two components. This is precisely what we observed.

From about 1100 K on, two frequency clusters become visible, which must be related to helium-filled vacancies because these are the only stable defects at such an elevated temperature. At still higher annealing temperatures the frequency clusters contract to rather sharp peaks at 122 Mrad/s and 210 Mrad/s, with asymmetry parameters $\eta = 0$ and $\eta = 1$, respectively (fig. 14). Both peaks disappear at 1300 K. This temperature corresponds with the so-called H-peak in THDS. On the basis of this correspondence and the observed symmetry properties we tentatively conclude that the above quadrupole frequencies may be assigned to HeV and He_2V_2. Since no divacancies were present after the preparative annealing at 750 K (*cfr.* fig. 7c), this would imply that trap mutation has indeed occurred.

ACKNOWLEDGEMENT. The experiments described in sect. 4.3 were carried out in collaboration with J.R. Fransens, M.S. Abd El Keriem and D.P. van der Werf. This work forms part of the research programme of the "Stichting voor Fundamenteel Onderzoek der Materie" (FOM) with financial support from the "Nederlandse Organisatie voor Wetenschappelijk Onderzoek" (NWO).

APPENDIX: STOCHASTIC HYPERFINE INTERACTIONS

If the Hamiltonian that describes the hyperfine interaction fluctuates at a rate λ that is comparable with the interaction frequency ω, the theory presented in sect. 2 and sect. 3 is no longer valid. There exists an extensive literature on this subject which, in general, is difficult to tackle (see, *e.g.*, Dattagupta [32]). However, when the fields are isotropically oriented, and the fluctuations are due to stochastic processes that alter the orientation but not the strength of the hyperfine field, an expression for the perturbation factor can be derived in the following simple and elegant manner which is due to Blume [33].

We assume that during the time interval [0, t] precisely s collisions at random instants $t_1, t_2, \ldots\ldots t_s$ take place. We assume further that in between the collisions the hyperfine field is time independent and gives rise to the static perturbation $G_k^o(t)$. Then the dynamic perturbation factor, which describes the momentary nuclear alignment, can be written as the product of s static factors, each accounting for the perturbation of the angular correlation in between two successive collisions:

$$G_k^s(t) = G_k^o(t-t_s)G_k^o(t_s-t_{s-1}) \ldots\ldots G_k^o(t_1) \tag{A1}$$

This expression has to be averaged over all possible collision sequences with $s = 0, 1, 2, \ldots\ldots$, and all possible distributions of $t_1, t_2, \ldots\ldots t_s$ over the time interval [0, t]. This yields

$$G_k(t) = \sum_s P_s(t)\overline{G_k^s(t)}$$

$$= \sum_s P_s(t)\int_0^t dt_s \int_0^{t_s} dt_{s-1} \ldots\ldots \int_0^{t_2} dt_1 W_s(t_1 t_2 \ldots\ldots t_s;t) G_k^s(t) \tag{A2}$$

For a stochastic process with a collision rate λ we may write

$$P_s(t) = (\lambda t)^s e^{-\lambda t}/s! \tag{A3}$$

$$W_s(t_1 t_2 \ldots\ldots t_s;t) = s!/t^s \tag{A4}$$

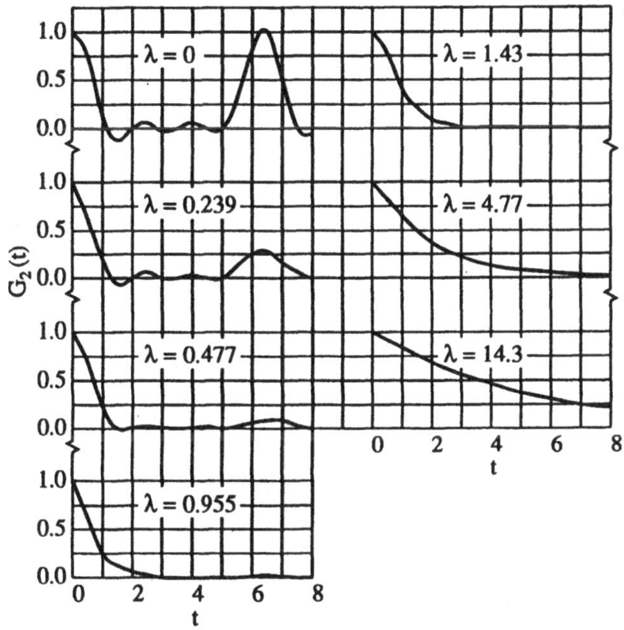

Fig. 15. The perturbation factor for an isotropically reorienting electric field gradient, calculated for different values of λ/ω_Q, *according to the Blume theory (from Ref.* [32]*).*

Inserting (A1), (A3) and (A4) into eq. (A2), one obtains the following s–fold convolution integral

$$
\begin{aligned}
G_k(t) &= \sum_s \lambda^s e^{-\lambda t} \int_0^t dt_s \int_0^{t_s} dt_{s-1} \,.....\, \int_0^{t_2} dt_1 G_k^0(t-t_s) G_k^0(t_s-t_{s-1}) \,.....\, G_k^0(t_1) \\
&= \sum_s \lambda^s \int_0^t dt_s \int_0^{t_s} dt_{s-1} \,.....\, \int_0^{t_2} dt_1 \times \\
&\quad \times e^{-\lambda(t-t_s)} G_k^0(t-t_s) e^{-\lambda(t_s-t_{s-1})} G_k^0(t_s-t_{s-1}) \,.....\, e^{-\lambda t_1} G_k^0(t_1)
\end{aligned}
$$

Introducing the Laplace transform of the perturbation factor

$$\tilde{G}_k(p) = \int_0^\infty e^{-pt} G_k(t)\, dt, \qquad p > 0$$

and remembering that the Laplace transformation transforms a convolution integral into a simple algebraic product, one finally obtains

$$\tilde{G}_k(p) = \sum_s [\lambda\tilde{G}_k^o(p+\lambda)]^s \tilde{G}_k^o(p+\lambda)$$

$$= \frac{\tilde{G}_k^o(p+\lambda)}{1 - \lambda\tilde{G}_k^o(p+\lambda)} \tag{A5a}$$

$$= \tilde{G}_k^o(p+\lambda) + \lambda\tilde{G}_k^o(p+\lambda)\tilde{G}_k^o(p) \tag{A5b}$$

The dynamic perturbation factor $G_k(t)$ is obtained either by a Laplace transformation of the right-hand side of eq. (A5a), or by a transformation of eq. (A5b) back into the time domain and solving the resulting inhomogeneous Volterra integral equation

$$G_k(t) = e^{-\lambda t}G_k^o(t) + \lambda\int_0^t e^{-\lambda t'}G_k^o(t')G_k(t-t')dt' \tag{A5c}$$

Fig. 15 shows some results for the case of an isotropically reorienting electric field gradient and a nuclear spin $I = 5/2$, for different values of λ/ω_Q.

REFERENCES

1 Arends, A.R., Hohenemser, C., Pleiter, F., Waard, H. de, Chow, L., and Suter, R.M., Hyp. Int. 8, 191 (1980).
2 Butz, T., Hyp. Int. 52, 189 (1989).
3 Fano, U., Revs. Modern Phys. 29, 74 (1957).
4 Frauenfelder, H. and Steffen, R.M., in: *Alpha-, beta- and gamma-ray spectroscopy*, ed. K. Siegbahn (North-Holland, Amsterdam, 1968) ch. XIXA.
5 Steffen, R.M. and Alder, K., in: *The electromagnetic interaction in nuclear spectroscopy*, ed. W.D. Hamilton (North-Holland, Amsterdam, 1975) ch. 12-13.
6 Krane, K.S., Steffen, R.M. and Wheeler, R.M., At. Data Tables 11, 351 (1973).
7 Rots, M., Claes, J. and Namavar, F., Hyp. Int. 7, 323 (1979).
8 Rots, M., Coussement, R., Claes, J., and Hermans, L., Hyp. Int. 11, 185 (1981).
9 Wegner, D., Hyp. Int. 23, 179 (1985).
10 Waard, H. de, and Drentje, S.A., Phys. Lett. 20, 38 (1966).
11 Hinman, G.W., Hoy, G.R., Lees, J.K., and Serio, J.C., Phys. Rev. 135, A218 (1964).
12 Pleiter, F. and Hohenemser, C., Phys. Rev. 25, 106 (1982).
13 Recknagel, E., Schatz, G. and Wichert, Th., in: *Hyperfine interactions of radioactive nuclei*, ed. J. Christiansen, Topics in Current Physics, Vol. 31 (Springer, Berlin, 1983) ch. 4.
14 Collins, G.S., Shropshire, S.L. and Fan, J., Hyp. Int. 62, 1 (1990).
15 Hohenemser, C., Arends, A.R., Waard, H. de, Devare, H.G., Pleiter, F., and Drentje, S.A., Hyp. Int. 3, 297 (1977).
16 Pütz, U., Hoffman, A., Rudolf, H.J., and Vianden, R., Z. Physik B 46, 107 (1982).
17 Kolk, G.J. van der, Post, K., Veen, A. van, Pleiter, F., and Hosson, J.Th.M. de, Rad. Effects 84, 131 (1985).
18 Gorkum, A.A. van, and Kornelsen, E.V., Rad. Effects 42, 93 (1979).
19 Veen, A. van, and Caspers, L.M., in: *Harwell symposium on inert gases in metals and ionic solids*, ed. S.F. Pugh, AERE Report 9733 (1980) p. 517.

20 Fransens, J.R., Pleiter, F. and Meinders, J., Solid State Commun. 71, 1155 (1989).
21 Schilling, W., Ehrhart, P. and Sonnenberg, K., in: *Fundamental aspects of radiation damage in metals*, eds. M.T. Robinson and F.W. Young Jr., CONF–751006–P1 (Gatlinburg, Tennessee, 1975) p. 470.
22 Seeger, A., *loc. cit.* Ref. [21)], p. 493.
23 Frank, W. and Seeger, A., Solid State Commun. 68, 495 (1988).
24 Fransens, J.R., Abd El Keriem, M.S. and Pleiter, F., to be published.
25 Myers, S.M., Richards, P.M., Wampler, W.R., and Besenbacher, F., J. Nucl. Mater. 165, 9 (1989).
26 Nordlander, P., Nørskov, J.K., Besenbacher, F., and Myers, S.M., Phys. Rev. B 40, 1990 (1989).
27 Nordlander, P., Nørskov, J.K. and Besenbacher, F., J. Phys. F 16, 1161 (1986).
28 Weidinger, A., J. Less Comm. Metals 103, 285 (1984).
29 Abd El Keriem, M.S., Werf, D.P. van der, and Pleiter, F., to be published.
30 Veen, A. van, in: *Vacancies and interstitials in metals and alloys*, eds. C. Abromeit and H. Wollenberger, Materials Science Forum 15–18, 3 (1987).
31 Veen, A. van, Evans, J.H., Buters, W.Th.M., and Caspers, L.M., Rad. Effects 78, 53 (1983).
32 Dattagupta, S., Hyp. Int. 11, 77 (1981).
33 Blume, M., Nucl. Phys. A167, 81 (1971).

PAC spectroscopy of electronic ceramics

John A. Gardner, Ruiping Wang, and Rainer Schwenker
Department of Physics, Oregon State University, Corvallis, Oregon 97331-6507 USA

William E. Evenson
Department of Physics, Brigham Young University, Provo, Utah 84602 USA

Robert L. Rasera
Department of Physics, University of Maryland, Baltimore County, Catonsville, Maryland 21228 USA

James A. Sommers
Teledyne Wah Chang Albany, Inc., Albany, Oregon 97321 USA

ABSTRACT

Dilute indium dopants in cerium oxides and $YBa_2Cu_3O_x$ have been studied by $^{111}In/Cd$ Perturbed Angular Correlation (PAC) spectroscopy. By controlling oxygen vacancy concentration in the cerium oxides through doping or high-temperature vacuum annealing, we have found that indium always forms a defect complex unless the sample is doped to reduce greatly the oxygen vacancy concentration. Three different vacancy-associated complexes are found with concentraions that depend on doping and oxygen stoichiometry. Another defect complex occurs in samples having negligible vacancy concentration. At low temperatures, evidence is found of interaction with an electronic hole trapped by ^{111}Cd after the radioactive decay of the ^{111}In parent. In $YBa_2Cu_3O_x$ the indium substitutes preferentially at the Y site but has measurable probability of substitution in at least one of the two copper sites. A symmetry change near 650 °C is consistent with the well-documented orthorhombic/tetragonal transition for samples in air or oxygen.

I. INTRODUCTION

Perturbed Angular Correlation (PAC) spectroscopy is an excellent technique for investigating a number of important properties of electronic ceramics.[1–4] We have used $^{111}In/Cd$ PAC to probe the structure and stability of indium donor-oxygen defect complexes in cerium oxide and have begun a study of microscopic structure of $YBa_2Cu_3O_x$. Cerium oxide has large ionic conductivity at high temperature and has great potential for technological applications. $YBa_2Cu_3O_x$ is a 90 K superconductor.

CeO_2 is the stable oxide of cerium in air. The oxygen concentration can be decreased by annealing in a reduced oxygen pressure at high temperature. Such material has a substantial electronic and ionic conductivity.[5–8] Ceria is often doped with trivalent (usually rare-earth) or divalent (Ca, Mg) cations. The cationic charge difference is compensated largely by oxygen vacancies, so such materials are high-temperature ionic conductors but have small electronic conductivity.[6–8] The interaction of dopants with oxygen defects is an important topic. Our work probes complexes of (trivalent) indium dopants with oxygen defects.

CeO_2 has the cubic fluorite structure, so the electric field gradient (efg) at the Ce position in a perfect crystal will be zero. If the ^{111}Cd nucleus is substituted at a Ce site with no nearby defects, then the ^{111}Cd PAC spectrum should be unperturbed. We have investigated a number of ceria samples in which the oxygen vacancy concentration is controlled by doping or high-temperature annealing. Measurements have been made on samples held at temperatures ranging from 10 K to 1400 K, and only near the highest temperatures in this range is the PAC spectrum unperturbed. At lower temperatures, we find a variety of interactions and can identify at least four well-defined defect-associated sites. At temperatures below 500 K, we always find evidence for "aftereffects" of the ^{111}In electron capture decay.

^{111}In/Cd PAC in $YBa_2Cu_3O_x$ has not been a straightforward experimental investigation. A number of groups have attempted PAC measurements[9–14] and have obtained non-reproducible spectra whose interpretation has been controversial.[15] Although the indium solubility in $YBa_2Cu_3O_x$ is approximately 3% per formula weight,[16] it has proven extraordinarily difficult to synthesize samples in which the majority of the ^{111}In tracer is dissolved in high-quality $YBa_2Cu_3O_x$. Most of the common synthesis methods produce material in which the majority of the ^{111}In is dissolved in phases other than $YBa_2Cu_3O_x$. We have recently developed a reproducible method for making samples in which the majority of the ^{111}In is dissolved in $YBa_2Cu_3O_x$. We find that indium substitutes predominantly at the Y site but also has a substantial probability of substitution in at least one of the copper sites. We have observed the orthorhombic/tetragonal phase transition in samples heated in flowing oxygen and report preliminary data here.

II. RESULTS OF CeO_2 EXPERIMENTS

We have performed measurements over a wide temperature range on pure and lightly-doped CeO_2 samples containing trace levels (of order 10^{-9} per formula weight) of ^{111}In. Samples were made by ammonium precipitation from cerium ammonium nitrate solutions of the desired cation stoichiometry. ^{111}In in HCl solution, obtained commercially from New England Nuclear Dupont, was stirred into the solution just prior to precipitation. Samples were calcined at 1500 °C or higher for 6 hours prior to measurement. This preparation method resulted in oxide grains of order 10 microns diameter. Samples calcined below 1200 °C had much smaller grains and a significant probability that the ^{111}In was not dissolved in the bulk.

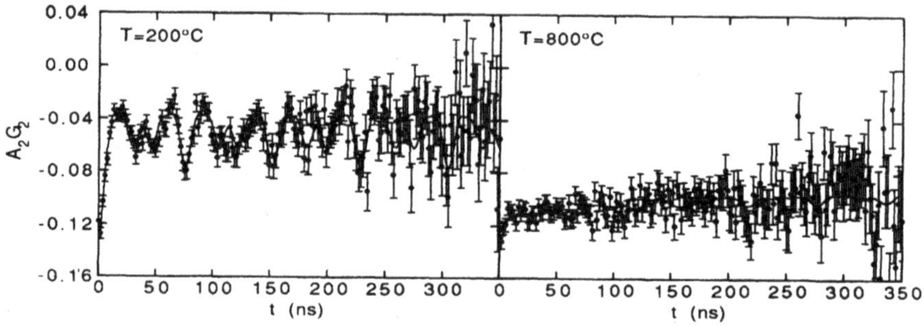

Figure 1: ^{111}Cd PAC experimental A_2G_2 spectra for cerium oxide doped with 500 ppm Nb per formula weight.

The PAC measurements were carried out using a 4-detector spectrometer with BaF_2 scintillator detectors. The apparatus is similar to one described previously.[17]

Figure 1 shows the experimental $A_2G_2(t)$ PAC spectra for a typical ceria sample doped with Nb, a pentavalent cation introduced to reduce the oxygen vacancy concentration to negligible levels. Below 175 K, the spectra are typical of an "aftereffects" interaction.[18] $G_2(t)$ decays rapidly with time due to interaction between the Cd nucleus and an electronic hole metastably trapped after the electron capture decay of the ^{111}In parent. This interaction is not understood in detail, and we model it by a single exponential. The model is heuristic and fits the data well but not perfectly. In all cases, the experimental $G_2(t)$ does not decay exactly to zero, although the "hard core" remaining at long time is small and may be an experimental artifact.

We have made only cursory measurements below room temperature, so this region remains to be explored in future work. At and above room temperature, we find that a small fraction of the Cd nuclei are subject to a well-defined interaction with a large, asymmetric electric field gradient (efg). The relative fraction of such sites decreases with temperature and becomes negligible at temperatures of order 1000 K. The interaction frequencies are reproducible from sample to sample, but the relative fraction is not. We do not presently understand how this static interaction arises. The PAC spectrum due to the remaining ^{111}Cd nuclei can be well fitted by an exponential and a constant. The fractional weights of the exponential, constant, and static interaction contributions to the spectrum are shown in Fig. 2. The damping rate of the exponential is shown in Fig. 3(a).

It is reasonable to assume that the exponential and constant terms arise from nuclei in substitutional cerium positions in which the escape time for the trapped "aftereffects" hole is of order nanoseconds. It is straightforward to show[19,20] that in such a case $G_2(t)$ is given by:

$$G_2(t) = \frac{r}{r+\lambda_a} + \frac{\lambda_a}{r+\lambda_a} e^{-(\lambda_a+r)t} \tag{1}$$

provided $G_2(t)$ is well-approximated by a single exponential when the hole escape

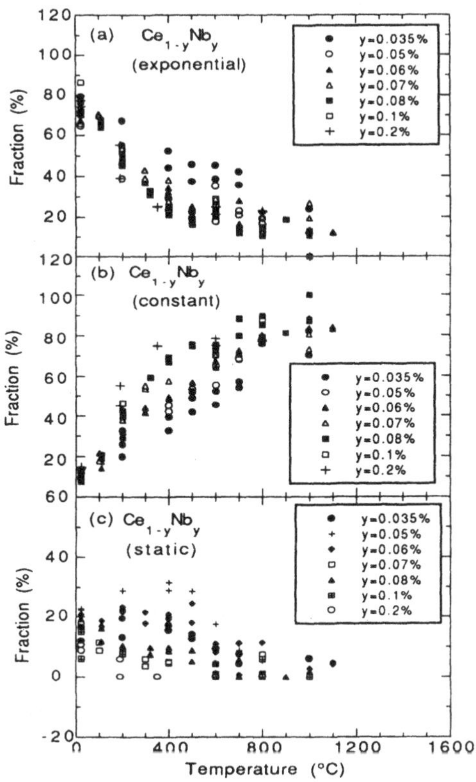

Figure 2: Fractions of (a) exponential, (b) constant, and (c) static interaction models used to fit the spectra of Nb-doped ceria at elevated temperatures.

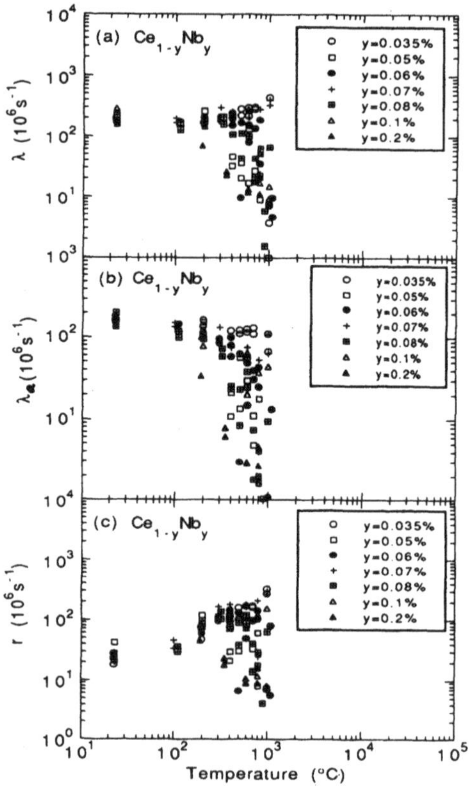

Figure 3: (a) Damping rate of the exponential model for Nb-doped ceria. (b) Exponential decay rate λ_a for ^{111}Cd PAC spectrum in presence of trapped "aftereffects" hole, and (c) escape rate r of the hole. Within this model $\lambda_a + r = \lambda$, where λ_a and r are defined by Eq. 1 in text.

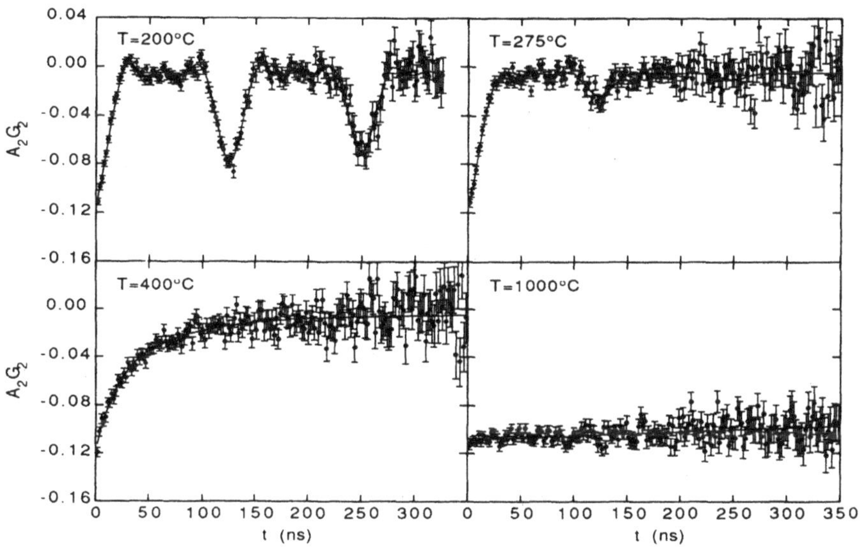

Figure 4: ^{111}Cd PAC experimental A_2G_2 spectra for undoped cerium oxide.

rate r becomes very small. The rates λ_a and r are shown in Figs. 3(b) and (c).

Spectra qualitatively similar to those of Fig. 1 are found for samples containing more than 300 ppm Nb. PAC spectra from undoped ceria are shown in Fig. 4. The spectrum at low temperature has a substantial aftereffects behavior, but almost half the ^{111}Cd nuclei are subject to a static interaction with a moderate-magnitude axially-symmetric efg. At 200 °C aftereffects are negligible, and the fraction of Cd nuclei subject to the static interaction is 100% within experimental uncertainty. We speculate that the ^{111}In parent is associated with a strongly-bound oxygen vacancy complex and that the probability of trapping a hole after ^{111}In decay is approximately 50%. Thus, at low temperature, approximately 50% of the spectrum is affected by the aftereffect interaction, but above 200 °C the hole escape rate is fast enough that its interaction with the ^{111}Cd nucleus has negligible effect on the PAC spectrum.

Figure 5 shows the Fourier transform of the 200 °C $A_2G_2(t)$ spectrum. This figure shows a frequency triplet with frequencies having ratios 1:2:3, the signature of a quadrupolar interaction with an axially-symmetric efg. At higher temperatures, the frequencies broaden, and the $A_2G_2(t)$ spectrum shows clearly that the efg is no longer static on the time scale of the intermediate state lifetime. We have examined two possibilities for the time-dependence. One possibility is that the oxygen vacancy escapes at high temperature. The second possibility is that the complex remains bound but that its symmetry axis direction changes due to thermal excitation. Detailed analysis, to be described in a later paper, clearly demonstrates that

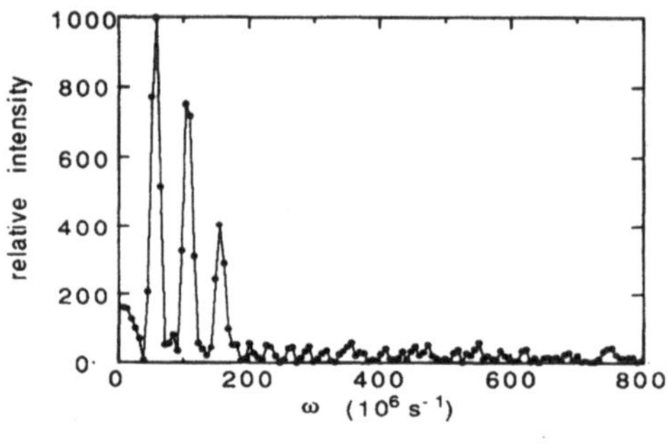

Figure 5: Fourier transform of 200 °C spectrum from Fig. 4.

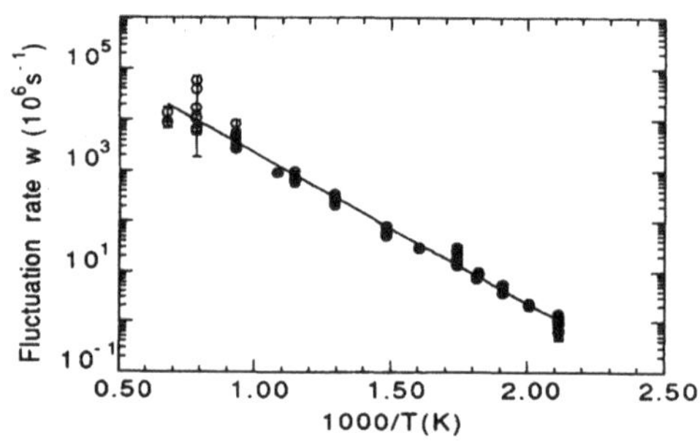

Figure 6: Single fitting parameter w vs. 1/T. The fitting model is described in Ref. 21.

the experimental data disagree qualitatively with the first possibility but can be quantitatively explained by the second.

The random axis-reorientation hypothesis can be described accurately by a model developed by Evenson *et al.*[21,22] We assume that the efg tensor remains constant apart from random axis reorientation at rate w. Data below 300 °C are well-represented by the "slow" limit (w/ω_Q small) of the model, whereas data above 400 °C are accurately described by the "fast" limit. The single fitting parameter w is shown vs. 1/T in Fig. 6. The activation energy for the axis-reorientation process is 0.6(0.2) eV. Figure 7 shows the (a) fractional weight, (b) the damping rate of the exponential decay used to approximate the "aftereffects" spectrum, and (c) the interaction frequencies ω_1 and ω_2.

In Fig. 8(a) we show the PAC spectrum at 23 °C for an undoped ceria sample that had been annealed for 30 minutes in a mechanical pump vacuum at 900 °C and in 8(b) for a sample annealed at 1000 °C. After annealing, these samples were sealed in order to keep the oxygen vacancy concentration constant. In addition to the site seen in undoped ceria (referred to as site A), two additional sites (B and C) are

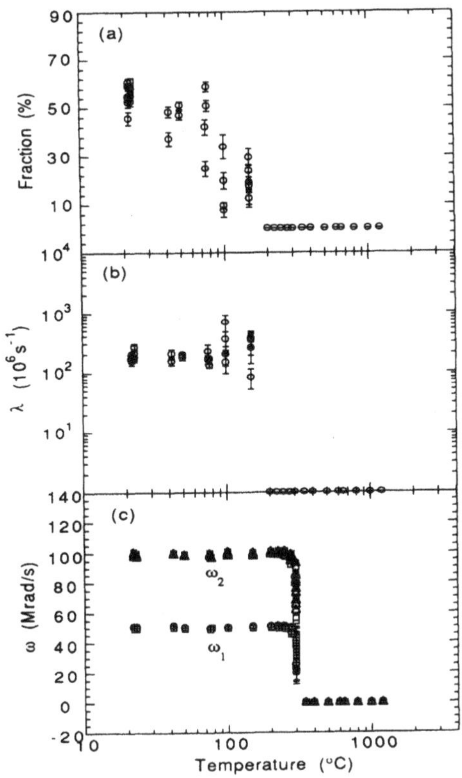

Figure 7: (a) Fraction and (b) damping rate of exponential interaction used to fit undoped ceria spectra. (c) Interaction frequencies for the A-site interaction. The quadrupole frequency $\omega_Q = 300 \times 10^6$ is obtained from fitting spectra in the static region below 300 °C and is held fixed at that value for higher temperatures.

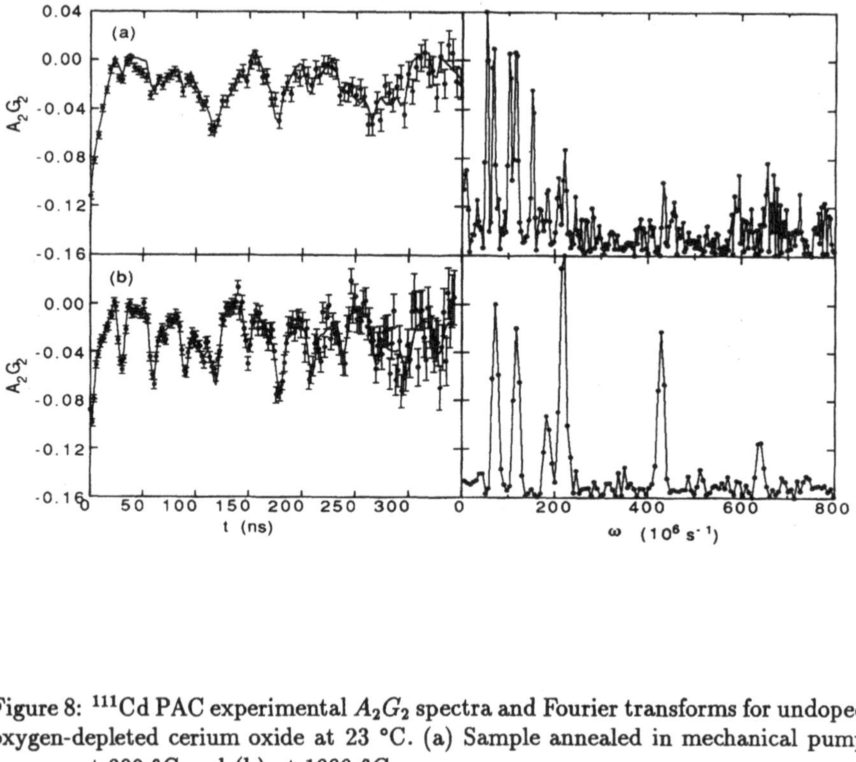

Figure 8: ^{111}Cd PAC experimental A_2G_2 spectra and Fourier transforms for undoped oxygen-depleted cerium oxide at 23 °C. (a) Sample annealed in mechanical pump vacuum at 900 °C and (b) at 1000 °C.

found. We tentatively attribute sites B and C to the presence of additional oxygen vacancies bound to the original site-A complex. The microscopic structure giving rise to these spectra has not been established unambiguously. We have speculated[4] that site A has a missing oxygen atom next to the indium atom and that the indium atom of site C has two neighboring oxygen vacancies forming a Bevan cluster.[23]

We find that PAC spectra are not strongly affected by minor contamination by rare-earth impurities that are commonly found in cerium oxides. We have used both a very high purity stock solution (cation impurities fewer than 20 ppm) and a commercial solution containing approximately 300 ppm rare-earth contaminants. No differences were found. It is interesting to note that approximately 300 ppm Nb is required to change the spectra from those of Fig. 4 to those of Fig. 1.

We have also carried out measurements of PAC spectra on samples doped with as much as 1% per formula weight of yttrium or indium. Samples doped with more than 0.4% yttrium have measureable contributions from sites of types B and C. All In-doped samples have spectra identical to those of pure ceria. These findings imply that indium does not introduce free oxygen vacancies and should therefore not yield materials with large ionic conductivity. To our knowledge, the ionic conductivity of indium-doped ceria has not been measured.

III. RESULTS OF $YBa_2Cu_3O_x$ EXPERIMENTS

We have tried a variety of sample-making techniques. Diffusion of indium into previously-synthesized $YBa_2Cu_3O_x$ was unsuccessful. The indium was apparently largely dissolved in non-representative surface layers or intergranular phases. We found very good spectra and extremely sharp Meissner flux exclusion for samples made by an evaporation method.[13,14] Unfortunately, little indium was dissolved in the majority phase. When samples were synthesized in flowing air or oxygen, most indium dissolved in Y_2BaCuO_5, although only a few percent of this phase was present in our samples. When samples were prepared in a restricted air flow, the indium dissolved preferentially in some other second phase whose composition we have not tried to identify. The PAC spectrum in this latter case[13] is characterized by frequencies $\omega_1 \approx \omega_2 \approx 240 \times 10^6 s^{-1}$. Uhrmacher[15] has presented credible evidence that this particular PAC signal arises from indium substituted in an yttrium-copper oxide phase.

We have achieved better success using a sol-gel sample preparation method. An EDTA solution containing Y, Ba, and Cu in ratio 1:2:3 was prepared from high purity metal nitrates. After stirring in the appropriate amount of ^{111}In solution, the mixture was dehydrated and converted to oxides in a ceramic casserole on a hotplate. Samples were synthesized under a variety of conditions at temperatures ranging from 900 to 940 °C for times of 15 minutes to 12 hours. Some were air-quenched. Others were cooled slowly over a number of hours before being placed in the PAC spectrometer furnace. In all cases, most of the indium was dissolved in the majority $YBa_2Cu_3O_x$ phase, but no synthesis condition was found that resulted in dissolving 100% of the indium tracer.

In Fig. 9 we show ^{111}Cd PAC spectra for such samples at four different temperatures. These data were taken with a spectrometer using four NaI(Tl) scintillator detectors but otherwise similar to the apparatus used for the ceria experiments. In Fig. 10 we show the frequencies and relative fractions of two sites that we identify tentatively as indium substituted at the Y site and at the Cu-1 site in $YBa_2Cu_3O_x$. Although the exact site identification is not certain, it is unambiguously clear that the two sites are associated with $YBa_2Cu_3O_x$ and not some second phase. We base this assertion on the behavior at the melting transition. At a temperature near 1000 °C in air and approximately 1030 °C in flowing oxygen, $YBa_2Cu_3O_x$ decomposes into Y_2BaCuO_5 and a melt. As seen in Fig. 9, the PAC spectrum changes substantially above this temperature, with the two $YBa_2Cu_3O_x$ sites disappearing and the Y_2BaCuO_5 site becoming substantially stronger. We have also observed that the 240 Mrad signal described above persists well above the $YBa_2Cu_3O_x$ decomposition temperature. In such high temperature spectra there is also a substantial contribution from indium atoms dissolved in the melt where their average efg is zero and their contribution to $G_2(t)$ is unperturbed.

The frequencies and relative fractions of the two $YBa_2Cu_3O_x$ sites shown in Fig. 10 both change substantially in the temperature range between 600 and 750 °C where the structure is known to change from orthorhombic to tetragonal. As expected, the site symmetries change from slightly asymmetric at low temperature to symmetric at high temperature. The transition region is still under investigation.

An interesting feature shown in Fig. 10 is that the site occupation probabilities change substantially above 900 °C. A third site having a very large ω_1 of order $450 \times 10^6 s^{-1}$ is found near 1000 °C. We suspect that this may be the Cu-2 site, but further measurements on a higher-resolution spectrometer are required to test this hypothesis.

IV. ACKNOWLEDGEMENTS

We thank Mr. John Griffith and Mr. Randy Lundquist for technical assistance in data acquisition and computer analysis. We also thank Mr. Mark Preddy for his typesetting efforts. The research on cerium oxide was sponsored in part by the US Department of Energy contract DE-FG06-85ER45191. The work on $YBa_2Cu_3O_x$ was supported in part by the US National Science Foundation under contract MSM-8717809.

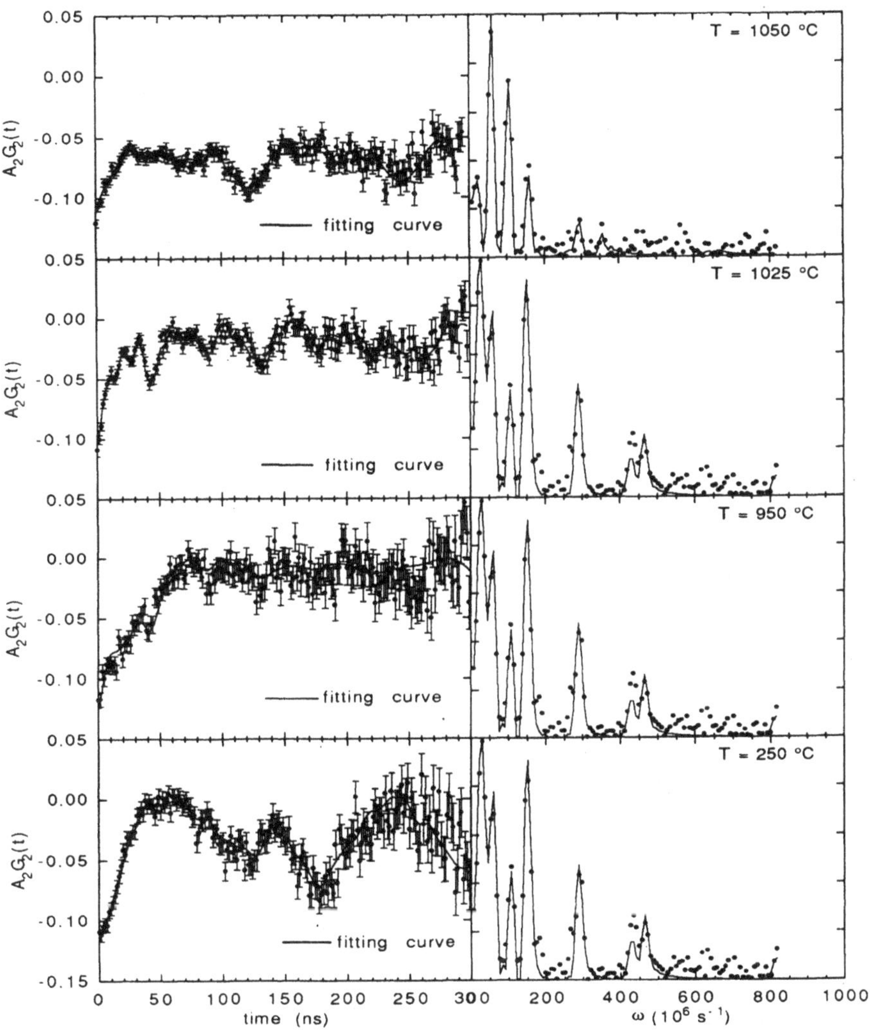

Figure 9: ^{111}Cd PAC experimental A_2G_2 spectra and Fourier transforms for sol-gel-synthesized $YBa_2Cu_3O_x$ samples held in flowing oxygen. At the lowest temperature, the material is in the orthorhombic phase, at the two higher temperatures, it is tetragonal. The highest temperature is above the decomposition temperature.

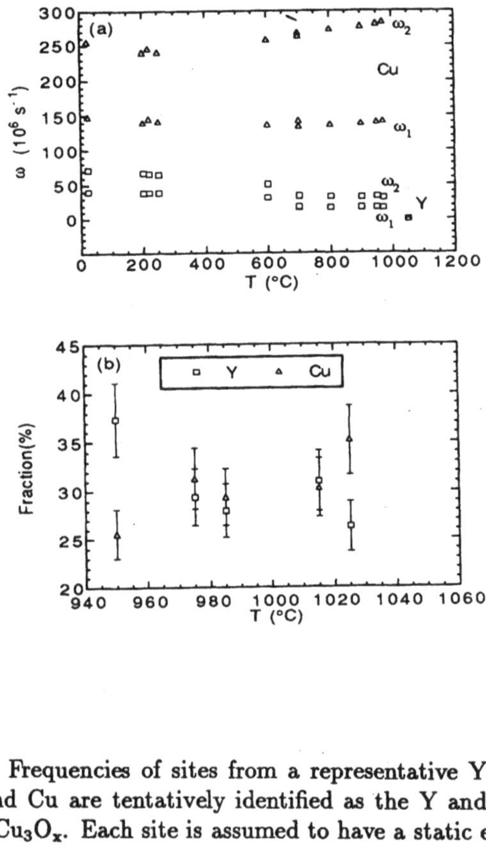

Figure 10: (a) Frequencies of sites from a representative $YBa_2Cu_3O_x$ sample. Sites labeled Y and Cu are tentatively identified as the Y and Cu-1 substitution positions in $YBa_2Cu_3O_x$. Each site is assumed to have a static efg, so spectra near the orthorhombic transition temperature were not properly fitted. (b) Fraction of the Y site and Cu-1 site in a sample for which the Y211 site fraction has a value of 20%.

References

[1] H. Jaeger, John A. Gardner, John C. Haygarth, and R. L. Rasera, J. Am. Ceram. Soc. 69, 458 (1986).

[2] J. A. Gardner, H. Jaeger, H. T. Su, and J. C. Haygarth, Physica B 150, 223 (1988).

[3] H.-T. Su, R. P. Wang, H. Fuchs, J. A. Gardner, W. E. Evenson, and J. A. Sommers, J. Am. Ceram. Soc. 73, 3215 (1990).

[4] R. P. Wang, J. A. Gardner, W. E. Evenson, and J. A. Sommers, paper presented at the 1991 annual meeting of the American Ceramics Society, and to be published in proceedings of the symposium on defects in oxides.

[5] H. L. Tuller and A. S. Nowick, J. Phys. Chem. Solids, 38, 859 (1977).

[6] R. Gerhardt-Anderson and A. S. Nowick, *Transport in Nonstoichiometric Compounds,* edited by G. Simkovich, (Plenum Press, New York, 1985).

[7] D. Y. Wang and A. S. Nowick, J. Phys. Chem. Solids, 77, 639 (1983).

[8] R. Gerhardt, W.-K. Lee, and A. S. Nowick, J. Phys. Chem. Solids 48, 563 (1987).

[9] M. Uhrmacher and A. Bartos, paper presented at the Eighth International Conference on Hyperfine Interactions, Prague, 1989, to be published in conference proceedings issue of Hyperfine Interactions.

[10] H. Plank, F. Meyer, and W. Witthuhn, Phys. Lett. A133, 451 (1988).

[11] P. Singh, M. N. Nyayate, S. H. Devare, and H. G. Devare, Phys. Rev. B 39, 2308 (1989).

[12] G. L. Catchen, M. Blaszkiewicz, A. J. Baratta and W. Huebner, Phys. Rev. B 38, 2824 (1988).

[13] H. T. Su, A. G. McKale, S. S. Kao, L. L. Peng, W. H. Warnes, John A. Gardner, and J. A. Sommers, *Ceramic Superconductors II,* (American Ceramic Society, Westerville, Ohio, 1988).

[14] J. A. Gardner, H. T. Su, A. G. McKale, S. S. Kao, L. L. Peng, W. H. Warnes, J. A. Sommers, K. Athreya, H. Franzen, and S.-J. Kim, Phys. Rev. B 38, 11317 (1988).

[15] M. Uhrmacher and A. Bartos, Hyperfine Interactions 61, 1073 (1990).

[16] G. Weidlich, M. Goelz, R. P. Wang, W. E. Evenson, J. A. Gardner, D. A. Keszler, J. A. Sommers, and J. E. Ostenson, J. Mater. Res. 6, 446 (1991).

[17] H. Jaeger, J. A. Gardner, H. T. Su, and R. L. Rasera, Review of Scientific Instruments, 58, 1694 (1987).

[18] A. G. Bibiloni, C. P. Massolo, J. Desimoni, L. A. Mendoza-Zélis, F. H. Sánchez, A. F. Pasquevich, L. Damonte, and A. R. López-García, Phys. Rev. B 32, 2393 (1985).

[19] A. G. Bibiloni, J. Desimoni, C. P. Massolo, L. A. Mendoza-Zélis, A. F. Pasquevich, F. H. Sánchez, and A. R. López-García, Phys. Rev. B 29, 1109 (1984).

[20] R. P. Wang, PhD thesis, Oregon State University, 1991, (unpublished).

[21] W. E. Evenson, A. G. McKale, H. T. Su, and J. A. Gardner, Hyperfine Interactions 61, 1379 (1990).

[22] W. E. Evenson, J. A. Gardner, R. P. Wang, H.-T. Su, and A. G. McKale, submitted to Hyperfine Interactions, July, 1990.

[23] D. J. M. Bevan and J. Kordis, J. Inorg. Nucl. Chem. 26, 1509 (1964).

PHASE TRANSITIONS STUDIED VIA PERTURBED ANGULAR CORRELATIONS: METAL OXIDES AND ION-MIXED METALLIC MULTILAYERS

K. P. Lieb, T. Weber, M. Uhrmacher, Z. Inglot and A. Bartos
II. Physikalisches Institut, Universität Göttingen,
D-3400 Göttingen, Fed. Rep. Germany

ABSTRACT
This lecture surveys PAC studies with implanted radioactive ^{111}In tracers on various structural and magnetic phase transitions in binary metal oxides. First results on the amorphisation of Ni-Al multilayers and the formation of the NiAl intermetallic phase after heavy-ion beam irradiations are described. It is shown that PAC is a very sensitive method to monitor phase transitions on a microscopic scale.

1. INTRODUCTION

Since many years, nuclear methods like Mössbauer spectroscopy (MS), perturbed angular correlations and distributions (PAC, PAD) and muon spin resonance (μSR) have been applied in condensed matter physics to extract information on the micro-environments of the probes, via their hyperfine interaction(s). The PAC method with ^{111}In tracers has been most widely used to measure electric field gradients (efg) and/or magnetic hyperfine fields in metals, semiconductors and compounds and to study in this way radiation induced defects [1], aggregates of the probe atoms with defects and gases (like He- and O-decorated defect clusters [2-4]) and many kinds of solid state reactions. The Zakopane Schools have witnessed challenging developments in this field and have established a lively tradition in presenting and discussing them.

*Supported by Deutsche Forschungsgemeinschaft (program Li325/2-1 and Sonderforschungsbereich 345, Universität Göttingen)

At Göttingen we have performed PAC studies with implanted ^{111}In tracer ions in a large number of binary and ternary metal oxides and more recently in high-temperature superconductors and similar ceramics [5-18]. The main motivation of these experiments has been to correlate the measured hf constants, i.e. the asymmetry parameter η and strength V_{zz} of the efg tensor, or the magnetic hf field, to the crystallographic structure of the lattice and the local oxygen coordination. Chapter 2 summarizes some recent results on the scaling of the quadrupole hf interaction of implanted ^{111}In in binary oxides. Having identified the efg parameters of substitutional ^{111}Cd in a defect-free environment, we then have been able to follow phase transitions (structural, magnetic and chemical phase transitions [5-7]). Some of the most illustrative examples which we have obtained in the last two years will be sketched in chapter 3.

More recently, Weber et al. [20,21] and Wodniecki et al. [22] have demonstrated that PAC also serves as a very sensitive method to study phase transitions induced by heavy ion bombardments of metallic (multi-)layers. We shall demonstrate in chapter 4 the potential of this method in the case of ^{111}In-doped Ni-Al multilayers irradiated with Ar and Xe ions. For this system the stoichiometry of the layers and the irradiation and annealing temperatures have been varied in order to test under which conditions an amorphous or cristalline intermetallic phase is being produced. A similar study has been performed for the Ni-Ti system [22].

2. THE EFG OF ^{111}Cd IMPURITIES IN SOME BINARY METAL OXIDES

2.1 Experimental Procedure

In a typical PAC experiment, some 10^{12} $^{111}In^+$-ions are implanted at 400 keV and at room temperature into the oxide powder samples pressed onto thin supporting Ag-foils. The samples are then annealed for 30-60 min at a temperature T_a and oxygen pressure p_a where no phase transformation is known to occur and ^{111}In does not diffuse out of the sample. Alternatively, the temperature and oxygen pressure during annealing can be chosen in order to induce a phase transition which most often results in large fractions of the ^{111}In probes on substitutional defect-free cation sites. The PAC data are taken with several four-detector apparatus, equipped with either NaI(*Tl*) or BaF_2 detectors. The measuring temperature T_m of the sample can be chosen in the range from 10 K (closed cycle He-cryostate) to ≈1000 K; the oxygen pressure at the sample can be varied from $p_m \approx 10^{-3}-10^{3}$ mbar, data can also be taken at 10^{-6} mbar.

The perturbation functions R(t) are generally fitted with up to five fractions f_i of static hf environments, where each quadrupolar fraction is characterized by its mean quadrupole frequency $\nu_Q \propto V_{zz}$, asymmetry parameter $\eta=(V_{xx}-V_{yy})/V_{zz}$ and width δ of the Lorentzian/Gaussian efg distribution. In some cases, a dynamic hf interaction component had to introduced which we relate to the relaxation of the ^{111}Cd atomic shell after the distortions caused by the EC-decay of ^{111}In mother nuclei. By either reducing the atomic relaxation time via an increased measuring temperature T_m or by taking PAC spectra with isomeric ^{111}Cd probes, Bartos et al. [15] have been able to show that, indeed, this relaxation mechanism causes ^{111}In-decay "after-effects". In the case of magnetic interaction, the PAC spectrum exhibits the Larmor frequency ω_L and its first harmonic $2\omega_L$. Only in the case of antiferromagnetic CuO and Cr_2O_3, we had to deal with combined interactions.

2.2 Scaling of the efg in Oxides of Bixbyite Structure

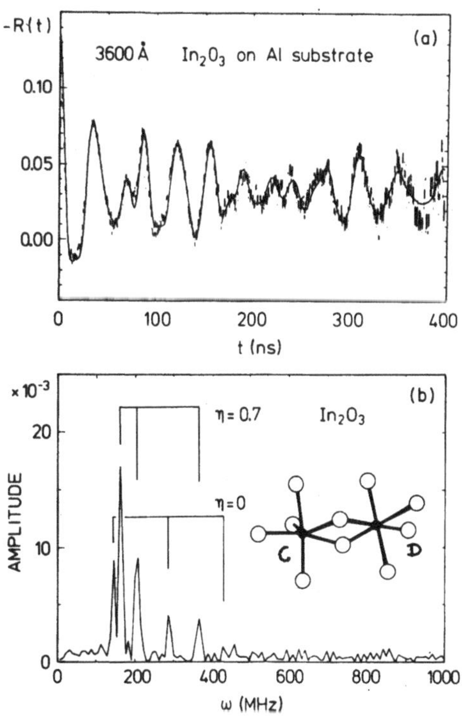

As most oxides studied are ionic semiconductors or insulators, one expects that the point charge model (PCM) gives a rather reliable estimate of the asymmetry parameter η of the efg. This parameter is very essential, as many oxides feature several non-equivalent ^{111}In cation sites as well as complex oxygen coordinations and structural defects in the cation sublattice, which often can only be distinguished by their η-values. A very good example is the In_2O_3 bixbyite-type lattice which in its cu-

Fig. 1: Perturbation function and Fourier transform of ^{111}In in In_2O_3 [6].

bic C-form [space group $Ia3(T_h^7)$] features two non-equivalent sites with fractions in the ratio $f_C/f_D=3:1$, quadrupole frequencies in the ratio $\nu_{QD}/\nu_{QC}=1.60$ and asymmetry parameters $\eta_C=0.81$ and $\eta_D=0$, according to the PCM and when adopting a normal population of both sites. The measured perturbation function of ^{111}Cd in In_2O_3 (see fig. 1) indeed shows these two fractions with the expected efg parameters [6]: $\eta_D=0.00(5)$, $\eta_C=0.69(5)$, $f_C/f_D=3.3(5)$ and $\nu_{QD}/\nu_{QC}= 1.29(3)$.

Bartos, Shitu, Wiarda et al. [15-17] have recently studied in detail the scaling of the efg in a number of bixbyite oxides differing in their lattice constant a as much as by 21%. Fig. 2 illustrates the perturbation functions and Fourier transforms for ^{111}In in Sc_2O_3, Y_2O_3, Dy_2O_3 and Yb_2O_3 [15]. On first inspection, these data nicely confirm the similarity among these four oxides and clearly exhibit the symmetric ($\eta_D=0$) and asymmetric ($\eta_C\approx0.8$) fractions. If we plot the quadrupole frequencies ν_{QC} and ν_{QD} on both cation sites versus a^{-3} (see fig. 3a), we do not find, however, an exact linear dependence as implied by the PCM, although both frequencies staedily increase for decreasing a: This small deviation from linearity can be traced to the varying degree of ionicity of the cation-oxygen bond in the various compounds. An expression for the ionicity is the bond valence $s(Cd) = \exp[(R-d_{Cd-O})/B]$ where R=0.1904 nm and B=0.037 nm are empirical constants and d_{Cd-O} is the Cd-O bond length in the various oxide lattices [19]. It is seen in fig. 3c that ν_{QC} strongly depends on s(Cd), while ν_{QD} does not. By plotting $\nu_Q/s(Cd)$ versus a^{-3}, we find that both the ionic and covalent bonding contribution for site C are taken into account. The constant value of $\nu_{QC}/s(Cd)=238(8)$ MHz/v.u. (fig. 3b) thus indicates that ν_{QC} scales with $s(Cd)/a^3$. For the efg at site D this conclusion cannot been drawn: ν_{QD} depends very weakly on the lattice constant. A similar result has been previously obtained for ^{111}Cd in Cu_2O and Ag_2O [6,14]. MSX-cluster calculations by Nagel have shown that the efg at ^{111}Cd in this linear $Cd-O_2$ complex is nearly independent on the Cd-O bond length.

2.3 Validity of the Point Charge Model

In spite of the partially covalent bonding of most metal oxides, it has been customary to compare the measured quadrupole hf interaction parameters with the predictions of the PCM. The calculated PCM parameters, ν_Q^{PCM} and η^{PCM}, strongly depend on the lattice sum, i.e. the number of ions contributing to the efg. This is illustrated in fig. 4 for the case of a dilute ^{111}Cd ion in a defect-free In_2O_3

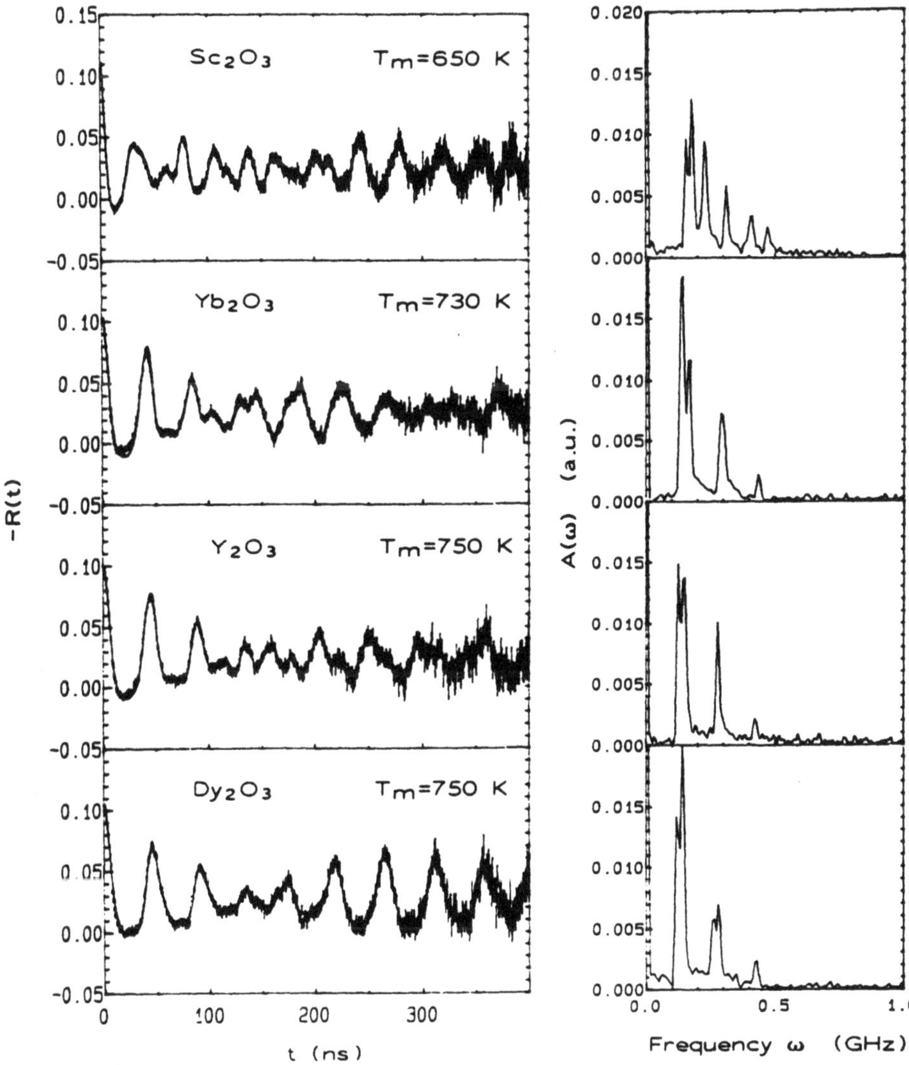

Fig. 2: Perturbation functions and Fourier transforms of ^{111}Cd in some bixbyite structure oxides. Note the similarities in efg-parameters [15].

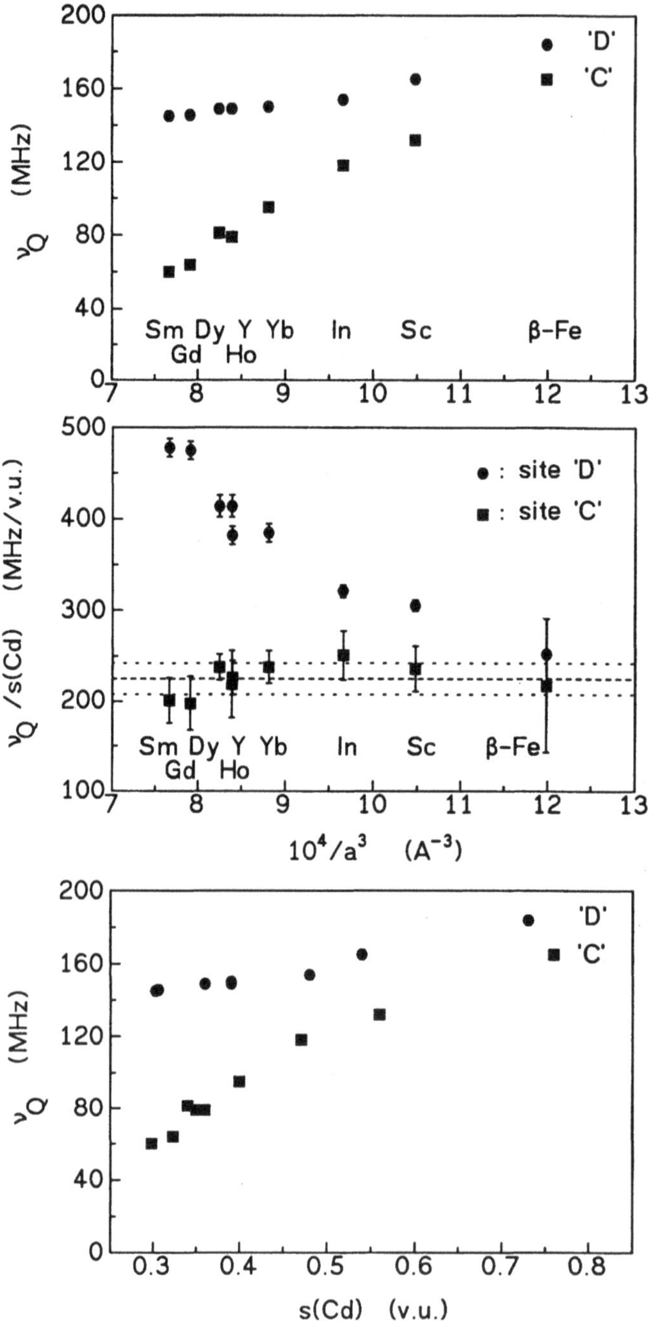

Fig. 3: Scaling of efg-parameters in bixbyite oxides;
a) ν_Q versus a^{-3};
b) $\nu_Q/s(Cd)$ vs. a^{-3};
c) ν_Q versus $s(Cd)$.

lattice: the parameters ν_Q^{PCM} and η^{PCM} stabilize if the radius R_L of the sphere of ions around the probe contributing to the lattice sum exceeds 15A and 30A, which is about 7-15 times the average In-O bond length. This calculations neglects the change in ion valence and size resulting from the ^{111}In → ^{111}Cd radioactive decay.

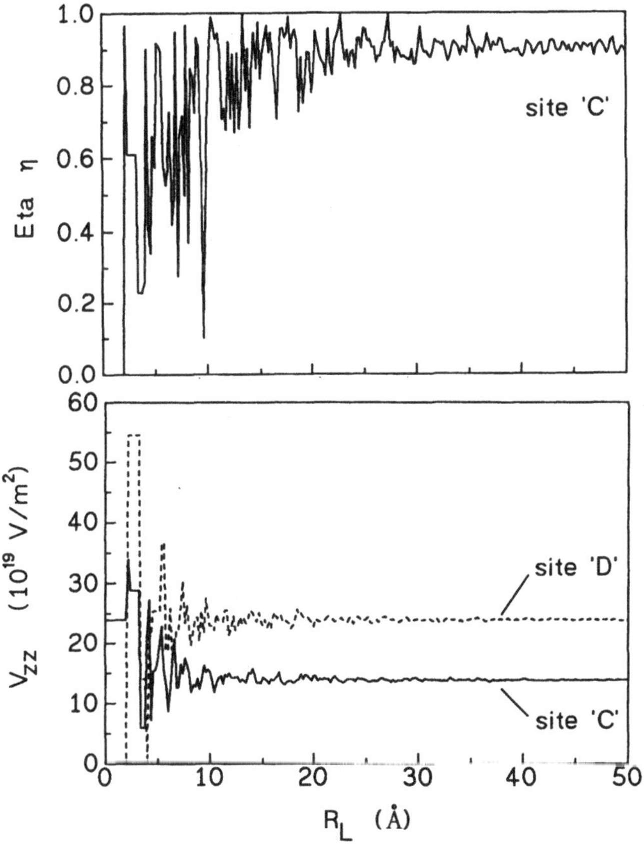

Fig. 4: Calculated η and V_{zz} parameters for In in In_2O_3 as function of the size of the sphere the ions of which contribute to the lattice sum, according to the PCM.

Fig. 5 illustrates the status of measured quadrupole frequencies for substitutional ^{111}Cd in binary metal oxides of cuprit, tenorit, bixbyite and corundum structure. Shown are the antishielding factors $\beta=\nu_Q/\nu_Q^{PCM}$ [13]. The data show a clear correlation of the antishielding factor β with the number n_o of nearest-neighbor oxygen ions and the oxide cristallographic structure, with β increasing for increasing n_o. The antishielding factor changes from β≈3-4 for Ag_2O and Cu_2O to β≈100 in corundum (six neighboring O^{--} ions). The origin of this correlation is under investigation.

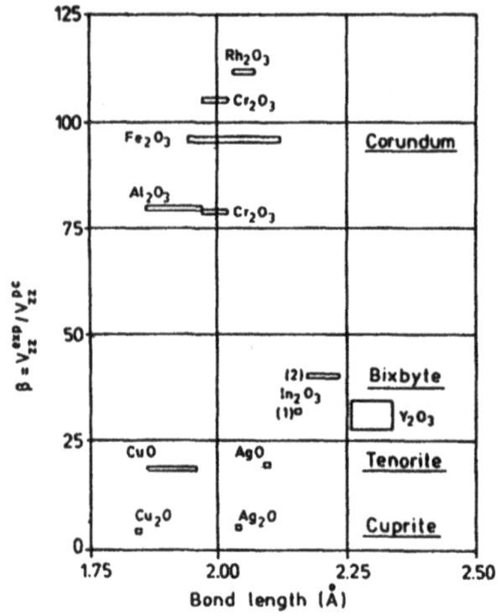

Fig. 5: Antishielding factor β for substitutional ^{111}Cd in various binary oxides. Note that the β-values for a given structure are very similar.

In summary, the scaling of the quadrupole frequencies mentioned above and the agreement of the measured asymmetry parameters with the predictions of the PCM are first (and still rather empirical) steps into an understanding of the efg's in oxides. These results should provide a good experimental basis for theoretical calculations. Recently, we have used these arguments not only to identify the formation of various oxide phases during REDOX transformations of some fcc-metals and various oxides [6,7,9-11,27], but also to distinguish among possible defect configurations associated with these phase transformations. In this context we mention our studies on cobalt, iron and manganese monoxides [11,18,27] where a large variety and concentration of structural defects have been proposed.

3. STRUCTURAL AND MAGNETIC PHASE TRANSITIONS IN METAL OXIDES

3.1 Structural Phase Transition in Rh_2O_3

In their investigation of oxides with corundum structure, Kesten et al. [12] have reported on the α-Rh_2O_3 → β-Rh_2O_3 phase transition. While α-Rh_2O_3 cristallizes in the rhombohedral (corumdum) phase, the structure of β-Rh_2O_3 is orthorhombic; the phase transition is first order and irreversible, it occurs rather slowly at 975-1325 K. The PCM predicts axial symmetry in the α-phase, but a large η-value for the β-phase. After ^{111}In implantation into polycristalline α-Rh_2O_3, PAC spectra were taken in an isochronal annealing cycle (t_a=30 min; T_a=475 and 675 K in vacuum; T_a=875 K, 1075 K, 1183 K and 1223 K at p_a=50 mbar oxygen to avoid desintegration at the high temperatures). Two fractions with well defined quadrupole parameters and a third one with a broader frequency distribution were observed:

1) ν_Q=131(1) MHz, $\eta\approx0.2$, $\delta\approx10$ MHz,
2) ν_Q=118(1) MHz, η=0.90(1), $\delta\approx$ 7 MHz,
3) ν_Q=165(9) MHz, η=0.55(4), δ=53 MHz.

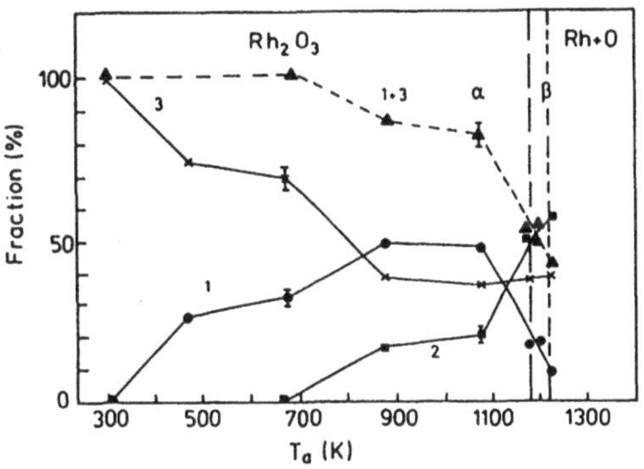

Fig. 6: Structural phase transition in Rh_2O_3 [13].

We interpret fraction 1 as due to substitutional ^{111}Cd in the α-phase, which then transform near 1100 K into the β-phase (fraction 2) as shown in fig. 6. The origin of fraction 3 is still unclear and may be related to defects trapped at the In probe in the α-phase.- Another good example are the PAC experiments in manganese oxides discussed by Wiarda, et al.[27].

3.2 Magnetic Phase Transitions

Magnetic phase transitions have been for a long time the subject of intense hf interaction studies [28]. Oxides offer a rich variety of magnetic ordering phenomena many of which can be made visible via PAC. The line of research which we pursue concentrates on the following topics:

i) Critical behaviour for substitutional ^{111}Cd impurities in comparison with bulk properties: determination of the critical temperature T_c and exponent β and range of the validity of the Curie-Weiss power law $B(t)/B_0=t^\beta$ with $t\equiv1-T/T_c$;

ii) Dependence of the supertransferred antiferromagnetic field on the oxygen coordination and bonding;

iii) Influence of defects and the role of magnetic ordering during REDOX phase transitions.

So far we have obtained results on ^{111}Cd in NiO, CoO, CuO, MnO, Cr_2O_3, Co_3O_4 and Fe_3O_4 [6,9-12,18,27]. Figs. 7 and 8 show the most recent results by Inglot et al. [18] for the ferrimagnetic system

^{111}CdFe_3O_4: fig. 8 illustrates the development of perturbation functions and their Fourier transforms as function of the measuring temperature T_m, while fig. 7 shows the critical behaviour of $\omega_L(t)/\omega_0$ from which the Neel temperature T_c=848(2) K and the critical exponent β=0.392(6) have been derived. In spite of the difficulty to measure small Larmor frequencies near T_c, it is estimated that the t^β-law is fulfilled over a quite broad temperature range: $3*10^{-3} \leq t \leq 0.6$. MS experiments in Fe_3O_4 by Häggström [29] have given β=0.39(1). A recent, less complete PAC study with ^{111}In-tracers by Asai et al. [30] has confirmed our results.

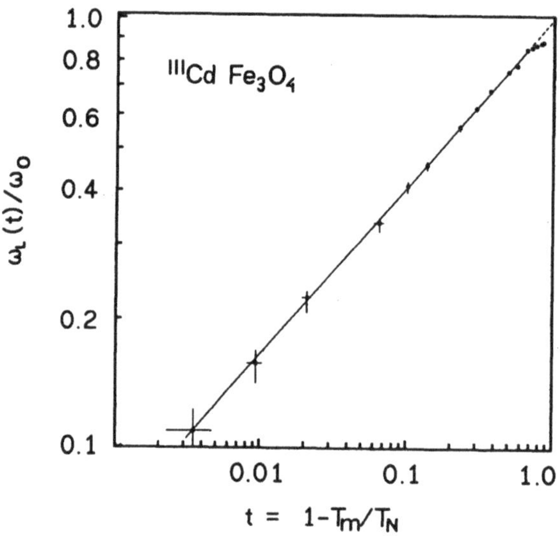

Fig. 7: Critical behaviour of the Larmor frequency for substitutional ^{111}Cd in ferrimagnetic Fe_3O_4.

(Below T≈170 K, i.e. outside the range of the t^β-validity, an additional 10-15% change of the Larmor frequency was noted in the PAC spectra which is associated with a Vervey phase transition [31].)

Table 1 summarizes information on the critical parameters for substitutional ^{111}Cd in the above mentioned oxides. Apart from CuO and Cr_2O_3, the magnetic sites have a cubic oxygen coordination and therefore no efg. In CuO and Cr_3O_3, the (antiferro-)magnetic frequencies are considerably smaller than the quadrupole frequencies leading to a line broadening and/or frequency splitting in the Fourier spectra. In all cases where β could be determined, it is close to the value β=0.380 typical for a 3D Heisenberg magnet. However, the very precise results in CoO and Fe_3O_4 indicate a 2-4% deviation from a pure Heisenberg behaviour.

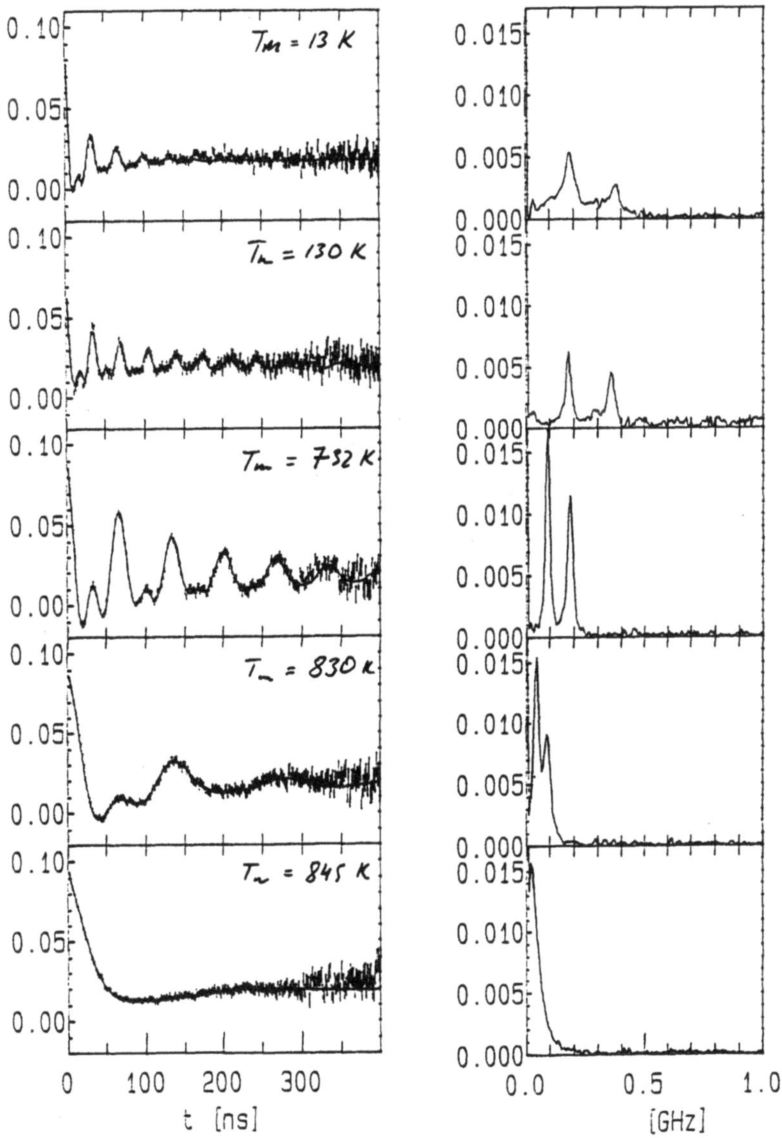

Fig. 8: PAC spectra of ^{111}Cd in ferrimagnetic Fe_3O_4 taken as function of the measuring temperature T_m. The critical parameters T_c, ß and ω_0 are listed in Table 1 (18).

Table 1: *Critical parameters of ^{111}Cd impurities in magnetically ordered oxides*

System	Structure^{a}	T_c (K)	β	B_0^{b} or B (T)	Ref.
NiO	NaCl; AF	543(2)	0.385(15)	22.2(4)	[6.9]
CoO	NaCl; AF	291.2(3)	0.393(5)	20.9(4)	[11]
Fe_3O_4	Spinell; FM	848(2)	0.392(3)	13.9(2)	[18]
CuO	monocline; AF	224(2)	0.365(5)	2.7(3)^{c}	[10]
α-Cr_2O_3	Corundum; AF	307	----	$\leq$0.41^{d}	[12]

a) FM=ferrimagnetic hf; AF=antiferromagnetic hf

b) B_0 extrapolated to 0K with Curie-Weiss law

c) At 10 K; d) At 60 K

4. PAC AS AN INDICATOR FOR ION-BEAM INDUCED PHASE TRANSITIONS IN METALLIC MULTILAYERS

4.1 Motivation

Ion-beam irradiation of metallic multilayers is a common method to produce intermetallic or amorphous phases which often cannot be generated in thermodynamic equilibrium conditions. Ion-beam mixing is a complicated process which depends on the layer composition and temperature as well as on the ion beam energy, species and fluence. We have recently demonstrated that PAC is a very elegant and profitable analyzing method in this field and can add detailed information, in particular to clarify the early stages of the mixing process in thin films which may be difficult to detect by Rutherford backscattering (RBS), TEM or X-ray diffraction.

Weber et al. [20,21] have collected rather complete information on the Ni-Al system which in this context offers various advantages: the mass contrast is high enough to follow the ion mixing process via RBS. Moreover, the known magnetic hf interaction at ^{111}In-dopants in Ni and the quadrupole interaction in the intermetallic NiAl phase are very clear markers which allow to carefully monitor the phase transition induced by ion irradiation. The Ni-Al phase diagram is shown in fig. 9. NiAl is a highly ordered B2 alloy with a wide composition range from c_{Ni}=45-60 at%. After room temperature Xe-mixing of Ni-Al multilayers, the formation of crystalline NiAl has been observed by Jaouen et al. [23]; according to Hung et al. [24], this phase is stable under RT Xe-bombardment. Other stable equilibrium phases are Ni_3Al, Ni_2Al_3 and $NiAl_3$.

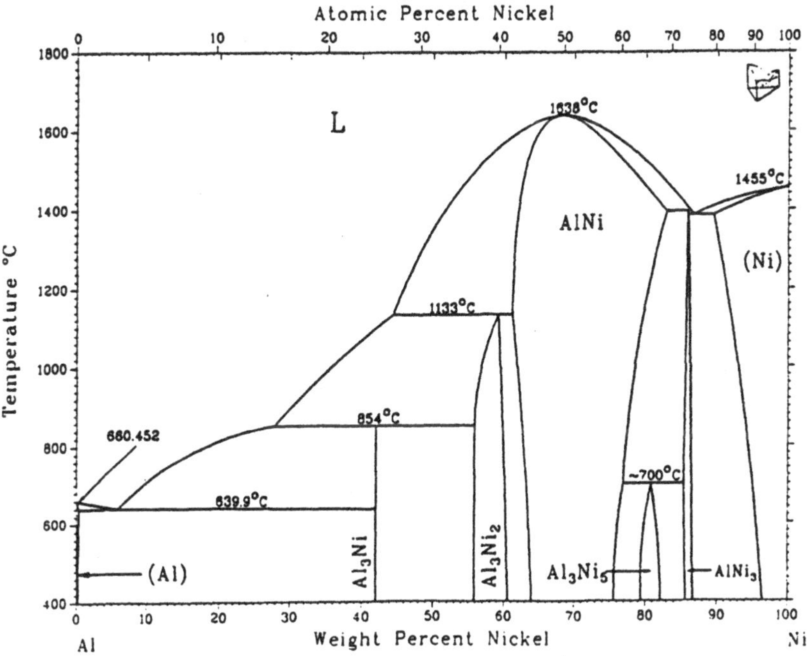

Fig. 9: Phase diagram of the Al-Ni system

4.2 Experimental Techniques

A typical experiment proceeds in the following way: A stack of 4-8 Ni and Al layers was deposited via electron-gun evaporation onto an Al or Al_2O_3 substrate at a pressure of $\sim 10^{-8}$ mbar in an UHV chamber which is connected to our IONAS implanter, via a differential pumping system. This arrangement allows layer production, RBS depth-profiling, ion mixing and ^{111}In implantation without breaking the vacuum (see fig. 10). So far PAC was performed with the ^{111}In-doped samples taken out of the chamber. The individual layers were 10-30 nm thick, with their relative thicknesses adjusted to produce the desired stoichiometry after mixing. In order to avoid oxide layers, the top layer of the stack was always Ni. The multilayer stack then was doped at 400 keV with 10^{12} $^{111}In^{+}$-ions so that the ^{111}In range profile covers most part of the stack. These (unannealed) samples were finally irradiated with 600 keV Ar^{+} or 900 keV Xe^{++} ions with fluences up to $2*10^{16}$ ions/cm^2, either at RT or LN temperature. The irradiated samples were RBS analyzed at RT with 900 keV α-particles and via PAC. Thermal annealings were also carried out.

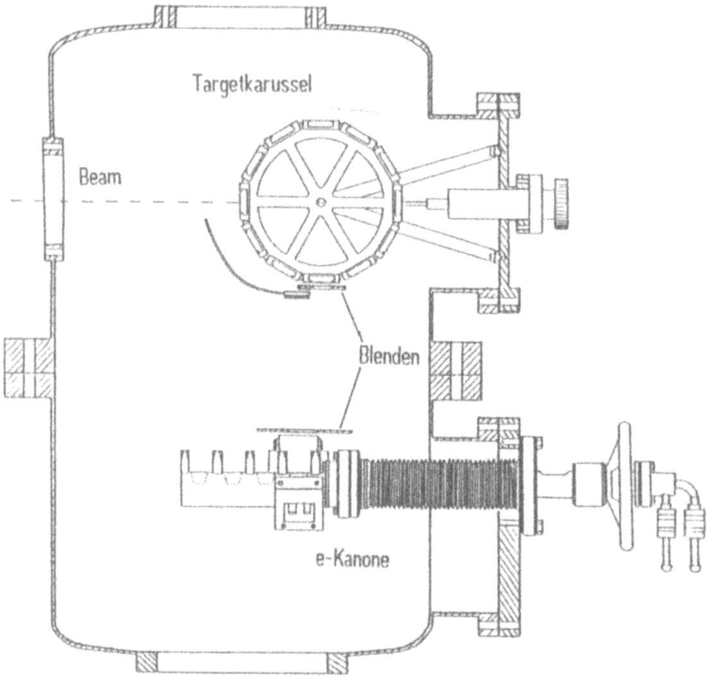

Fig. 10: UHV chamber with electron gun and target carousel connected to the IONAS ion implanter

4.3 Results

Figs. 11 and 12 illustrate how the RBS and PAC spectra of a 4+4 stack develop as function of the 900 keV Xe^{++} ion irradiation at RT, up to a fluence of $12*10^{15}$ ions/cm^2. In the RBS spectra of the Ni component (fig.11), one notes how the "valleys" representing the Al-layers are successively filled with Ni indicating interface mixing. In the PAC spectra and their Fourier transforms (fig. 12), we observed the disappearance of the Ni Larmor frequency (ω_L=93 MHz [25]) and the appearance of the quadrupole frequency typical for ^{111}Cd in crystalline NiAl [26]. The RBS spectra revealed full intermixing at about 10^{16} ions/cm^2, while the PAC spectra indicate the formation of the NiAl phase already at $4*10^{15}$ions/cm^2. It thus appears that the early stages of phase formation can easily be made visible via PAC.

A second series of PAC measurements in the Ni-Al system was devoted to study the effects of changing the relative layer thicknesses and

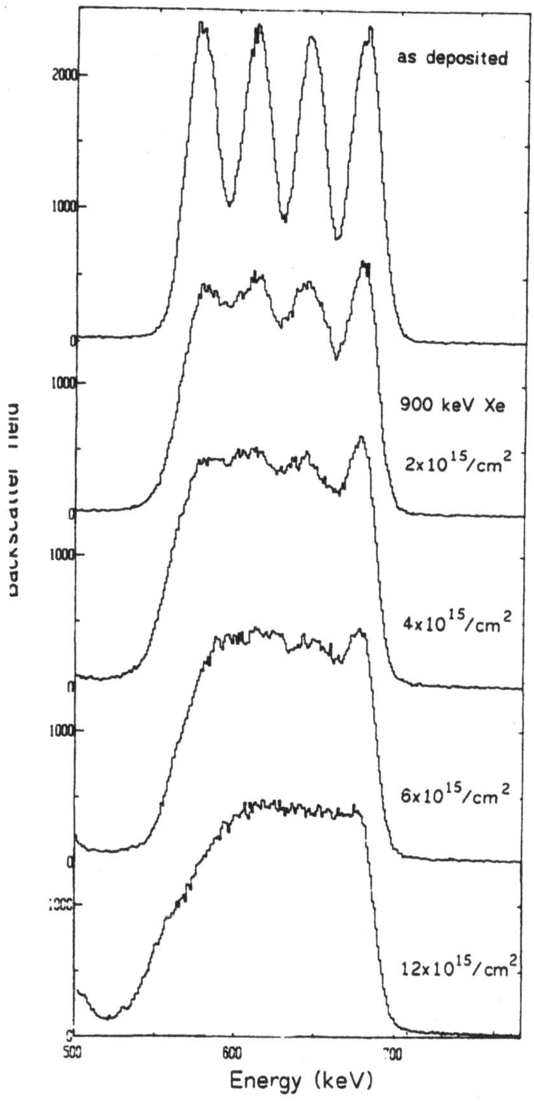

irradiation temperature [21]. 900 keV Xe^{++}-ion irradiations at RT of 4+4 stacks prepared in such a way as to produce the stoichiometries of the Ni_3Al or $NiAl_3$ phases after ion mixing, resulted in broad frequency distributions typical for ^{111}Cd in amorphous samples (see fig. 13). It is, however, noteworthy that the centers of these distributions do not coincide, being ≈85 MHz in the case of Ni_3Al and ≈200 MHz in the case of $NiAl_3$.
Finally, LN irradiations of 4+4 multilayer stacks in the 50:50 atomic ratio gave a broad frequency distribution peaking around 250 MHz and possibly indicating an amorphous phase. We also noted a small component typical for crystalline AlNi. Xe post-irradiation of this sample at RT enhanced re-cristallisation and generated a PAC pattern typical for the intermetallic NiAl alloy.

Fig. 11: RBS spectra from a Ni-Al 4+4 stack irradiated with 900 keV Xe^{++}-ions to the doses indicated [20].

Acknowledgements

Some of the results presented here have been accumulated in collaboration with A.F. Pasquevich, J. Shitu, T. Wenzel and D. Wiarda. We are most grateful to them for allowing us to quote from unpublished work. The help of L. Ziegeler and D. Purschke is also acknowledged.

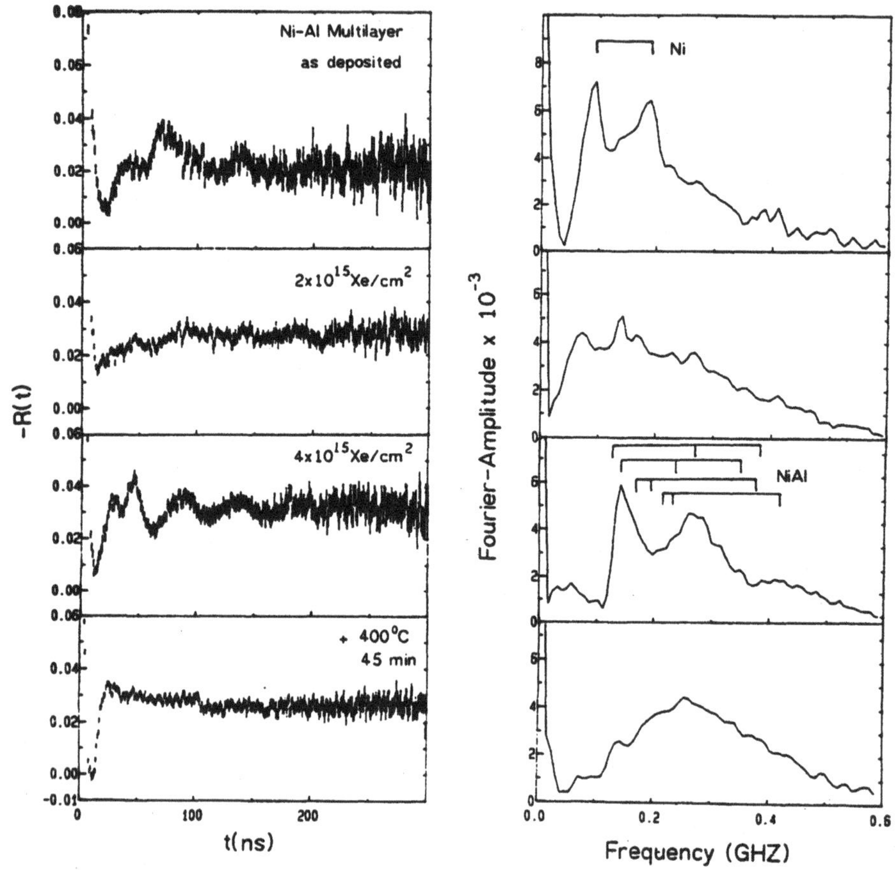

Fig. 12: PAC spectra obtained from a 4+4 Ni-Al multilayer stack doped with ^{111}In and irradiated at RT with 900 keV Xe^{++} ions [20]. The typical frequencies of ^{111}Cd in Ni and intermetallic NiAl are indicated.

REFERENCES

[1] Recknagel, E., Schatz, G., Wichert, T., In "Hyperfine Interactions of Radioactive Nuclei, ed. by Christiansen, J., Topics of Current Physics, vol. 31 (Springer, Heidelberg, 1983) p. 133

[2] Bolse, W., Wodniecki, P., Schröder, H., Uhrmacher, M., Lieb, K.P., Phys. Lett. 93A,429 (1983); Wegner, D., Schröder, H., Lieb, K.P., Rad. Eff. Lett. 85,13 (1984)

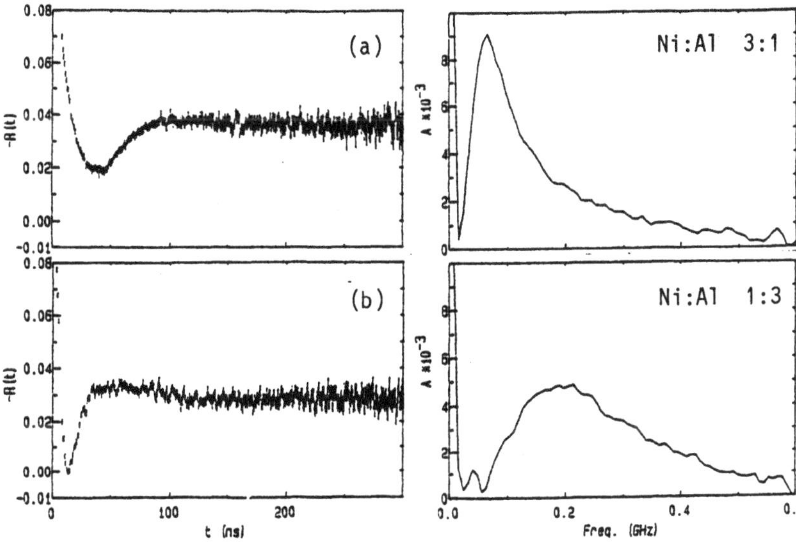

Fig. 13: PAC spectra of two 4+4 Ni-Al multilayer stacks doped with ^{111}In and irradiated at RT with $10^{16}/cm^2$ 900 kev Xe-ions. The Ni-Al stoichiometries were 3:1 (a) and 1:3 (b). Ion irradiation does not form the intermetallic phases [21].

[3] Schröder, H., Bolse, W., Uhrmacher, M., Wodniecki, P., Lieb, K.P., Z. Phys. B65, 163 (1986); Wegner, D., Uhrmacher, M., Lieb, K.P., Z. Phys. B68, 461 (1987)

[4] Pleiter, F., in "Solid State Reactions After Ion Implantation Detected with Nuclear Methods" (Göttingen, 1986) p. 125

[5] Uhrmacher, M., Schröder, H., Bolse, W., Wodniecki, P., Lieb, K.P., XX Winter School on Physics, Krakow (1985) p. 100; Bolse, W., Bartos, A., Kesten, J., Uhrmacher, M., Lieb, K.P., XXIII Zakopane School of Physics, Krakow (1988) p. 127

[6] Bolse, W., Uhrmacher, M., Lieb, K.P., Phys. Rev. B36, 1818 (1987)

[7] Bolse, W., Lieb, K.P., Uhrmacher, M., Wegner, D., ELAF '87, La Plata, (World Scientific, Singapore, 1988) p. 665.

[8] Uhrmacher, M., Bartos, A., Hyp. Int. 61, 1073 (1990); XXV Zakopane School of Physics (World Scientific, Singapore, 1990) p. 25
[9] Bolse, W., Uhrmacher, M., Lieb, K.P., Hyp. Int. 35, 631 (1987)
[10] Bartos, A., Bolse, W., Lieb, K.P., Uhrmacher, M., Phys. Lett. A130, 177 (1988); Hyp. Int. 50, 619 (1989)
[11] Inglot, Z., Wegner, D., Lieb, K.P., Hyp. Int. 50, 785 (1989); Wegner, D., Inglot, Z., Lieb, K.P., Ber. Bunsen-Ges. Phys. Chem. 94, 1 (1990)
[12] Kesten, J., Uhrmacher, M., Lieb, K.P., Hyp. Int. 59, 309 (1990)
[13] Kesten, J., Bolse, W., Lieb, K.P., Uhrmacher, M., Hyp. Int. 60, 683 (1990)
[14] Bolse, W., Uhrmacher, M., Kesten, J., Hyp. Int. 35, 931 (1987)
[15] Bartos, A., Lieb, K.P., Pasquevich, A.F., Uhrmacher, M., submitted
[16] Shitu, J., Bartos, A., Lieb, K.P., Uhrmacher, M., Wenzel, T., Wiarda, D., to be publ.
[17] Wiarda, D., Wenzel, T., Uhrmacher, M., Lieb, K.P., to be publ.
[18] Inglot, Z., Wiarda, D., Lieb, K.P., Wenzel, T., Uhrmacher, M., J. Phys. Cond. Matter, in press
[19] Brown, I.D., Shannon, R.D., Acta Cryst. A29, 266 (1973); Brown, I.D., Aldermatt, D., Acta Cryst. B41, 244 (1985)
[20] Weber, T., Lieb, K.P., Uhrmacher, M., Surf. Interf. Anal., in press
[21] Weber, T., Lieb, K.P., contribution to IBA-10, Eindhoven (1991)
[22] Wodniecki, P., Corts, T., Lieb, K.P., Uhrmacher, M., contribution to E-MRS, Strassbourg (1991)
[23] Jaouen, C., Riviere, J.P., Delafond, J., Nucl. Instr, Meth. B7/8, 591 (1985); *ibid.* B19/20, 549 (1987)
[24] Hung, L.S., Nastasi, M., Gyalai, J., Mayer, J.W., Appl. Phys. Lett. 42, 672 (1983)
[25] Hohenemser, C., Kachnowski, T., Bergstresser, T.K., Phys. Rev. B13 3154 (1976); Pleiter, F., Hohenemser, C., Phys. Rev. B25, 106 (1982)
[26] Fan, J., Collins, G., Hyp. Int. 60, 655 (1990)
[27] Wiarda, D., Bartos, A., Lieb, K.P., Uhrmacher, M., Wenzel, T., XXV Zakopane School of Physics (World Scientific, Singapore, 1990)
[28] Rinneberg, H.H., Shirley, D.A., Phys. Rev. B13, 2138 (1976)
[29] Häggström, L., Annersten, H., Ericsson, T., Wäppling. R., Karner, W., Bjarman, S., Hyp. Int. 5, 201 (1978)
[30] Asai, K., Okada, T., Sekizawa, H., J. Phys. Soc. Jap. 54, 4325 (1985)
[31] Evans, B.J., Westrum, E.F., Phys. Rev. B5, 3791 (1972); Buckwald, R.A., Hirsch, A.A., Sol. State Comm. 17, 621 (1975)

Electron-Gamma PAC Studies at CERN/ISOLDE: present and future

J.G. Correia*, A.A. Melo* M.F. da Silva**, J.C. Soares* and H. Haas***

* Centro de Física Nuclear da Universidade de Lisboa,
Av. Prof. Gama Pinto nº 2, 1699 - Lisboa, Portugal

** Departamento de Física, ICEN/LNETI,
Estrada nacional nº 10 - 2685 Sacavém, Portugal

*** Hahn-Meitner Institut and Freie Universität Berlin, Germany

and

ISOLDE-Collaboration, CERN, Switzerland

Abstract:

The e^--γ PAC Technique has been applied to Materials Science Studies in the seventies. One limitation of the technique is its strong dependence of a flexible radioactive isotope production facility with the possibility of ion implantation. In the present work, the new development of the e^--γ PAC technique as intalled at ISOLDE/CERN is introduced. Results from the first measurements show the advantage in using this facility on line with the ISOLDE separators at CERN for Materials Science and basic Nuclear Physics studies. Short lived isotopes produced in the decay of nuclei very far from the stability line with low energy and highly converted cascades are appropriate candidates for Nuclear Spectroscopy studies and Materials Science Applications.

I. Introduction

The γ - γ PAC technique has been developed in the middle of the seventies and has been extensively applied to Materials Science studies in the last decade. Its contribution to the understanding of the microscopic characterization of defects in materials is very clear[1,2]. However, some limitations, like in other methods, result from the availability of isotopes with halflives and nuclear magnetic or quadrupole moments appropriate to carry on the respective measurements. In many cases γ - γ cascades can be alternatively substituted with advantage by e^-- γ cascades. In fact, low energy E1, E2, M2 and M4 transitions are highly converted and the particle parameter in the K or L-shell is sometimes greater or equal than one. In this case, the well known γ - γ PAC theory is extensive to the e^-- γ PAC studies and new isotopes can be used, particularly, in the case where one of the transitions is completely converted. This is the case in decays of high spin metastable states like in the decay of $^{119m}Sn \rightarrow {}^{119}Sn$ or $^{197m}Hg \rightarrow {}^{197}Hg$ where the M4 conversion electron coefficient is very high[3,4]. The e^-- γ cascade can also be used alternatively in the decay of $^{111m}Cd \rightarrow {}^{111}Cd$ where the conversion electron coefficient is 1.5. The advantage of the method results from the fact that much higher activities can be used with a clear improvement in the statistics and, particularly, in the time range of the analysis due to the suppression of random coincidences. This advantage makes the method attractive, particularly, for studies of long lived levels with lifetimes of few microseconds populated or depopulated by transitions with a very high conversion electron coefficient. The decay of $^{73}As \rightarrow {}^{73}Ge$ is an ideal case to illustrate this sentence.

Details are given in the following sections on the method and on the results, starting to show that the experimental set up fits well near or on line with the ISOLDE facility at CERN. At the end it is shown that basic Nuclear Phyics studies on short lived isotops lying very far from the stability line can open a new and unique field of research for this equipment. New cascades will be prepared for PAC studies applications in the field of Material Science.

II. Experimental Details

In an electron-gamma angular coincidence experiment a compromise has to be done between solid angle integration, energy resolution of the electrons and ratio of true to random coincidences. In order to maximize the true to random coincidence ratio we have to increase the solid admission angle of the magnetic spectrometer, degrading, therefore, the energy and time resolution performance of the set up. The magnetic spectrometer from the Siegbahn type[5] with two magnetic lenses at 90° offers the advantages of increasing the solid admission angle without degrading energy and time resolution. If the system is coupled with two fixed NaI(Tℓ) - detectors at 90° it is equivalent to a 4-detectors PAC γ - γ set up. This situation is very important in order to optimize the statistics and to eliminate effects of misalignments in the sample installation.

The e^- - γ or γ - e^- PAC coincidences have to be performed with the sample in vacuum due to the fact that electron scattering has to be minimized. This disadvantage, relative to the γ - γ PAC technique, offers flexibility, if the sample holder is coupled to the head of a cryopump system and if the equipment is mounted on line with an ion implanter to perform the experiment at any temperature and to carry on temperature dependence experiments in a large temperature range. Ultra-high-vacuum development is possible and in this case the way is open to study, for the first time, surface or near surface phenomena using the e^- - γ PAC technique. The electron detection in the magnetic lenses is done using plastic scintillators in order to optimize the time resolution of the system. This has a disadvantage when detecting low energy electrons if the amplitude of the signals are lower than the photomultiplier noise. In the present situation, using XP1021 photomultipliers the cutting energy is about 11 keV. In order to eliminate this limitation an electron pre-acceleration system can easily be developed which enables to detect electrons with only a few keV energy as if they had the energy of 20 keV. In this way the signal is completely separated from the noise of the photomultiplier. Possible effects of the pre-acceleration in the attenuation of the angular correlation anisotropy coefficient are now being studied. This is very important to use the e^{-}-γ or γ-e^{-} technique for the determination of electromagnetic transition properties in Nuclear Physics studies or to use low anisotropy cascades in Materials Science applications.

The time resolution of the equipment depends strongly on the type of gamma detectors used and on the energy of the electrons detected. Using NaI(Tl) detectors a time resolution of about 2 ns is achieved without big effort for the usual cascades of ^{111m}Cd and ^{197m}Hg. Half of this value can be obtained using BaF_2 detectors.

A routing system has been constructed and coupled to an external Camberra ADC and an IBM AT compatible 20 MHz computer which allow to store the data and to evaluate at the same time the experimental anisotropy function R(t). This is very important if decisions have to be taken when short lived isotopes are measured. The computer system is equipped with all the software needed to analyse the data and to fit the experimental results to the theoretically expected functions.

III. Results

III.1. Magnetic hyperfine fields and electric field gradients measured at Hg and Cd implanted in hcp cobalt single crystals.

The main aim of this experiment is the precise measurement of the absolute values of the magnetic hyperfine fields and electric field gradients of Cd and Hg in hcp Co single crystals. For the first time we have carefully derived the electric field gradients at Hg and Cd in the hcp cobalt lattice. The high performance of the technique is clearly shown in the case of an unavoidable mixed implantation of ^{111m}Cd and ^{111}In, both decaying to the same level in ^{111}Cd through two different γ transitions. In this case the γ-γ PAC technique, cannot be used because the anisotropies are of the same magnitude but with opposite signals. On the contrary, using the e^--γ PAC technique, first, the activity of ^{111m}Cd has been used to measure the hyperfine interaction of Cd in Co and, later, the activity of ^{111}In has been used to measure the temperature dependence of the same hyperfine interaction using the γ-γ PAC technique.

Figs.1 and 2 show representative results obtained with the e^--γ and γ-γ PAC techniques. The cobalt single crystal had the <0001> axis in the plane of the detectors at 45° between the magnetic lens and the gamma detector in the e^--γ set up. In the γ-γ set up, the axis was oriented between detectors (1,2) and (3,4) also at 45°. Using this geometry, it can be shown that only the ω_L Larmor frequency is observed, being the $2\omega_L$ amplitude completely attenuated. The $2\omega_L$ component appears if the measurement is performed with the <0001> axis of the Co single crystal set perpendicular to the detectors plane as is shown in Fig. 3. From the data shown in Figs. 1 to 3 it is clear that the magnetic dipole and electric quadrupole interaction constants can be derived very accurately. Taking into account the values of the magnetic and quadrupole moments[6,7], μ = -0.7656(25) μ_N and Q = +0.83(13)b of the $5/2^+$, 247 keV level in ^{111}Cd we derive from our data for Cd in Co

$$B_{hf}\,(Cd\ \underline{Co}) = 28.82(30)T$$

and

$$V_{zz}\,(Cd\ \underline{Co}) = 3.19(50) \times 10^{16}\ V/cm^2$$

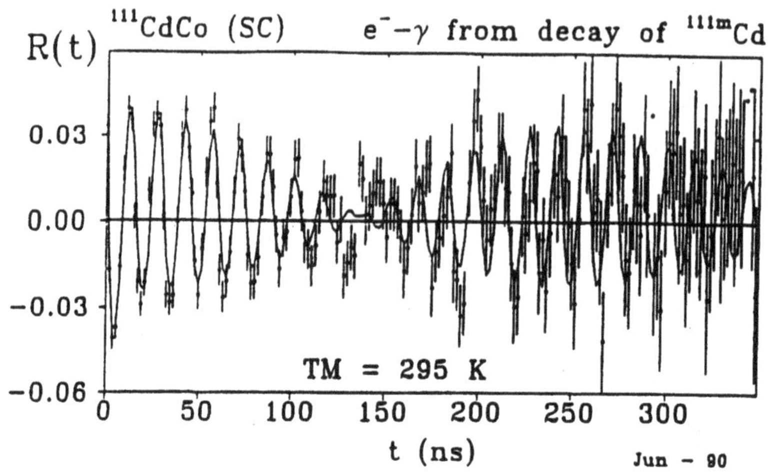

Fig.1 - Experimental spin precession curve, R(t), for ^{111}Cd implanted at ISOLDE with 60 keV in Cobalt single crystal measured using the e$^-$-γ PAC technique with the $\vec{c}$-axis of the single crystal in the detectors plane at 45°. The solid curve is a fit of the expected function to the experimental data.

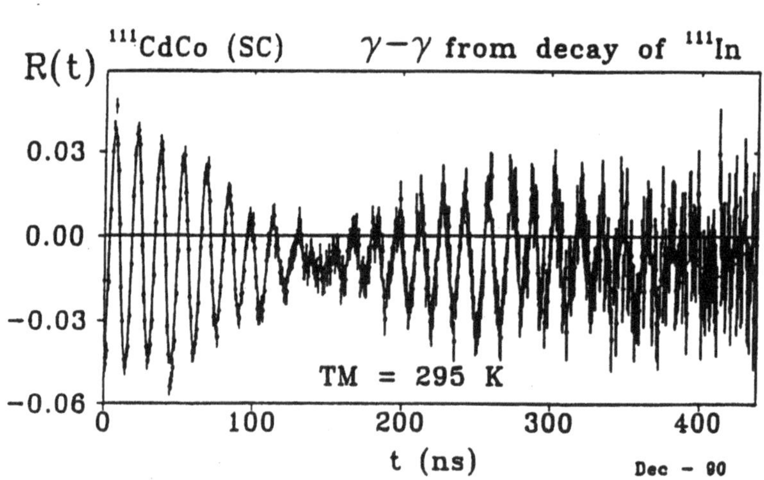

Fig.2 - The same as in Fig. 1 but the measurement has been performed using the γ - γ PAC technique.

For the $5/2^+$ 134 keV level populated in the decay of the ^{197m}Hg isotope taking into account the respective values of the magnetic[8] and quadrupole moments[7], $\mu = 0.965(65)\mu_N$ and $Q = -0.081(6)b$, we derive for the two different fractions of Hg ions, respectively,

$$B_{hf}(Hg\ \underline{Co}) = 42.5(2.9)T \text{ and } 50.1(3.4)T$$

and

$$V_{zz}(Hg\ \underline{Co}) = 153(29) \times 10^{16}\ V/cm^2 \text{ and } 49(44) \times 10^{16}\ V/cm2$$

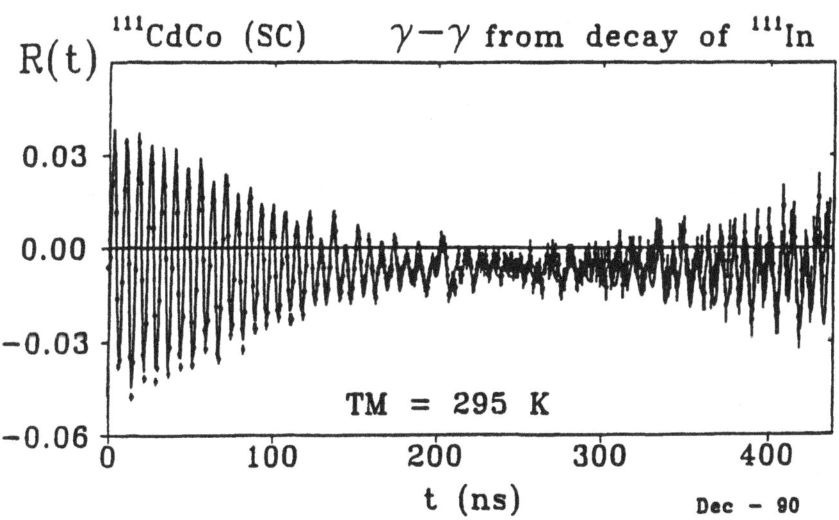

Fig.3 - The same as in Fig. 2 but the measurement has been performed with the $\vec{c}$-axis of the crystal perpendicular to the detectors plane.

III.2. Lifetime measurement using pre-accelerated electrons in the decay of ^{73}As. It is shown the feasibility of using the γ -e$^-$ PAC technique in the decay of the long lived ^{73}As.

The ^{73}As isotope has been selected in order to show that new low energy cascades can be studied and used in Materials Science applications. The ^{73}As isotope decaying to ^{73}Ge is also a very important candidate for studies of defects in semiconductors.

In Fig. 4 electron spectra are compared with and without pre-acceleration. The big advantage of using the pre-acceleration in order to detect the L-electron line of the 13.3 keV transition is clear. The L-electron line is well separated from an electron line due to Auger electrons: These electrons are not in coincidence with the 53.4 keV γ transition. At a much lower electron energy, only a few keV, a very intense peak has been also observed well separated from the detector noise. For comparison the electron spectrum obtained with the same discriminator low energy level set up but without pre-acceleration is overlapped. In this case we can see only the 53 keV conversion electrons.

Fig. 5 shows the lifetime measurement performed during the short time interval of two days using an weak ^{73}As 80 d halflife source. The $T_{1/2}$ period of 2.862(30) μ s has been measured being the error bars after six halflifes very small.

Further studies are in progress in order to measure the anisotropy of the cascade and the influence of the pre-acceleration and electron scattering in the attenuation anisotropy factor.

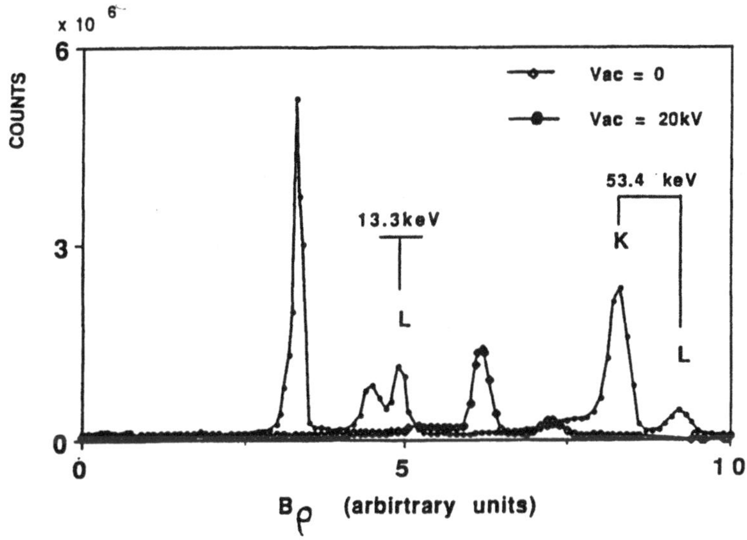

Fig.4 - Electron spectrum of the ^{73}As isotope implanted at ISOLDE in a Ni-foil. The pre-acceleration of 20 kV makes possible the use of the γ - e⁻ cascade for lifetime measurements and PAC Studies.

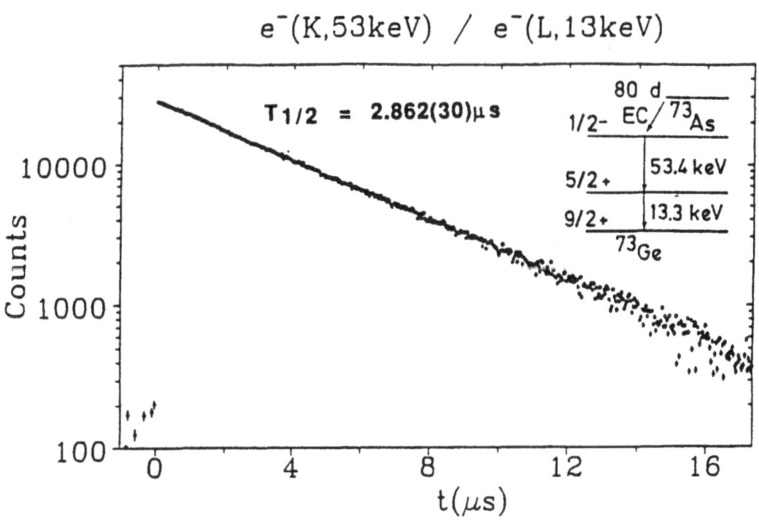

Fig. 5 - Delayed coincidences of the 53.4 keV e_K^- - 13.3 keV e_L^- cascade measured in the decay of ^{73}As.

III.3. Lifetime measurements in the decay of the 4.3 min. ^{75}Kr isotope: a test experiment for the performance of the e^--γ and γ-e^- PAC technique for Nuclear Physics and Materials Science studies.

Very strong beams of short lived isotopes are available at CERN/ISOLDE. The ^{75}Kr isotope with a lifetime of 4.3 min has been selected in this work in order to test, both, the feasibility to use this isotope on line in Nuclear Physics experiments and to prepare a new experiment for studying the spectroscopy of this isotope. The spin and parity of most of its levels are not yet known. The 4.3 min activity of ^{75}Kr has been collected in the isotope separator ISOLDE 2 on beryllium and nickel foils and it has been introduced in the chamber of the magnetic spectrometer through a pre-evacuated vacuum chamber so that the experiment could start between 1.5 and 2 min after the implantation was finished. The collected activity was optimized so that the gamma detectors could stand the counting rate.

Both lenses of the spectrometer selected the K 88 keV conversion electron line. In both γ-detectors a wide window selected the 132 keV γ line which could not be separated from the 154 keV γ-line.

The prompt curve for this set up has been shifted to the middle of the 500 ns range of the time to amplitude converter in order to observe the possibility of measuring lifetimes of levels populating the 220.9 keV level. Due to the very high activities and low energies of the transitions the effective time resolution of this experiment has been estimated to be about 3.5 ns. In fig 6 the results of all the measurements performed are shown after the subtraction of random coincidences. The lifetime of the 132.5 keV level corresponding to the K 88.4 keV e^--132.5 keV γ cascade appears on the right side of the prompt curve and has been calculated, taking into account the 3.5 ns time resolution of the set up, to be $T_{1/2}$ = 5.7(3)ns.

On the left side of the prompt curve a 5 times longer lifetime could be observed which has not yet been measured in other experiments. Since in the decay of ^{75}Br there are lifetimes of this order of magnitude, experiments have been performed in order to identify the level to which this lifetime could be atributed. However, it has been immediately shown that the activity feeding the γ -e$^-$ cascade responsible for this lifetime had also the period of about 4.3 min which excludes the possibility to atribute this lifetime to any level of the daughter activity of ^{75}Br.

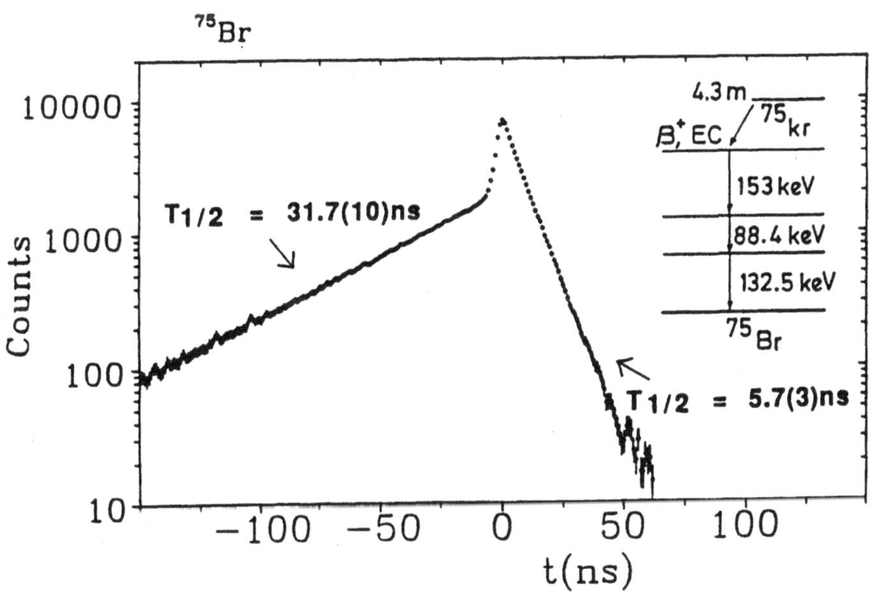

Fig.6 - The halflife of the 132.5 keV level in ^{75}Br using the 88.4 keV e_k^- - 132.5 γ cascade in the decay of the 4.3 min. ^{75}Kr isotope. A new halflife of 31.7 ns has been measured.

Taking into account the results of ref.[9] and our data, the observed lifetime can be attributed to the 220.9 keV level considered a possible candidate of a $9/2^+$ level. In this case we looked for the theoretical anisotropy of the cascade and we analysed the data of the different detectors forming the function

$$R(t) = 2\,\frac{W(180,t) - W(90,t)}{W(180,t) + 2W(90,t)}$$

which should be proportional to $A_{22}\ G_{22}(t)$ for polycrystalline samples and cascades where the A_{44} coefficient is small. In both set of measurements using beryllium and niquel foils it turns out that the effective measured anisotropy at time zero is negative and has the magnitude of about 10% for the 153 γ - K 88 keV e^- cascade and about 6% for the K 88 keV e^- - 132.5 keV γ cascade respectively for Be and Ni foils. These values, a factor of two lower than the values expected taking into account that the cascades have pure multipolarities, are quite resonable since in our set up in order to optimize the statistics of the measurements the admission angle in the electron lenses and the solid angle of the gamma detectors were quite big. Electron scattering in the beryllium and nickel foils and very fast fields seen by a fraction of ions can also be partially responsible for the attenuation.

IV. Conclusions

Electron-gamma and gamma-electron PAC studies using short and long lived isotopes are now possible at CERN/ISOLDE and the excellence of the technique has been demonstrated using the ^{111m}Cd 40 min, the ^{197m}Hg 24 h and the ^{75}Kr 4.3 min activities. For long lived isotopes a new set up is in development in Lisbon consisting of one electromagnetic lens and five BaF_2 gamma detectors. Big improvements in the CERN set up will be introduced replacing NaI(Tℓ) by BaF_2 detectors. The equipment has, also, to be prepared to be coupled with Ge detectors for level scheme studies in order to be more useful for basic Nuclear Physics experiments.

In the near future, some improvements have to be introduced in the experimental set up in order to lower the vacuum pressure and to run temperature dependent experiments.

Acknowledgements

The authors are grateful to Prof. E. Bodenstedt from the University of Bonn who supplied the magnetic spectrometer, to JNICT and CERN comission in Portugal who funded the project. One of us (J.G.C.) acknowledges JNICT for the master and Ph D. fellowships under the Science Programme, T. Butz and D. Forkel for help in using the γ -γ equipment.

References

[1] J. Christiansen: Hyperfine Interactions of Radioactive Nuclei (Springer Verlag, Berlin Heidelberg 1983).

[2] Th. Wichert and E. Recknagel, in: Microscopic Methods in Metals ed U. Gonser (Springer-Verlag, Berling Heidelberg, 1986) p. 317.

[3] J.C. Soares, K. Krien, A.G. Bibiloni, K. Freitag and R. Vianden, Phy. Lett. 45A (1973) 465 .

[4] K. Krien, A. G. Bibiloni, K. Freitag, J. C. Soares and R. Vianden, Phys. Rev. B8 (1973) 2248.

[5] P. Kleinheinz, L. Samuelsson, R. Vukanovic and K. Siegban, Nucl. Inst. and Meth. 32 (1965) 1.

[6] H. Bertschat, H. Haas, F. Pleiter, E. Recknagel, E. Schlodder, and B. Spellmeyer, A. Phys. 270 (1974) 203.

[7] R. Vianden, Hyp. Int. 35 (1987) 1079.

[8] E. Gerdau, D. Rüter and J. Braunsfurth, Z. Phys. 230 (1970) 72.

[9] A.J. Kreiner, M.A.J. Mariscotti, C. Backtash, E. der Mateosian and P. Thieberger Phys. Rev. C 24 (1981) 148.

ON THE EQUIVALENCE OF THE AFANAS'EV-TSYMBAL METHOD AND OF THE FOURIER TECHNIQUE OF MÖSSBAUER LINE SHARPENING*

D.L. Nagy+ and U. Röhlich
Physikalisches Institut der Universität Erlangen-Nürnberg,
Erwin-Rommel-Straβe 1, D-8520 Erlangen, F.R.G.

ABSTRACT: We show that the integral transformation method of Mössbauer line sharpening recently suggested by Afanas'ev and Tsymbal[1)] is exactly equivalent to the conventional Fourier method if the appropriate filter function is used.

1. INTRODUCTION

Any Mössbauer transmission spectrum $y(\nu)$ of thin absorbers in the case of static hyperfine interactions is the convolution of the distribution function of the hyperfine energies $p(\nu)$ with a Lorentzian $l(\nu)$ of full width at half maximum (FWHM) 2Γ:

$$y(\nu) = \int_{-\infty}^{+\infty} p(w)\, l(\nu - w)\, dw \tag{1}$$

with

$$l(\nu) = \frac{\Gamma}{\pi} \frac{1}{\nu^2 + \Gamma^2} \tag{2}$$

Γ being the natural linewidth (0.0973 mm/s for ^{57}Fe). The physical information about the system is contained in $p(\nu)$ which, obviously, consists of sharper peaks than $y(\nu)$ does. In principle, if we have a small number of well-defined environments of the resonant nucleus in the sample, $p(\nu)$ is the sum of delta-functions. In disordered or amorphous systems, however, the hyperfine interaction may have a distribution and, accordingly, $p(\nu)$ is the sum of somewhat broadened peaks which are still narrower than the corresponding peaks of the spectrum. There exist several procedures[2)] to extract the distribution function from the spectrum. All these line sharpening procedures face the problem of finding a reasonable compromise between increased noise and false oscillations and are basically equivalent to each other and to the Fourier technique (see below) which is the most transparent of them.

A new method of Mössbauer line sharpening has recently been proposed by Afanas'ev and

*Part of the lecture "Model-independent Data Reduction Methods in Nuclear Solid State Physics". Another part of this lecture is being published in *Hyperfine Interactions*.
+On leave from the Central Research Institute for Physics, P.O.B. 49, H-1525 Budapest, Hungary

Tsymbal[1]. This latter technique is based on a numerical integral transformation of the experimental spectrum and has efficiently been used to process the spectra of the high-T_C superconductor $YBa_2(Cu_{1-x}Fe_x)_3O_{6.8}$ [3]. In the present paper we are going to show that also the method of Afanas'ev and Tsymbal is equivalent to the Fourier technique and we give the corresponding filter function.

2. THE FOURIER TECHNIQUE

It is, in principle, very easy to solve Eq. (1) for $p(\nu)$. Since $y(\nu)$ is a convolution of $p(\nu)$ and $l(\nu)$:

$$y(\nu) = p(\nu) * l(\nu) \tag{3}$$

one obtains:

$$p(\nu) = (2\pi)^{-1/2}\, \mathbf{F}^{-1} \frac{\mathbf{F}\, y(\nu)}{\mathbf{F}\, l(\nu)} = (2\pi)^{-1/2}\, \mathbf{F}^{-1} \frac{Y(k)}{L(k)} \tag{4}$$

where $\mathbf{F}$ denotes the Fourier transformation while $Y(k)$ and $L(k)$ stand for the Fourier transforms of $y(\nu)$ and $l(\nu)$, respectively, k being the variable of the Fourier transformation. Hereafter the Fourier transforms will be denoted by the corresponding capitals:

$$Y(k) = (2\pi)^{-1/2} \int_{-\infty}^{+\infty} y(\nu) \exp(ik\nu)\, d\nu \tag{5}$$

and

$$y(\nu) = (2\pi)^{-1/2} \int_{-\infty}^{+\infty} Y(k) \exp(-ik\nu)\, dk \tag{6}$$

etc.

If the *measured* data $y^{\sim}(\nu)$ are noisy, which is always the case, the deconvolution of $p(\nu)$ from Eq. (3) as applied to $y^{\sim}(\nu)$ rather than to $y(\nu)$ belongs to the class of ill-posed problems[4] which can only be solved by finding some compromise between enhanced noise and distortion of $p(\nu)$. Indeed, denoting the noise (or statistical fluctuation) of the measured data in addition to the ideal case by $h(\nu)$, Eq. (1) reads:

$$y^{\sim}(\nu) = \int_{-\infty}^{+\infty} p(w)\, l(\nu - w)\, dw + h(\nu) \tag{7}$$

Eq. (4) is not applicable to $y^{\sim}(\nu)$ since white noise contains all frequencies with the same probability. When its Fourier transform is divided by

$$L(k) = (2\pi)^{-1/2} \exp(-|k|\Gamma) \tag{8}$$

the high frequency ($|k| >> 1/\Gamma$) contribution from $h(\nu)$ will be considerably enhanced.

The inconvenience of the high noise can be avoided by using some filter function. Before performing the inverse Fourier transformation in Eq. (4), $Y^{\sim}(k)/L(k)$ is multiplied by this filter function $D(k)$ resulting in the filtered experimental distribution

$$p_D^{\sim}(\nu) = (2\pi)^{-1/2}\, \mathbf{F}^{-1} \frac{Y^{\sim}(k)}{L(k)} D(k) \tag{9}$$

If the value of $D(k)$ is small enough at high frequencies the high frequency part of the noise is effectively suppressed.

The application of a filter $D(k)$ also distorts the noise-free distribution $p(\nu)$:

$$p_D(\nu) = (2\pi)^{-1/2}\, \mathbf{F}^{-1} \frac{Y(k)}{L(k)} D(k) \tag{10}$$

A comparison of Eqs. (4) and (10) yields:

$$P_D(k) = (2\pi)^{-1/2} \frac{Y(k)}{L(k)} D(k) = P(k)\, D(k) \tag{11}$$

i.e.

$$p_D(\nu) = (2\pi)^{-1/2} p(\nu) * d(\nu) = (2\pi)^{-1/2} \int_{-\infty}^{+\infty} p(w)\, d(\nu - w)\, dw \tag{12}$$

The effect of the filtering on $p(\nu)$ is, apart from a factor $(2\pi)^{-1/2}$, equivalent to smoothing with $d(\nu)$ the inverse Fourier transform of $D(k)$. A Fourier transformation - filtering - inverse Fourier transformation procedure is often used, even without deconvolution, to smooth measured data *in large*[5]) which again, of course, means the suppression of the high frequency contribution from the noise. From this point of view it is obviously immaterial whether the measured spectrum is smoothed with $(2\pi)^{-1/2} d(\nu)$ and the deconvolution is performed on the smoothed spectrum or, as above, $p(\nu)$ is first extracted from $y(\nu)$ and the smoothing is done afterwards. Indeed, the smoothed noise-free spectrum is

$$y_D(\nu) = (2\pi)^{-1/2} \int_{-\infty}^{+\infty} y(w)\, d(\nu - w)\, dw \tag{13}$$

The Fourier transform of $y_D(\nu)$ is

$$Y_D(k) = Y(k)\, D(k) \tag{14}$$

The application of Eq. (4) to $Y_D(k)$ rather than to $Y(k)$ obviously results in $P_D(k)$ of Eq. (11).

The basic problem of filtering is to find a filter function $D(k)$ so that it reasonably suppresses the enhanced noise but still does not significantly modify the shape of $p(\nu)$ (*cf.* Eq. (12)). The shape of $D(k)$ may depend on several parameters the most important of those being the cutoff frequency k_o which is defined so that - at least approximately -

$$\begin{aligned} D(k) &> 1/2 \quad \text{for } |k| < k_o \\ D(k) &< 1/2 \quad \text{for } |k| > k_o \end{aligned} \tag{15}$$

A filtering (*i.e.* a smoothing) should not change the normalization of $p(\nu)$:

$$\int_{-\infty}^{+\infty} p_D(\nu)\, d\nu = \int_{-\infty}^{+\infty} p(\nu)\, d\nu \tag{16}$$

Combining Eqs. (12) and (16) we obtain that

$$\int_{-\infty}^{+\infty} d(\nu) = (2\pi)^{-1/2} \tag{17}$$

Taking the Fourier transform of Eq. (17) we see that the normalization of $p(\nu)$ will not be affected by the filtering/smoothing if and only if $D(0) = 1$.

A good filter $D(k)$ should, obviously, fulfil two contradictory requirements: *a)* its value should be close to unity for $|k| << k_o$ and close to zero for $|k| >> k_o$ where k_o is chosen so that $k_o \approx 1/\Gamma_p$ where Γ_p is the width of the narrowest detail of $p(\nu)$ and *b)* its inverse Fourier transform should not contain oscillations which, when convoluted with $p(\nu)$, result in oscillations of higher amplitude than the noise of $p_D{}^{\sim}(\nu)$. A Gaussian filter

$$D(k) = \exp\,[-\ln 2\,(k/k_o)^2] \tag{18}$$

fulfils condition *b)* its inverse Fourier transform being a Gaussian again but it only very poorly meets condition *a)* resulting in a strong distortion of $p(\nu)$. An important feature of the Gaussian filter is that it asymptotically vanishes even when divided by $L(k)$ of Eq. (8), at frequencies $|k| >> k_o$. On the other hand, since $D(k) \approx 1$ and $L(k) \approx 1$ for $k \approx 0$ and $L(k)$ is monotonously decreasing, $D(k)/L(k)$ always has a more or less sharp maximum around k_o. Thus, even if k_o was correctly chosen from the point of view of $p(\nu)$, the $k \approx k_o$ part of the originally white noise will be overemphasized resulting in oscillations of frequency $k \approx k_o$ on $p_D^{\sim}(\nu)$ again. Therefore, a further requirement *c)* should also be fulfilled by $D(k)$: the function $D(k)/L(k)$ should remain asymptotically finite constant rather than vanish or diverge at high frequencies. Such a behaviour well reflects the essence of the resolution improvement procedure being quite natural that when the resolution of a spectroscopic method is increased the noise (both its low and high frequency part) is increased as well.

The filter suggested by Nagy and Kulcsár[6])

$$D(k) = (1 + C/L(k))^{-1} \tag{19}$$

with a constant $C << 1$ fulfils requirements *a)* and *c)* since $D(0) \approx 1$ and at high frequencies $D(k)/L(k) \approx 1/C$. Although the Fourier transform of (19) is not free from oscillations C can still always be chosen so that the oscillations and the noise of $p_D^{\sim}(\nu)$ are of the same order of magnitude. Substituting (8) into (19) and choosing k_o so that $C = (2\pi)^{-1/2}\exp(-k_o\Gamma)$ we get the Fermi-Dirac function as a filter:

$$D(k) = \{1 + \exp[(|k| - k_o)\Gamma]\}^{-1} \tag{20}$$

The condition $C << 1$ means that $k_o >> 1/\Gamma$ which can only be fulfilled if the noise is low. Otherwise neither requirement *a)* is fulfilled nor the asymptotic region of $D(k)$ meeting requirement *c)* is used. In order to fulfil at least the normalization condition $D(0) = 1$ usually the filter

$$D(k) = \{1 + \exp[(-k_o)\Gamma]\}\,\{1 + \exp[(|k| - k_o)\Gamma]\}^{-1} \tag{21}$$

is used. We shall refer to $D(k)$ of Eq. (21) as *modified Fermi-Dirac filter*.

3. THE AFANAS'EV-TSYMBAL TECHNIQUE

The line sharpening technique recently proposed by Afanas'ev and Tsymbal[1]) is based on subsequent integral transformations of the experimental spectrum which can easily be preformed numerically. We denote the original spectrum with

$$y_1(\nu) = y(\nu) \tag{22}$$

From $y_1(\nu)$ we obtain $y_2(\nu)$ by the integral transformation

$$y_2(\nu) = y_1(\nu) + \frac{\Gamma}{\pi} \int_{-\infty}^{+\infty} \frac{y_1(\nu) - y_1(w)}{(\nu - w)^2}\, dw \tag{23}$$

From $y_2(\nu)$ the next function $y_3(\nu)$ can be calculated:

$$y_3(\nu) = y_1(\nu) + 2y_2(\nu) + \frac{\Gamma}{\pi} \int_{-\infty}^{+\infty} \frac{y_2(\nu) - y_2(w)}{(\nu - w)^2}\, dw \tag{24}$$

etc. For numerical applications a more symmetric form of Eqs. (23) and (24) was used[3].

In their paper[1] Afanas'ev and Tsymbal give an elegant proof that

$$y_n(\nu) = \int_{-\infty}^{+\infty} p(w)\, l_n(\nu - w)\, dw \tag{25}$$

with

$$l_n(\nu) = C_n \frac{1}{(\nu^2 + \Gamma^2)^n} \tag{26}$$

where the normalization factor C_n is chosen so that

$$\int_{-\infty}^{+\infty} l_n(\nu)\, d\nu = 1 \tag{27}$$

i.e.

$$C_n = \frac{2^{n-1}\,(n-1)!}{(2n-3)!!} \frac{\Gamma^{2n-1}}{\pi} \tag{28}$$

The full width of $l_n(\nu)$ at half maximum is

$$W_n = \Gamma\,(2^{1/n} - 1)^{1/2} \tag{29}$$

which tends towards zero if $n \to +\infty$. Therefore, taking into account Eq. (27)

$$\lim_{n \to +\infty} l_n(\nu) = \delta(\nu) \tag{30}$$

and so, combining Eqs. (25) and (30):

$$\lim_{n \to +\infty} y_n(\nu) = p(\nu) \tag{31}$$

which means that the procedure asymptotically gives an exact solution to Eq. (1).

The Afanas'ev-Tsymbal technique enables one to easily estimate the error in the sharpened spectrum[1].

4. COMPARISON OF THE TWO METHODS

Since both Eq. (4) and Eq. (31) give the exact solution to Eq. (1), the Afanas'ev-Tsymbal method is asymptotically equivalent to the Fourier technique without applying a filter. In the following we show that the n-th step of the Afanas'ev-Tsymbal method is also equivalent to the Fourier technique with an appropriate filter function $D_n(k)$. Indeed, Eq. (25) says that $y_n(\nu)$ is nothing but the distribution function $p(\nu)$ smoothed with $l_n(\nu)$. According to Eqs. (10-12), however, this is equivalent to the application of the filter in the Fourier procedure

$$D_n(k) = (2\pi)^{1/2}\, \mathsf{F}\, l_n(\nu) \tag{32}$$

which can be given in closed form:

$$D_n(k) = \frac{1}{(2n-3)!!} (2/\pi)^{1/2} (\Gamma\,|\,k\,|\,)^{n-1/2} K_{n-1/2}(\Gamma\,|\,k\,|\,) \tag{33}$$

where $K_m(z)$ is the Bessel function of second kind (Macdonald function).

The Afanas'ev-Tsymbal filter function $D_n(k)$ fulfils requirement *a)* of Section 2 (the normalization condition) being $D_n(0) = 1$ for any n. Moreover, the low-k components of $p(\nu)$ are only slightly distorted by this filtering since the derivative of the filter $D_n'(0) = 0$ for $n > 1$. From this point of view both the Gaussian filter (18) and the Afanas'ev-Tsymbal filter (33) are superior to the modified Fermi-Dirac filter (21) which does not have this property. Clearly, like the Gaussian filter the Afanas'ev-Tsymbal filter also meets requirement *b)* its inverse Fourier transform $(2\pi)^{1/2}\, l_n(\nu)$ being completely free from oscillations.

For $|\,k\,| \to +\infty$ the asymptotics of $D_n(k)$ is:

$$D_n(k) \approx \frac{1}{(2n-3)!!} (\Gamma\,|\,k\,|\,)^{n-1} \exp(-\Gamma\,|\,k\,|\,) \tag{34}$$

which, unfortunately, does not meet requirement *c)* of Section 2. Indeed,

$$\frac{D_n(k)}{L(k)} \approx \frac{(2/\pi)^{1/2}}{(2n-3)!!} (\Gamma \mid k \mid)^{n-1} \tag{35}$$

which is strongly divergent if $|k| \to +\infty$ for $n >> 1$. In other words, the Afanas'ev-Tsymbal procedure unnecessarily enhances the noise. This property of the technique has been realized by the authors themselves who suggested to perform a proper smoothing of the experimental spectrum prior to the integral transformation[1]. However, as shown is Section 2, such a smoothing is again equivalent to an additional filtering in the Fourier procedure.

In the Afanas'ev-Tsymbal procedure the degree of line sharpening is determined by the number of steps n. In the Fourier technique the cutoff frequency k_o plays a similar role. It is easy to find the relation of these two parameters. The finest relevant detail of the resolved distribution function $p(v)$ is $1/k_o$ if the Fourier procedure is applied and W_n of Eq. (29) if the Afanas'ev-Tsymbal technique is used. For $n >> 1$ following approximation of Eq. (29) holds:

$$W_n \approx (\ln 2)^{1/2} n^{-1/2} \Gamma \tag{36}$$

and so

$$1/k_o \approx (\ln 2)^{1/2} n^{-1/2} \Gamma \tag{37}$$

i.e.

$$k_o \approx (\ln 2)^{-1/2} n^{1/2} \Gamma^{-1} = 1.201\, n^{1/2} \Gamma^{-1} \tag{38}$$

Since k_o is proportional to $n^{1/2}$ quite many steps may be necessary if a line sharpening corresponding to a high k_o value in the Fourier procedure is desired.

The properties of the Gaussian, modified Fermi-Dirac and the Afanas'ev-Tsymbal filter can be compared in Figs. 1 and 2.

5. CONCLUSIONS

The Afanas'ev-Tsymbal procedure of Mössbauer line sharpening is exactly equivalent to the usual Fourier technique if the filter function of Eq. (33) is used. While this filter has some advantageous properties at low frequencies it unnecessarily enhances the noise in the high frequency region. This problem can only be overcome by smoothing the raw data which is equivalent to an additional filtering in the Fourier procedure. For numerical applications the Fourier technique should preferably be used especially for high cutoff frequencies since to reach the effect of line sharpening which can be achieved in one step with the Fourier technique many steps of the Afanas'ev-Tsymbal procedure are necessary.

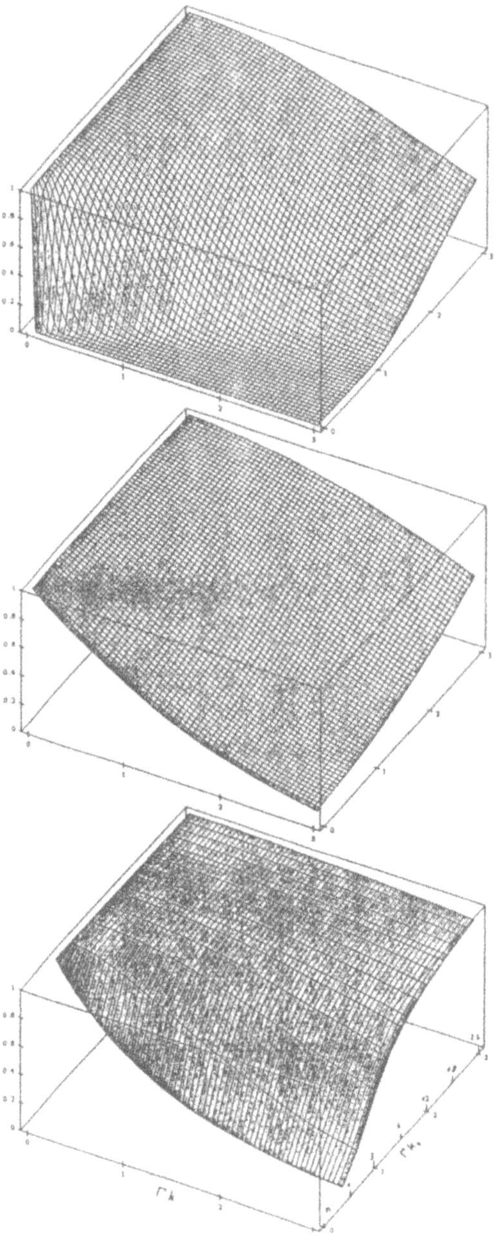

Fig. 1. The Gaussian, modified Fermi-Dirac and the Afanas'ev-Tsymbal filters $D(k)$ (from top to down).

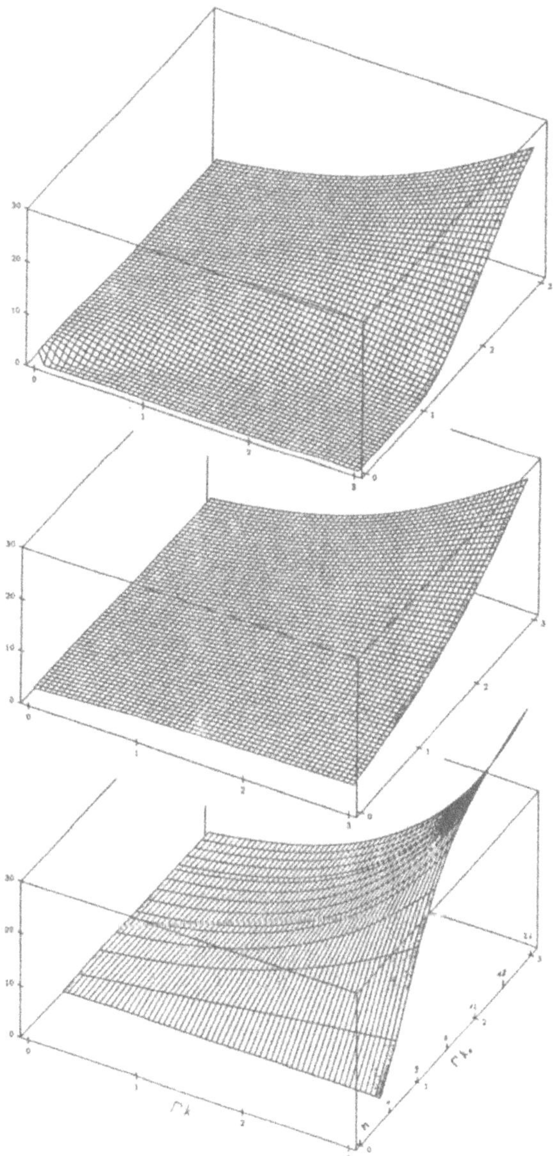

Fig. 2. The Gaussian, modified Fermi-Dirac and the Afanas'ev-Tsymbal filters $D(k)$ divided by the Fourier transform $L(k)$ of a Lorentzian (from top to down).

REFERENCES

1) Afanas'ev, A.M. and Tsymbal, E.Yu, Report IAE-4890/10, Atominform, Moscow, 1989.
2) Nagy, D.L. and Röhlich, U., Hyp. Int., in the press and references therein.
3) Afanas'ev, A.M., Tsymbal, E.Yu., Cherepanov, V.M., Chuev, M.A., Yakimov, S.S., Zinn, W., Sauer, Ch. and Bush, A.A., Solid State Commun., 76, 1099 (1990).
4) Tichonov, A.N., "Solutions of Ill-Posed Problems", Winston, Washington DC, 1977.
5) Lánczos, C., "Applied Analysis", Pitman, London, 1957, p. 217.
6) Nagy, D.L. and Kulcsár, K., in: "Proc. Int. Conf. Mössbauer Spectrometry", Dresden, 1971, ed. Schnorr, H. and Kautz, M., Physikalische Gesellschaft der DDR, 1971, p. 618.

E.F.G. ASPECTS IN AMORPHOUS FERRIC FLUORIDES a-FeF_3

JEAN-MARC GRENECHE

Laboratoire de Physique des Matériaux, U.A. CNRS n°807
Université du Maine B.P 535
72017 LE MANS Cedex, FRANCE

ABSTRACT

We report first the different kinds of continuous random networks of corner-sharing octahedra simulating the structure of ideal amorphous ferric fluorides a-FeF_3. Then, the electric field gradient calculations developed in these structural models using a point charge model are discussed. The results are compared to experiment data deduced from Mössbauer Spectroscopy and to those predicted from previous theoretical developments established on dense random packings of spheres.

INTRODUCTION

During the last decennies, much effort has been devoted to elucidate the structural properties of amorphous materials, in other words the chemical and positional orders at short and medium ranges (see for example /1/. The diffraction techniques (X-rays, neutrons) widely used to investigate these materials are unfortunately limited to radial information (pair-correlation function) because of their isotropic behaviour from a macroscopic point of view.

In opposite, the local techniques such as extended X-ray absorption fine structure (EXAFS), nuclear magnetic resonance (NMR), Mössbauer effect (ME) and electron paramagnetic resonance (EPR) are excellent tools to approach higher order correlation functions describing local environments. So, the NMR and ME techniques provide useful information concerning electric quadrupolar interactions through the electric field gradient [EFG] tensor related to the bond angle distribution. In this paper, we focused our attention on the Mössbauer spectroscopy.

In the case of some nuclear spin Mössbauer resonance (for example ^{57}Fe with I=3/2), one can establish the following relation,

$$\Delta = \frac{eQ|V_{zz}|}{2}\left(1+\frac{\eta^2}{3}\right)^{1/2}$$

where Δ is the quadrupolar splitting, V_{zz} the main component of the E.F.G. tensor, $\eta = \left(V_{xx} - V_{yy}\right)/V_{zz}$ the asymmetry parameter (with $o\leq\eta\leq1$ using the usual convention $|V_{xx}|\leq|V_{yy}|\leq|V_{zz}|$) and Q the nuclear quadrupole moment, OXYZ being the set of principal axes.

The ^{57}Fe Mössbauer resonance applied to amorphous systems leads to paramagnetic spectra with broad and non-lorentzian lines, due to the local distribution of charges around the probe nuclei. One can only estimate the quadrupolar splitting distribution $P(\Delta)$ while the joint distribution function $P(V_{zz}, \eta)$ cannot be directly observed from Mössbauer experiments because of the small nuclear spin value I = 3/2.

Nevertheless, complementary information can derive from High Field Mössbauer Spectroscopy [H.F.M.S.] spectra recorded in the paramagnetic state (see some review papers as /2, 3/). The E.F.G. sign is unambiguously determined in the case of powdered crystalline samples. On the contrary, the situation is less clear for disordered systems (yet discussed in /4/) because of (i) the complexity of the in-field ^{57}Fe M.E. spectra, (ii) the superimposition of both positive and negative V_{zz} distributions and (iii) the presence of random induced hyperfine fields, if any, which together prevent to a correct information. One can conclude that the iron Mössbauer spectroscopy is *a priori* rather limited in investigating local structure in amorphous systems.

Let us note that the distribution of E.F.G. could be potentially determined from out-of field spectra for nucleus values $I \geq 5/2$ but the analyses of these spectra become very complex on disordered systems, because of the presence of overlapped and broadened lines.

The electric field gradient distributions have been extensively investigated from an experimental point of view on disordered systems as amorphous alloys (metglasses based on transition metal and or rare-earth elements) /5, 6/, magnetic amorphous insulators /7/, on icosahedral and decagonal quasi-crystalline systems /8/. Furthermore some theoretical developments based on random coordination models have been proposed : the dense random packing model or the "shell model" /9/, the random gaussian model /10/ and the gaussian isotropic model /11/.

In the case of amorphous ferric fluorides a-FeF_3, we have developed electric field gradient calculations using a point charge model with atomic positions given by different kinds of structural computed networks resulting from continuous packing of octahedral units. The estimated quadrupolar splitting distribution $P(\Delta)$ was compared to the experimental one obtained by Mossbauer Spectroscopy at room temperature and the calculated joint distribution function $P(V_{zz}, \eta)$ were then compared to the expected distributions deduced from analytic developments established from environment models based on random ionic coordinations /9/.

In this synthesis paper, the expected distributions $P(V_{zz}, \eta)$ given by the shell model and the random gaussian model are briefly presented in a first section.

Then we present the cristallographic data of the three different crystalline forms of FeF_3 and the evolution of their magnetic characteristics is discussed in terms of frustrating topology. The structural results obtained by usual diffraction techniques and the Mössbauer studies on the amorphous varieties of FeF_3 are given in the following section.

The different existing structural random networks of corner-sharing octahedra, describing the ideal amorphous ferric fluoride a-FeF_3 are discussed in section 3.

After a brief review of the different electric field gradient calculation methods developed in crystalline systems, we discuss in section 4 the E.F.G. calculations adapted to amorphous structures. The results are then presented and compared to those established by Mössbauer spectroscopy and by theoretical developments.

1 - RANDOM COORDINATION MODELS

1-1 The Shell Model

A theoretical approach has been proposed by Czjzek et al /9/ to described the E.F.G. distributions in disordered systems assuming random ionic coordinations.

In the most general case, the components of the E.F.G. induced by a discrete distribution of charges q_n located on x_i^n (with polar coordinates r_n, θ_n, ϕ_n) can be described analytically in terms of spherical harmonics with five-independent parameters U_m (with $-2 \le m \le 2$) as following

$$\left| \begin{array}{ll} U_{-2} & = \frac{1}{2}\sqrt{3}\sum_n q_n \sin^2\theta_n \cos^2\phi_n\, r_n^{-3} \\ U_{-1} & = \sqrt{3}\sum_n q_n \sin\theta_n \cos\theta_n \sin\phi_n\, r_n^{-3} \\ U_o & = \frac{1}{2}\sum_n q_n (3\cos^2\theta_n - 1)\, r_n^{-3} \\ U_{+1} & = \sqrt{3}\sum_n q_n \sin\theta_n \cos\theta_n \cos\phi_n\, r_n^{-3} \\ U_{+2} & = \frac{1}{2}\sqrt{3}\sum_n q_n \sin^2\theta_n \sin^2\phi_n\, r_n^{-3} \end{array} \right.$$

The distribution functions $P(U_m)$ become calculable and independent, and exhibit equal width gaussian shapes centered at the origin, assuming spherical shells far from atomic probes and containing randomly distributed atoms with a weak density. The expression of $P(U_m)$ is given by

$$P_m (U_m) \propto \frac{1}{\sigma} \exp (- U_m^2/2\sigma^2)$$

With such considerations, the joint distribution function $P(V_{zz},\eta)$ was established as

$$P(V_{zz},\ \eta) = 2\ V_{zz}\ \eta\ (1-\frac{\eta^2}{9})\ \int d\alpha \int d\beta \int d\gamma\ \sin\beta\ \pi_m\ P_m\ (U_m)$$

and leads after the successive integrations to

$$P(V_{zz},\eta) = \frac{1}{2\pi^{1/2}}\ \frac{1}{\sigma^5}\ V_{zz}^{\ 4}\ \eta\ (1-\ \frac{\eta^2}{9})\ \exp\ (-\frac{V_{zz}^2\ (1+\frac{\eta^2}{3})}{2\sigma^2})$$

The main features observed on this function $P(V_{zz},\ \eta)$ which characterizes the "shell model" are (i) a null probability for η=0 and V_{zz}=0, not for η=1, (ii) a close equiprobability for finding negative and positive values of V_{zz}. The marginal functions $Q(V_{zz}) = \int_0^1 d\eta\ P(V_{zz},\ \eta)$ and $R(\eta) = \int_{-\infty}^{+\infty} dV_{zz}\ P(V_{zz},\ \eta)$ are presented in /9/ and compared to those resulted from directsimulation for randomly occupied spherical shells).

Because of the structural constraints (polyhedral coordination) of the local environment of the probe nuclei in amorphous solids, most of Mossbauer experiments did not agree with the present shell model /6, 11-13/.

Let us remark that in the case of isotropic solids (on the average), the distribution function of the V_{ij} components depends only on two invariant functions of the E.F.G. tensor:

$$S = V_{zz}^{\ 2}\ (\ 1 + \eta^2/3\ )\ \text{and}\ D = V_{zz}^{\ 3}\ (\ 1 - \eta^2\ )$$

In the case of a 3/2 nuclear spin as iron probe nuclei, the quadrupolar splitting $P(\Delta)$ (with $\Delta = e\ Q\ \sqrt{S}\ /\ 2$) was reevaluated and

$$P(\Delta) = \frac{\Delta^{d-1}}{\sigma^d}\ \exp\ -\ \frac{\Delta^2}{2\sigma^2}$$

where d represents the number of degrees of freedom of the multinormal distribution and $d < 5$ /14/.

In section 3, the d-value will be discussed in the case of amorphous ferric fluorides where the iron atoms are octahedrally surrounded : in spite of weak distortions, such a structural unit can be described through two independent random variables, which leads to the following form

$$P(\Delta) = \frac{\Delta}{\sigma^2} \exp - \frac{\Delta^2}{2\sigma^2}$$

1-2 The Random Gaussian Model

The random gaussian model proposed by Coey /10/ was established on the basis of independent distribution functions $P(V_{ii})$ coupled by the Laplace's equation ($\sum_i V_{ii} = 0$) : it is assumed that in the plane (111) containing the origin in the space coordinates V_{xx}, V_{yy} and V_{zz}, the density of probability is only dependent of the distance r to the origin, as it is shown on figure 1.

In the plane (r,ψ), one can estimate

$$\Delta = eQr \sqrt{\frac{2}{3}} \text{ and } \eta = \sqrt{3} \text{ tg } \psi \text{ , /15/}$$

and then $P(\Delta)$, assumed to be independent of ψ, with a gaussian shape centered at the origin, is given by the following expression.

$$P(\Delta) = \frac{\Delta}{\sigma^2} \exp - \frac{\Delta^2}{2\sigma^2}$$

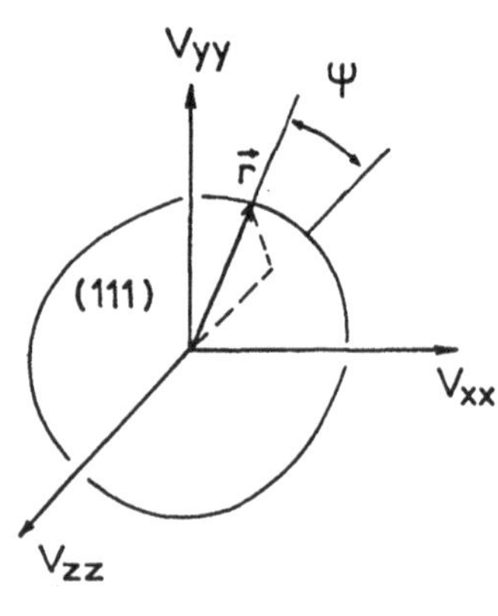

Fig.1: V_{xx} , V_{yy} and V_{zz} coordinates: each point of the (111) plane containing the origin corresponds to an E.F.G. tensor.

2 - THE DIFFERENT CRYSTALLINE AND AMORPHOUS FORMS OF FeF_3

2-1 The Crystalline Phases Of FeF_3

Up to now, three crystalline forms of FeF_3 have been synthesized: the rhombohedral r-FeF_3 /16/, the hexagonal tungsten bronze H.T.B. FeF_3 /17/ and the pyrochlore Pyr-FeF_3 /18/. Their crystallographic structures are respectively presented on figure 2.

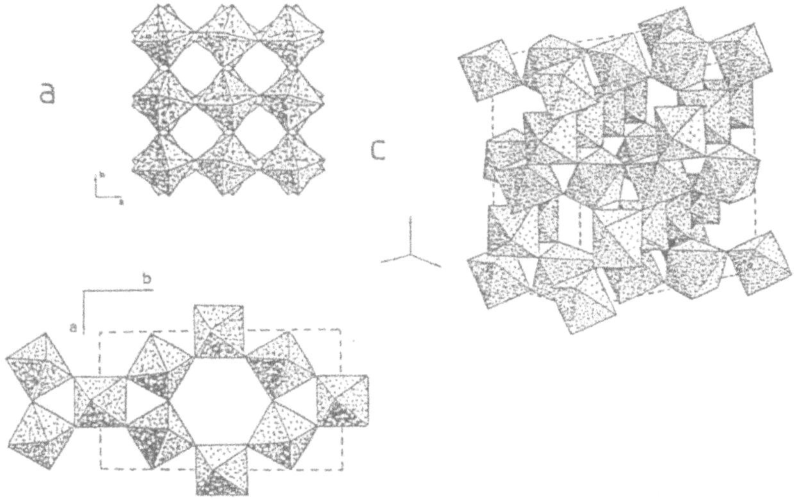

Fig.2: Crystal structure of the three crystalline forms of FeF_3: a) r-FeF_3, b) H.T.B.-FeF_3 and c) Pyr-FeF_3 (reprinted from /21/ with permission).

The original structure work on the stable rhombohedral form of FeF_3 established on powdered sample /16/ was recently confirmed by single crystal refinement /19/. The structure consists of corner-sharing octahedral units centered on a cubic lattice : the angle of tilting of the octahedra is 10.1° from cubic set of axes.

The structure of H.T.B.-FeF_3 is built up from corner-sharing octahedra and results from the superimposition of hexagonal tungsten bronzes planes in which the cationic network draws triangles and hexagones.

The cationic subnetwork of the pyrochlore phase of FeF_3 consists of corner-sharing tetrahedra.

	r-FeF_3	HTB - FeF_3	Pyr-FeF_3
Space group	$R\bar{3}c$	P6/mmm	Fd3m
Fe-F (Å)	1.923	1.916^a - 1.906^b	1.926
Fe-F-Fe (°)	152.15	142.3^a - 150.7^b	141.65
T_N (K)	363	~ 97	~ 20
Θ_p (K)	~- 1100 |22|	~ - 600	
μ (μ_B)	4.45	4.07	3.32
H_{hf} (kOe)	618	560	523
I.S.^{c} (mm s^{-1})	0.49	0.47	0.48
Δ^d (mm s^{-1})	0.05	- 0.20	- 0.22
Spin structure	antiferromagnet weak ferromagnet	triangular 120°	tetrahedral 109°
Percentage of triangular ferric rings	0	16.7	100

(a) (b) : values corresponding respectively to distances and and angles in the (001) plane and along [001]
(c) : isomer shift values relative to metallic iron at 300K
(d) : quadrupolar splitting values in the paramagnetic state.

TABLE 1 : Structural and magnetic characteristics on the three crystalline phases of FeF_3 (the structural parameters values were respectively refined from 300K, 4.2K and 2.5K diffraction experiments for r-FeF_3, HTB-FeF_3 and Pyr-FeF_3).

Susceptibility and magnetization measurements indicate the presence of antiferromagnetic exchange coupling for the three crystalline phases. Their magnetic characteristics are compared in table 1 while their magnetic structures are given on figure 3 : r-FeF_3 is a canted antiferromagnet i.e. a weak ferromagnet below 363 K where the canting is mainly due to the Dzialoshinsky-Moriya interaction /20/, (ii) the magnetic structure of H.T.B.-FeF_3 is described as three ferromagnetic sublattices at 120° one from each other in each (001) plane ; these planes are antiferromagnetically coupled /19/, (iii) the magnetic structure of Pyr-FeF_3 consists of four ferromagnetic sublattices oriented at 109° from each other /21/. One can explain these magnetic structures in terms of frustration: (i) r-FeF_3 contains only square platelets, (ii) in H.T.B.-FeF_3 four Fe-Fe interactions are involved in triangles and two in squares, whereas (iii) all the Fe-Fe interactions occur in triangular platelets in Pyr-FeF_3.

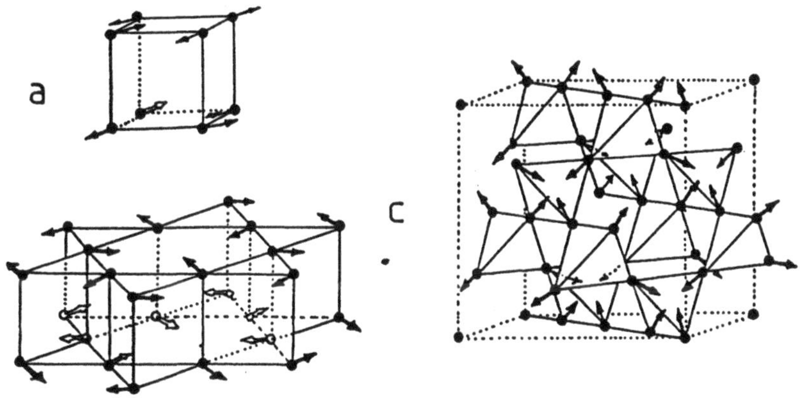

Fig.3: Magnetic structures of the three crystalline forms of FeF_3: a) r-FeF_3, b) H.T.B.-FeF_3 and c) Pyr-FeF_3 (reprinted from /21/ with permission)

The reduction of the magnetic ordering temperatures was attributed to the frustration of superexchange bonds implied by the cationic topology.

This ferric crystalline series provides a nice illustration of the influence of the cationic topology on magnetic behaviour in systems with antiferromagnetic interactions.

The concept of frustration has been first introduced to explain the magnetic behaviour of spin-glasses in which random distributions of exchange interactions of both signs are present and then in some crystalline lattices. None ground state was evidenced where all interactions are satisfied simultaneously /23-26/. One can note the presence of triangular platelets which induces frustrated interactions in H.T.B.-FeF_3 and Pyr-FeF_3, contrarily to the non-frustrated r-FeF_3 : the corresponding percentages of triangular frustrated interactions are reported in table 1.

2-2 The Amorphous Ferric Fluorides a-FeF_3

Two varieties of amorphous ferric fluorides a-FeF_3 have been prepared either by the vapour-deposition technique /27/ on a cold substrate or by fluorination technique /28/. Let us note that these compounds are highly hygroscopic and the samples are never pure : either a small quantity of divalent iron phase FeF_2 is observed after the deposition process or an HF contribution is induced during the fluorination.

In addition, the reproducibility of these amorphous materials is very dependent of the experimental conditions of the synthesis : purity of the initial crystalline powder, pressure, temperature, deposition rate...

In order to investigate the structural properties, diffraction techniques (neutron, X-rays, EXAFS, XANES and LAXS) have been performed /29/. The main results on structural data are reported in table 2 and are compared to those obtained on the different crystalline forms of FeF_3.

The ^{57}Fe Mössbauer spectroscopy has probably brought the main contribution to the determination of the chemical short range order in these amorphous fluorides /30, 31/.

The paramagnetic spectra recorded at room temperature on different samples are reported on figure 4. The mean isomer shift values (reported in table 2) are consistent with those measured on the crystalline phases and are significative of trivalent iron state in octahedral surroundings.

The quadrupolar splitting distribution P(Δ) can be characterised by the factor $q = \langle\Delta^2\rangle/\langle\Delta\rangle^2$ /12/ for an easier comparison to those deduced from the analytic development proposed by Czjzek $P(\Delta) = \Delta^{d-1}/\sigma^d \exp(-\Delta^2/2\sigma^2)$ /9/: one can conclude from the values of d reported in table 2 that the *iron ions are surrounded by octahedra weakly distorted in both amorphous forms of* FeF_3.

Over the temperature range 25-40K, the spectra are composed of non-resolved lines associated to the progressive spin freezing of the magnetic moments. The analysis of the spectra required correlated quadrupolar splitting P(Δ) and hyperfine field P(H) distributions /32/.

	Vapour - deposited a-FeF_3	Fluorinated a-FeF_3, x HF[a]
FeF_3 (Å)	1.95	1.95
H_{hf}[b] (KOe)	560	550-560
I.S[c] ($mm\ s^{-1}$)	0.46 0.58	0.46-0.48 0.56-0.58
Δ[d] ($mm\ s^{-1}$)	0.58	0.51-0.56
$q = \langle\Delta^2\rangle / \langle\Delta\rangle^2$	1.25	~ 1.30
d[e]	2.0	2.7
E.F.G.[f] (%)	45;55	48;52
T_G[g] (K)	~ 36	~ 36
Θ_p (K)	~ - 490	~ -1200
Spin structure	Speromagnet	Speromagnet

(a) for the present result x = 0.4 ; 0.8
(b) mean hyperfine field value at 4.2K
(c) isomer shift values at 300K and 4.2K relative to metallic iron at 300K
(d) mean quadrupolar splitting values at room temperature
(e) exponent value in the sense of Czjzek's analytic development /9/
(f) probability for finding positive and negative V_{zz} values at iron site
(g) spin-freezing temperature.

TABLE 2 :Structural, magnetic and hyperfine characteristics of both amorphous varieties of FeF_3.

Below 25K, the Mössbauer spectra consist in six broad but resolved lines indicating the presence of E.F.G. and hyperfine fields distributions. Hyperfine characteristics of some ferric amorphous fluorides obtained at liquid helium temperature are given in table 3.

The main conclusions are the following : (i) *the isomer shift values agree with the octahedral coordination of ferric iron,* (ii) *the hyperfine field mean values are weakly smaller than those observed on the analoguous crystalline forms* (see table 3 and 1), (iii) *the hyperfine field distribution is close to a symmetric Gaussian shape with a small low-fields tail,* and (iiii) *the hyperfine parameters are weakly correlated.*

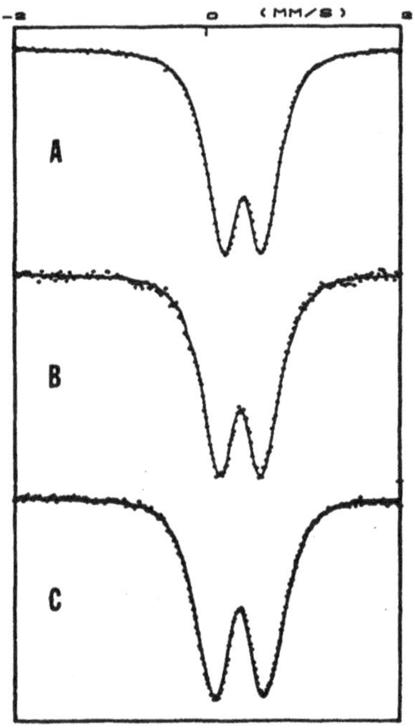

Fig.4: Mossbauer Spectra recorded at 300K on a-FeF_3,0.8HF (A), a-FeF_3,0.4HF (B) and deposited a-FeF_3 /33/.

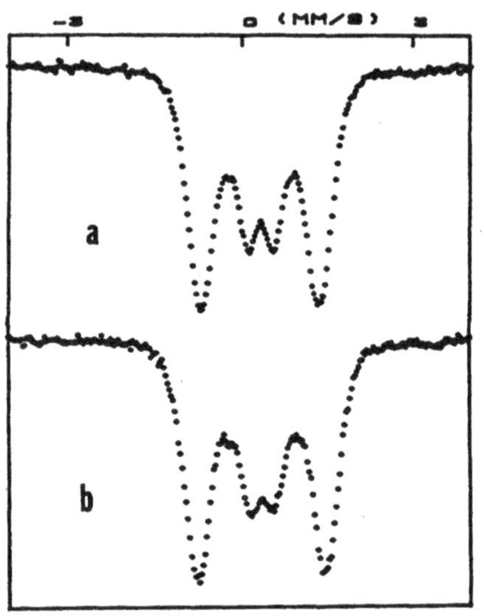

Fig.5: Mossbauer spectra recorded at 200K under 70 kOe applied field parallel to the γ-beam on FeF_3,0.4HF(a) and a-FeF_3(b) /33/

By applying external magnetic field parallel to the γ-beam in the paramagnetic state at high temperatures, one can establish from the symmetric shape of the spectra (as shown fig. 5) that the *E.F.G. values at iron sites are positive and negative in equal proportions* (see table 2). Such proportions were estimated by fitting the superimposition of quadrupolar spectra composed of pure positive or pure negative axial E.F.G. /33/. The slight misfit is probably due to a small contribution of induced hyperfine fields.

Below the spin-freezing temperatures, the external field cannot align the magnetic moments : the broadening of the outerlines (as shown fig. 6) is characteristic of a speromagnetic structure. Such a broadening was previously observed on the two frustrated **crystalline** forms of FeF_3. In addition, antiferromagnetic interactions are evidenced by susceptibility measurements. One can conclude that *the speromagnetic behaviour is in agreement with a highly frustrated cationic topology,* responsible for the low spin-freezing temperature value.

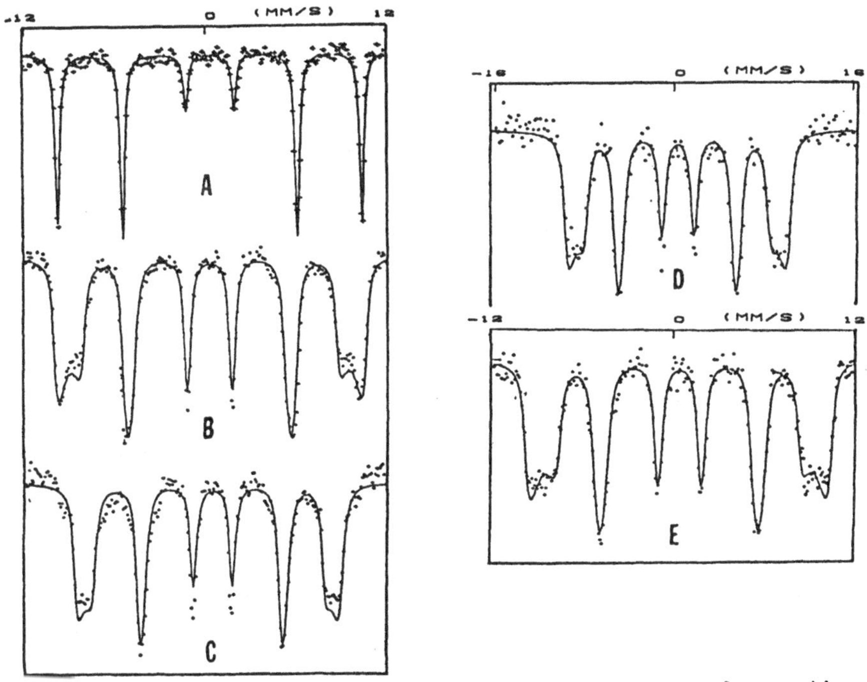

Fig.6:Mossbauer spectra recorded at 4.2K under an external magnetic field parallel to the γ-beam on r-FeF_3, H.T.B.-FeF_3, Pyr-FeF_3, a-FeF_3 and a-FeF_3,0.8HF /29/.

In the intermediate temperature range, the competition between molecular field distribution and applied field leads to a progressive canting of the magnetic moments /34/.

Recent E.P.R. experiments have been performed on amorphous AlF_3 doped with Cr^{3+} /35/. As the asymmetry parameter is given by $\lambda = b_2^2/b_2^0$ with $0 \le \lambda \le 1$, the E.P.R. spectra have been successfully taken into account by using the joint distribution function $P(b_2^0, \lambda)$ derived from random ionic arrangements with $2 \le d \le 3$.

All these structural and magnetic results gave evidence that *the "ideal" amorphous ferric fluoride a-FeF_3 can be described as a random network of corner-sharing octahedra made up of odd- and even-membered cationic rings.*

3 - REVIEW OF THE AMORPHOUS STRUCTURAL NETWORKS REPRESENTING a-FeF_3

3-1 A First Computer Modelling

A first approach of modelling the structure of a-FeF_3 has been proposed by Lines /36, 37/. A dense random packing model containing 858 ions has been computed using an algorithm based on the method pioneered by Bennett /38/: it consists in the sequential addition of hard spheres of two different radii ; in addition, the electrostatic forces are introduced to separate the ions of like charge as far as possible within the basic building scheme.

The E.F.G. values at iron sites were calculated using a point charge model : the E.F.G. distribution well agrees with Mössbauer data and exhibits an equal probability for finding positive and negative values of V_{zz} /39/. Let us remark that such a structural description does not respect the expected six-fold coordination.

3-2 Hand-Built Models

Two spherical continuous random networks composed of corner-sharing octahedra with no dangling bonds were then built by hand : the first one consists in 516 ball and spoke units (2064 atoms) /40/ and the second one results in a random packing of 300 plastic rigid octahedra (1200 atoms) /41/.

In order to minimise the elastic energy, a computer relaxation was carried out on these structural models assuming harmonic pair potentials, the topology being unchanged.

Structural parameters and ring statistics were calculated from these relaxed networks and discussed by comparison to available experimental data obtained on amorphous or crystalline phases.

3-3 A Random Simulated Model

A few years ago, a larger octahedral continuous random network containing 4396 atoms (1251 Fe atoms and 3145 F atoms) was successfully generated by computer /42/. The model starts from a nucleus made of nodes which represent cationic positions : the algorithm consists in growing this nucleus by adding new positions and connecting bonds between them. The physical criteria were the following : (i) the 6-fold coordination, (ii) the steric encumbrance of the atoms and (iii) the presence of odd-and even membered cationic rings. By such a way, a spherical random network composed of six-fold coordinated 1251 cationic positions with no dangling bonds was obtained, and 3145 anions were then included. The definitive model was refined after applying the relaxation procedure assuming harmonic pair potentials.

The structural and topological data deduced from the computed network are consistent with experimental results, like the E.F.G. calculations which are reported in section 4.

3-4 The Topologically Disordered Models

More recently, the topological defect method was developped to octahedrally-bonded systems : it consists in modifying the topology of a crystalline supercell by breaking and reforming bonds and then in applying relaxation with periodic boundary conditions (allowing more accurate structural data calculated after juxtaposition of these supercells).

Such an approach has been yet discussed in 2-dimensional case /43/ while the last refinements for 3-dimensional topological disordered systems are in progress /44/ but the E.F.G. calculations were carried out.

4- E.F.G. CALCULATIONS IN IONIC SYSTEMS.

4-1 Crystalline Ionic Structures

Up to now, E.F.G. calculations at iron sites have been carried out using a polarizable point charge model on several kinds of crystalline ionics : spinels /45, 46/, garnets /47-49/, pyrochlores

/50/ and fluorides ($AFeF_4$, compounds) /51/.

The experimental data given by the quadrupolar splitting value, were well reproduced in most cases by taking into account both monopolar and dipolar contributions.

The E.F.G. components are calculated in most ionic structures using a polarizable point charge model : an ionic compound is assumed to consist of discrete deformable cationic and anionic positions located on a three-dimensional network. Consequently, the local electric field E which derives from the potential V, acting on a site, induces a dipolar moment $p = [\alpha_d]\,E$ where $[\alpha_d]$ represents the dipolar polarizability tensor. In the same way, the E.F.G. tensor $[V_{ij}]$ induces a quadrupolar moment tensor $[q_{ij}] = [\alpha_q]\,[V_{ij}]$ where $[\alpha_q]$ is the quadrupolar polarizability tensor.

The potential V resulting from the charge distribution $\rho(r)$ located at the distance R is given by

$$V = \frac{1}{|R|} \int \rho(r)\, d\tau \ .$$

It can be expanded into a multipolar contribution series as following

$$V = \frac{q}{|r|} + \sum_i P_i \frac{\partial}{\partial x_i} \left(\frac{1}{r}\right)_{r=R} + \frac{1}{2} \sum_{i,j} q_{ij} \frac{\partial^2}{\partial x_i\, \partial x_j} \left(\frac{1}{r}\right)_{r=R} + \ldots$$

where q, P_i, and q_{ij} are respectively components of the monopolar, dipolar and quadrupolar moments.

Consequently, the electric field and the E.F.G. components can be calculated by successive derivations of the potential.

By such a way, one can find a set of equations which are linked together : so, the total system has to be self-consistently solved. The quadrupolar contribution is generally neglected. Consequently, the monopolar and the dipolar contribution of the E.F.G. components are given by the following expressions

$$V_{ij} = \frac{\partial^2 V}{\partial x_i \partial x_j} \simeq V^m_{ij} + V^d_{ij} \text{ with}$$

$$V^m_{ij} = \sum_k q_k \frac{(3\, x_{ik}\, x_{jk} - \delta_{ij}\, r_k^{\,2})}{r_k^{\,5}}$$

$$V^d_{ij} = \sum_k - 3 \left[\frac{x_{ik}\, P_{ik}\, (5x_{ik}\, x_{jk} - \delta_{ij}\, r_k^{\,2})}{r_k^7} - \frac{(P_{ik}\, x_{ik} + P_{jk}\, x_{jk})}{r_k^5} \right]$$

where q_k is the charge and x_{ik}, x_{jk} are the cartesian coordinates of the kth ion at a distance r_k from the origin located at a given site and P_{ik} is the ith component of the induced dipole moment on the kth ion.

In general, any E.F.G. tensor at nuclear sites is expressed as

$$[\, V_{ij} \,] = (\, 1 - \gamma \,)[\, V_{ij}^{Latt} \,] + (\, 1 - R \,)\, [\, V_{ij}^{Val} \,]$$

where V_{ij}^{Latt} is the lattice contribution resulting from the charge distribution of the surrounding ions in the network, V_{ij}^{Val} the valence term due to the electron distribution in incompletely filled electronic shells of the probe atom, γ and R the antishielding and shielding factors respectively. One can note that for ferric ions, the valence term is null.

One can calculate the electric field components and the E.F.G. components by lattice summations, solving respectively 3n self-consistent equations and 5n equations where n represents the number of unequivalent iron sites : consequently, such calculations included monopolar and quadrupolar contributions are restricted to the crystalline systems.

To obtain absolute convergence in lattice summations which are generally developped either in direct lattice or in reciprocal lattice, Coogan has established some rules as following /52/: the unit cell has (i) no monopolar moment for the E.F.G.; (ii) no monopolar or dipolar moment for the electric field ; (iii) no monopolar, dipolar or quadrupolar moment for the potential calculations to converge.

From a computational point of view, the elementary cell is composed of fractional charges according to the location (1/8 at corners, 1/4 at edges and 1/2 at faces) and to the space group symmetries /53/. Then, the calculations are carried out on volumes which are multiples of the elementary cell with $(2L-1)^3$ unit cells arranged symmetrically around the origin cell. The Coogan's rules previously described are well satisfied under this calculation method. By such a way, an absolute convergence on the E.F.G. components is rapidly observed as a function of L (for instance after L=4 /51/).

The calculations require accurate crystallographic data. An other alternative consists in calculating over spherical volumes /49/: this method is less efficient in the estimation of the values of E.F.G. components due to a slower convergence, because Coogan's criteria are satisfied only on the average.

Once the E.F.G. components are calculated over all the unequivalent iron sites, the main components V_{xx}, V_{yy} and V_{zz} and to the set of principal axes are determined after diagonalisation, according to the usual convention $|V_{xx}| \le |V_{yy}| \le |V_{zz}|$ with respect to the crystallographic reference. One can deduce the relevant iron Mössbauer parameters : the asymmetry parameter $\eta = (V_{xx} - V_{yy})/V_{zz}$ and the quadrupolar splitting $\Delta = \frac{e^2Q\,V_{zz}}{2}\left(1+\frac{\eta^2}{3}\right)^{1/2}(1-\gamma_\infty)$ where

Q is the quadrupolar moment, assumed to be 0.20 barns for ^{57}Fe /54/ and $(1 - \gamma_\infty)$ is the Sternheimer antishielding factor assumed to be +10.14 /55/ for ferric ions and finally the polar angles (θ, ϕ) of the principal E.F.G. axis OZ with respect to the spatial reference.

Monopolar and dipolar E.F.G. calculations have been carried out over the crystalline phases of FeF_3 : the results are reported in table 3. Except for an opposite sign in H.T.B.-FeF_3 phase, a quantitative agreement with the room temperature quadrupolar splitting value would require a fluorine polarizability of 0.5 - 0.6 $Å^3$ /56/, which is lower than classical values (≃ 1.0 $Å^3$) /57, 58/ (see table 1).

	r-FeF_3	H.T.B. - FeF_3	pyr - FeF_3
$\Delta(mm.s^{-1})$	$- 0.07^a$ /57/ $- 0.03^b$	$+ 0.33^a$ /57/ $+ 0.29^b$	$-.60^a$ /56/ $-.25^b$

(a,b) the fluorine polarizablility value is assumed to be 1.0 $Å^3$ and 0.5 $Å^3$ respectively.

TABLE 3 : E.F.G. including monopolar and dipolar contributions calculated in the three crystalline phases of FeF_3.

4-2 E.F.G. Calculation Method In Amorphous Structure /59/

In amorphous systems, the periodicity is lost : consequently the calculations of E.F.G. components have to be performed on spherical volumes. Furthermore, the dipolar contribution has to be dropped because of the large number of required equations ; fortunately, this term should decrease very rapidly with the integrating sphere radius.

The method was tested on a crystalline fluoride, and the computed monopolar contribution of E.F.G. components were compared with direct lattice summations calculations. As it is shown on fig. 7, the V_{ij} limit values were in excellent agreement ; one can note a slower convergence clearly due to large values of monopolar moments of the successive spheres. To speed up the convergence and to reduce the monopolar moments, we have introduced a transitional shell in which the ionic charges are progressively weighted from the inner to the outer surface as it is shown on fig. 8.

The improved method was optimized after testing the influence of the following parameters ; number and thickness of successive shells, thickness of the transitional shell, deviation of the quadrupolar splitting value which should be smaller than the typical step using in the procedure fitting of quadrupolar Mössbauer spectra (typically $\delta\,\Delta \sim 0.10\ mm.s^{-1}$).

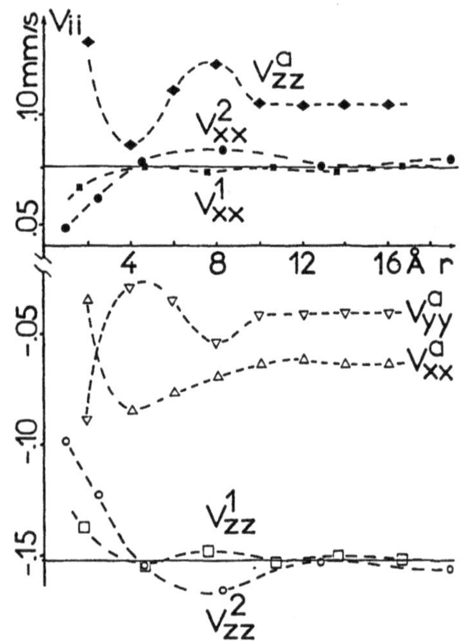

Fig.7: Evolution as function of the integrating sphere radius of the E.F.G. components for crystalline $CsFeF_4$ without the transitional shell (V_{ii}^2), with such a shell (V_{ii}^1) and for the amorphous case (V_{ii}^a) /59/.

Accurate E.F.G. calculations are obtained using an integration sphere of 10 Å radius at least, and radius step of 1 Å: this leads to define an E.F.G. sphere located at the centre of the structural network, in which E.F.G. components can be calculated over iron sites (see fig. 9).

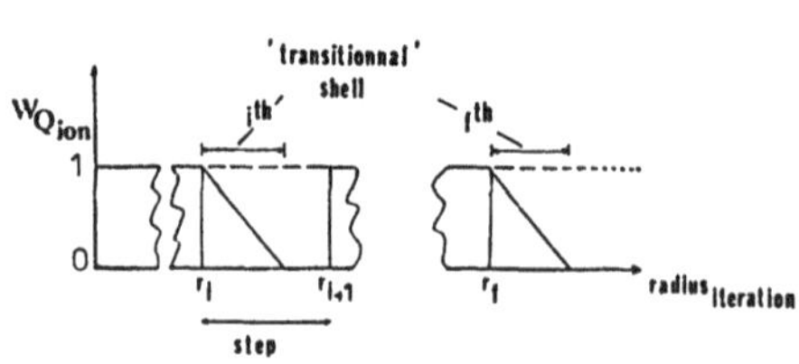

Fig.8: Weighting factor of the ionic charges as a function of the sphere radius /59/.

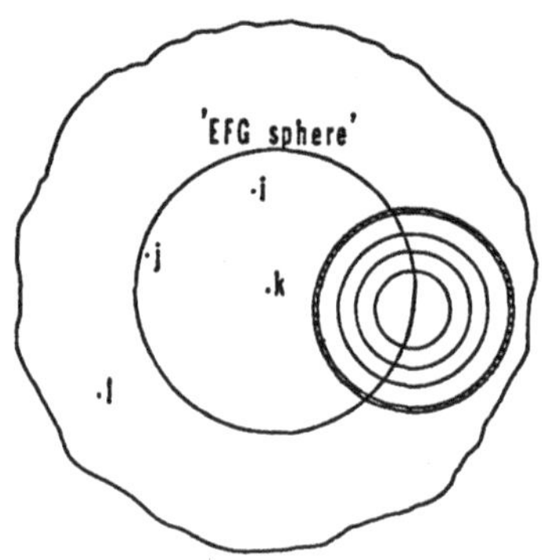

Fig.9: Schematic view of the E.F.G. sphere /59/.

4-3 E.F.G. Calculation Results On a-FeF_3 Structural Networks.

The first continuous random network composed of 2048 atoms was considered to investigate monopolar E.F.G. distributions. The quadrupolar splitting distribution was obtained from an E.F.G. sphere including 50 iron sites only : in spite of this rather small number of probes, the mean value of the quadrupolar splitting, the q factor value (reported in table 4) and the distribution shape were in good agreement with room temperature Mössbauer data.

More accurate E.F.G. data were obtained from the simulated model for which a larger E.F.G. sphere containing 495 iron sites was considered. The quadrupolar splitting mean value (reported in table 4) is weakly smaller than that observed at room temperature for both varieties of a-FeF_3. The quadrupolar splitting distribution well agrees with the experimental distribution (see fig. 10). Furthermore, both negative and positive values of Δ are obtained in equal amounts, as it is shown on figure 11, in agreement with experimental data reported in table 2 .

Fig.10: Quadrupolar splitting distribution calulated from the simulated model and compared to the experimental $P(\Delta)$.

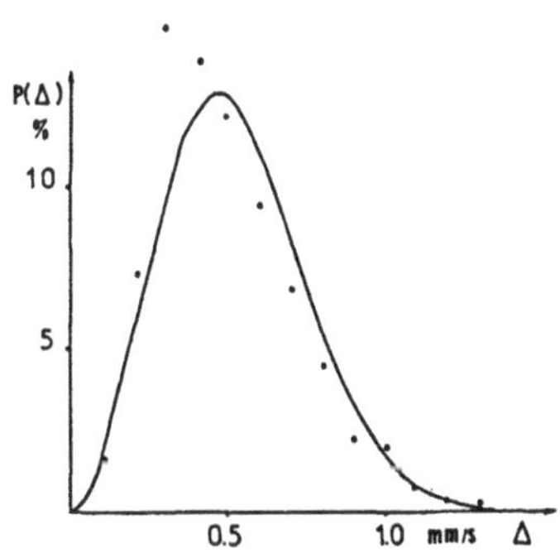

Fig.11: Negative and positive contributions of Δ values calculated from the simulated model /59/

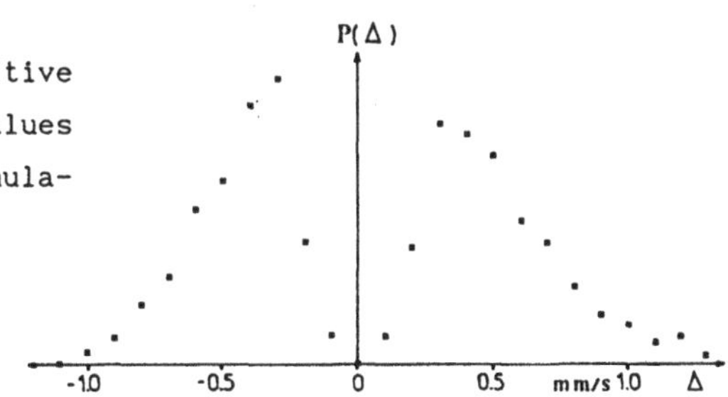

	Hand-built model 2064 atoms	Simulated random network 4396 atoms	Topologically disordered networks 4000 atoms
E.F.G. Sphere[a]	50	495	1000
$<\Delta>$ ($mm.s^{-1}$)	0.79	0.46	0.55-0.70
$<\eta>$	0.63	0.6-0.7	0.53-0.58
d^b	2	2.9	$2<d<3$
E.F.G.[c] (%)	46 ; 54	52 ; 48	~ 50 ; 50

(a) number of iron sites
(b) exponent value in the sense of Czjzek's analytic development /9/
(c) probability for finding positive and negative V_{zz} values

TABLE 4 : E.F.G. characteristics deduced from the different networks representing a-FeF_3.

The quadrupolar splitting distribution and asymmetry parameter distribution are compared respectively on figure 12 and 13 to those theoretically predicted by Czjzek et al with $2 \leq d \leq 3$ / 9 / in the case of dense random packings. One can note null probability for $\Delta = 0$ or $\eta = 0$, in agreement with theoretical predictions. Moreover, the marginal η distributions for a given quadrupolar splitting value Δ do not sizeably depend on Δ values but differ from dense random packings predictions : such a difference is well reflected by the most probable η value estimated to $\simeq$ 0.6-0.7, smaller than the predicted value ($\simeq$ 0.8-0.9) in the case of dense random packing models.

At last, E.F.G. distributions were calculated on some random computed packings of octahedra obtained according to the topological defect method /60/. As the crystalline cubic supercell was composed of 10x10x10 = 1000 octahedral units on one hand, and as the topologically disordered networks were relaxed with periodic boundary conditions, the monopolar contributions of the E.F.G. components were calculated over 1000 iron sites.

The quadrupolar splitting mean values were found in the range 0.55-0.70 $mm.s^{-1}$, in agreement with Mössbauer data, like the quadrupolar splitting distribution. One can also observe an equiprobability between positive and negative values of V_{zz}.

The same conclusions as those established from the simulated network were derived except that the asymmetry parameter mean value was sizeably reduced in the range 0.53-0.58.

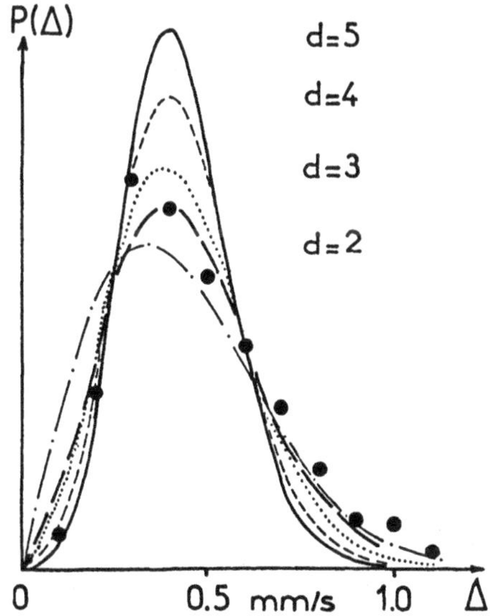

Fig.12: Calculated P(Δ) from the simulated model compared to those deduced from Czjzek's analytic developments /59/

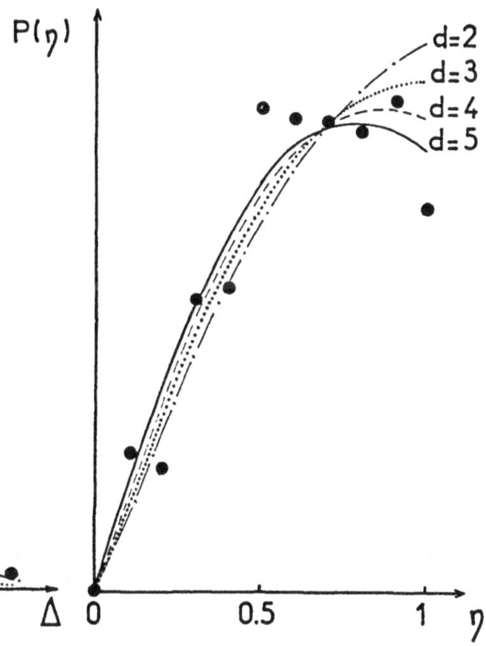

Fig.13: Calculated P(η) from the simulated model compared to those deduced from Czjzek's analytic developments /59/

4-4 Topology Of The Distribution $P(V_{zz},\eta)$

One can describe the topology of the joint distribution function $P(V_{zz}, \eta)$ once the E.F.G. parameters V_{zz} and η have been expressed in terms of polar coordinates (r,φ) as following /61/.

$$\begin{cases} V_{zz} = -(+)\ r\cos\left(\frac{\pi}{6} +(-)\ \varphi\right) \\ \eta = \sqrt{3}\ \tan\left(\frac{\pi}{6} +(-)\ \varphi\right) \qquad \text{when } V_{zz} < (>)\ 0 \end{cases}$$

Consequently, the function $P(V_{zz},\eta)$ is transformed into an $F(r,\varphi)$ distribution : in the case of the dense random packing model, the joint distribution function takes the following form :

$$F(r,\varphi) = \frac{r^4 \cos 3\varphi}{(2\pi)^{1/2}\sigma^5} \exp\left(-r^2/2\sigma^2\right)$$

By such a reference transformation, the parameter values defined in the range $-\infty < V_{zz} < +\infty$ and $0 \leq \eta \leq 1$ are now contained in a 60° angular sector : figure 14a and 14b shown contour maps of $P(r,\varphi)$ estimated from the simulated random network (before smoothing) and from topologically disordered networks respectively.

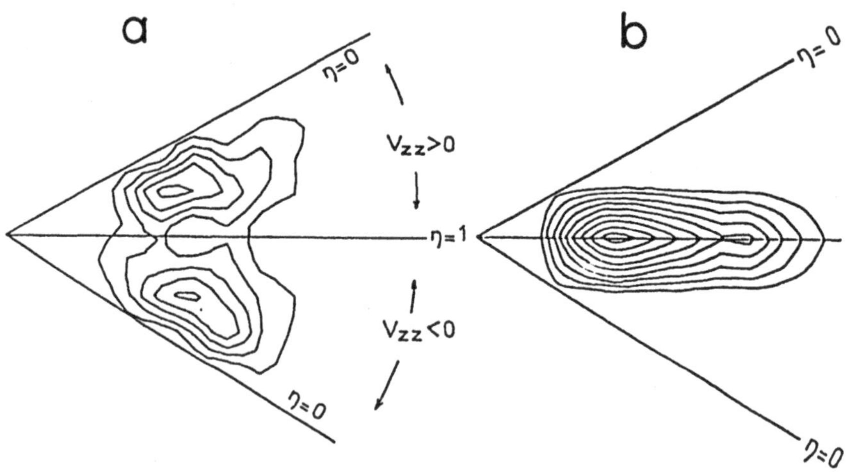

Fig.14: Contour maps of $P(r,\varphi)$ figuring the topology of $P(V_{zz},\eta)$ calculated on the simulated model before smoothing (a) /59/ and on the topologically disordered network (b) /60/.

In both cases, the contour maps vizualize the equiprobability for finding positive and negative values of V_{zz}, and the absence of correlation between η and V_{zz}.The major difference which appears between the joint distributions, is the presence of a valley centered around $\eta = 1$ only in the simulated random network.

Furthermore, theoretical contours of P(r, φ) established either from the analysis independent of φ proposed by Coey (random gaussian model) or from the development based on a dense random packing developed by Czjzek (shell model) are respectively illustrated on fig. 15. One can remark that the joint distribution function P(r, φ) calculated on the topologically disordered networks better agree with that predicted by the shell model than that estimated from the random simulated network, in spite of similar structural characteristics. The difference might purely originate from different size of the E.F.G. spheres.

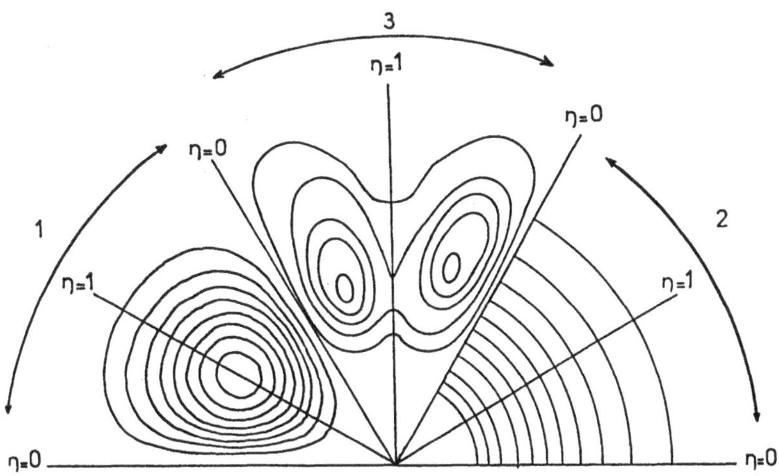

Fig.15: Contour maps of P (r,φ) calculated from the shell model (1) /61/ and from the random gaussian model (2) and from the simulated model after smoothing (3) /59/.

ACKNOWLEGMENTS

The author wishes to express his gratitude to Prof. F. Varret, J.M.D. Coey, J. Teillet, G. Ferey and M. Leblanc for their kind contribution over recent years. Acknowledgements are due to Prof. F. Varret for a critical reading of the manuscript.

REFERENCES

/ 1/ 3rd International Conference on the structure of Non-Crystalline Material, Grenoble, eds JANOT, C. and WRIGHT, A. (Editions de Physique) (1985)

/ 2/ Chappert, J., J de Physique C6 35 71-88 (1974)

/ 3/ Chappert, J., Hyp. Int 13 25. (1983)

/ 4/ Lines, M.E., J. Phys. Chem. Solids 43 723-730 (1982)

/ 5/ Czjzek, G., Nuc. Inst. Methods 199, 37-44 (1982)

/ 6/ Maurer, M. and Friedt, J.M., Hyp. Int. 27, 135-146 (1986)

/ 7/ Eibschutz, M., Lines, M.E., Hyp. Int. 27, 47-58 (1986)

/ 8/ Brand, R.A., le Caer, G. and Dubois, J.M., J. Phys. : Condens Matter 2 6413-6431 (1990)

/ 9/ Czjzek, G., Fink, J., Gotz, F., Schmidt, H., Coey, J.M.D., Rebouillat, J.P. and Lienard, A., Phys. Rev B 23, 2513-2530(1981)

/10/ Coey, J.M.D. and Czjzek, G., unpublished results

/11/ Le Caer, G., Dubois, J.M. and Brand, R.A., Amorphous Metals and Non-equilibrium processing, Von Allmen M.Ed. (les Editions de Physique) 249.(1984)

/12/ Lopez-Herrera, M.E., Greneche, J.M. and Varret, F., Phys.Rev B 28, 4944-4948 (1983)

/13/ Eibschütz, M., Lines M.E., Van Uitert, L.G., Guggenheim, H.J. and Zydzik, G.J., Phys. Rev. B 24, 2343-2348 (1981)

/14/ Czjzek, G., Phys. Rev. B 25 4908-4910 (1982)

/15/ Varret, F. and Henry, M., Revue de Physique Appliquée 15, 1057-1061 (1980)

/16/ Hepworth, M.A., Jack, K.H., Peacock, R.D. and Westland, G.J., Acta. Cryst. 10 63 (1957)

/17/ Leblanc, M., De Pape, R., Ferey, G. and Pannetier, J., Solid State Comm. 58 171-176 (1986)

/18/ De Pape, R. and Ferey, G., Mat. Res. Bull. 21 971-978 (1986)

/19/ Leblanc, M., Pannetier, J., Ferey, G. and De Pape, R., Revue de Chimie Minérale 22, 107-114 (1985)

/20/ Wertheim, G.K., Guggenheim, H.J. and Buchanan, D.N.E., Sol. State Comm. 5 537 (1967)

/21/ Ferey, G., De Pape, R., Leblanc, M. and Pannetier, J., Revue de Chimie Minérale 23 474-484 (1986)

/22/ Lahlou-Mimi, M., Pennec, Y., Bassat, J.M., Leblanc, M. and Greneche, J.M., to be published

/23/ Toulouse, G., Commun. Phys. 2 115-118 (1977)

/24/ Vannimenus, J. and Toulouse, G., J. Phys. C.: Solid State Physics 10L 537-542 (1977)

/25/ Kickpatrick, S., Phys. Rev. B 16 4630-4641 (1977)

/26/ Ferey, G., Leblanc, M., De Pape, R. and Pannetier, J. in Inorganic Solid Fluorides, Hagenmuller, P. Ed, Academic Press, 395-414 (1985)

/27/ Ferey, G., Leclerc, A.M., De Pape, R., Mariot, J.P. and Varret, F., Solid State Comm. 29 477-480 (1979)

/28/ Leblanc, M., Ferey, G., Greneche, J.M., Le Bail, A., Varret, F., De Pape, R. and Pannetier, J., J. de Physique C8 46 175-179 (1985)

/29/ Greneche, J.M., Le Bail, A., Leblanc., M., Mosset, A., Varret, F., Galy, J. and Ferey, G., J. Phys. C : Solid State Physics 21 1351-1361 (1988)

/30/ Ferey, G., Varret, F. and Coey J.M.D., J. Phys. C : Solid State Physics 12 L531-537 (1979)

/31/ Eibschütz, M., Lines, M.E., Van Uitert, L.G., Guggenheim, H.J. and Zydzik, G.J., Phys.Rev. B 29 3843-3851 (1984)

/32/ Teillet, J., Hauet, A. and Greneche, J.M., J. de Physique C8 49 935-938 (1988)

/33/ Greneche, J.M., Varret, F., Leblanc, M. and Ferey, G., Solid.State Comm. 63 435-438 (1987)

/34/ Greneche, J.M., and Varret, F., Hyp. Int. 42 935-938 (1988)

/35/ Latelli, H., Emery, J., Buzaré, J.Y., Duroy, H., Fourquet, J.L., Leblanc, M. and Ferey, G., to be published

/36/ Lines, M.E., Phys.Rev. B 21 5793-5801 (1980)

/37/ Lines, M.E., J. Mag. Mag. Mat. 36 1-25 (1983)

/38/ Bennett, C.H., J. Appl. Phys. 43 2727 (1972)

/39/ Lines, M.E., J. Non-Cryst. Solids 46 1-12 (1981)

/40/ Coey, J.M.D. and Murphy, P.J.K., J. Non-Cryst. Solids 50 125-129 (1982)

/41/ Greneche, J.M., Teillet, J. and Coey, J.M.D., J. Non-Cryst. Solids 83 27-34 (1986)

/42/ Greneche, J.M., Teillet, J. and Coey, J.M.D., J. Physique 48 1709-1714 (1987)

/43/ Greneche, J.M. and Coey, J.M.D., J. Physique 51 231-242(1990)

/44/ Greneche, J.M. and Coey, J.M.D., to be published

/45/ Hudson, A. and Whitfield, H.J., Molec. Phys. 12 165 (1967)

/46/ Yadnik, C.M. and Mathur, H.B., Molec. Phys. 16 625 (1969)

/47/ Burns, G., Phys. Rev. 124 524 (1961)

/48/ Nickolson, W.J. and Burns, G., Phys. Rev. A 133 1568 (1964)

/49/ Stadnik, Z.M., J. Phys. Chem. Solids 45 11-318 (1984)

/50/ Calage, Y. and Pannetier, J., J. Phys. Chem. Solids 38 711-718 (1977)

/51/ Teillet, J., Calage, Y. and Varret, F., J. Phys. Chem. Solids 43 863-869 (1982)

/52/ Coogan, C.K., Aust. J. Chem. 20,2551-2565 (1967)

/53/ Stevens, J.G. et al, J. Phys. Chem. 5 1093 (1976)

/54/ Sternheimer, R.M., Phys. Rev. 130 1423 (1963)

/55/ Evjen, H.M., Phys.Rev. 39 675 (1932)

/56/ Calage, Y., Zemirli, M., Greneche, J.M., Varret, F., De Pape, R. and Ferey, G., J. Sol. State. Chemistry 69 197-201 (1987)

/57/ Calage, Y., unpublished results

/58/ Wertheim, G.K. and Buchanan, D.N.E., Phys. Rev. 161 478 (1967)

/59/ Greneche, J.M., Varret, F. and Teillet, J., J. Physique 49 243-248 (1988)

/60/ Greneche, J.M., Hyp. Int. 55 953-956 (1990)

/61/ Czjzek, G., Hyp. Int. 14 189-194 (1983)

EPITAXIAL $CoSi_2$ ON Si<111>

André Vantomme and Guido Langouche*
Instituut voor Kern- en Stralingsfysika
University of Leuven
B-3001 Leuven, Belgium

* Research Assistant, N.F.W.O. (National Fund for Scientific Research, Belgium).

1. INTRODUCTION

In recent years, Co–silicides have been the object of very intensive studies [1–11]. A substantial amount of effort has especially been devoted to the investigation of $CoSi_2$. Both surface and buried layers of $CoSi_2$ have been of interest for applications in microelectronics, due to their high thermal stability, their very low resistivity (15–20 $\mu\Omega$cm at room temperature) and their small lattice mismatch to Si (only 1.2 %). Thin surface layers are very interesting as low resistance contact layers to Si. Buried layers, on the other hand, being semiconductor–metal–semiconductor (SMS) structures, are attractive for novel device applications, such as metal base transistors and three–dimensional devices [12–14]

Surface silicide layers can be obtained by evaporation techniques (furnace annealing of Co–covered Si single crystals or Molecular Beam Epitaxy [MBE]) or by implantation techniques (annealing of high dose Co–implanted Si single crystals [IBS : Ion Beam Synthesis] or by ion beam mixing [IBM]). The first example of a buried Co–silicide layer was realized by a two step epitaxial growth of a $CoSi_2$ and a Si layer sequentially on a Si substrate with MBE [4]. White et al [5] first successfully synthesized a heteroepitaxial Si/$CoSi_2$/Si structure by high dose Co implantation in a Si substrate (IBS : ion beam synthesis), resulting in monocrystalline buried $CoSi_2$ layers with sharp interfaces and with better electrical transport properties than those grown by more conventional techniques. In most cases, implantation energies around 200 keV are used. The lowest energy used is 100 keV [6,7]. From the point of view of applications, it is more interesting to use thinner buried $CoSi_2$ layers [12]. Thinner films can be obtained by lowering the implantation energy. However, owing to the higher sputtering yield and the smaller projected range when using lower implantation energies, one might form surface $CoSi_2$ layers instead of buried layers. Moreover, to be suitable for applications, the thermal stability of these layers should be investigated in detail.

In principle, if the thickness of the epitaxial $CoSi_2$ layer is below a critical value t_C, the lattice mismatch will be completely accommodated by elastic strain and the epitaxial growth will be pseudomorphic. Otherwise, the lattice mismatch will be partially accommodated by the strain and partially by misfit dislocations near the interface. We used channeling spectroscopy and X–ray rocking curve measurements to determine the strain in our samples.

Another interesting phenomenon to be studied is the orientation of epitaxial silicide layers. Two orientations can exist on Si<111> : type A, which has exactly the same orientation as the substrate, and type B, which is rotated by 180° about the surface normal <111> axis. It has been checked what type of orientation can be obtained by implantation technique, and how this orientation can depend on the $CoSi_2$ layer thickness.

To investigate these silicide/silicon structures, several techniques have been combined, such as Rutherford backscattering spectroscopy – RBS – (relative composition, interface abruptness, layer thicknesses), channeling spectroscopy (crystalline quality, strain, orientation), cross section transmission electron microscopy – TEM – (crystalline quality, interface abruptness, layer thicknesses, orientation), electrical measurements (resistivity, thermal stability), X–ray rocking curve measurements (strain, orientation). These techniques, which give a more macroscopic (averaged in space) information, were combined with Mössbauer spectroscopy, a method which gives a more local (microscopic) information, and thus can distinguish between different chemical phases.

2. EXPERIMENTAL

All Mössbauer spectra were measured at room temperature, using a single line absorber of 0.5 or 1.0 mg/cm^2 ^{57}Fe SFC (sodium ferrocyanide). The isomer shifts values were converted to absorber isomer shifts relative to α–iron.

A 2 MeV He^+ beam produced by a 5SHD–2 Pelletron is used for the RBS and channeling measurements with normal scattering geometry. To improve the depth resolution, glancing–angle geometry was sometimes used as well.

The resistivity measurements are carried out with a standard four–point–probe technique. The probes are formed by thin (30 μm diameter) Al alloy (99% Al + 1% Si) wires which are mounted on the sample with an Ultrasonic Wire Bounder. Measurements are made down to 0.5 K in a conventional ^3He resistivity cryostat. For the high–temperature measurements ($10 < T < 300$ K), a ^4He flow cryostat is used.

The X–ray rocking curve measurements were done with a computer–controlled 12 kW rotating anode Rigaku DMax II system, equipped with a flat pyrolytic carbon monochromator, and using a CuK_α ($\lambda = 1.5418$ Å) source. All the diffraction spectra were measured with a divergence slit of 1/6°, a receiving slit of 0.3 mm and a scattering slit of 1.2°. Typical counting times were between 1 and 20 seconds, stepping intervals were always 0.01° 2θ.

3 RESULTS AND DISCUSSION

3.1 Surface silicide layers by annealing of Co–covered Si.

Surface Co–silicide layers were formed by furnace annealing (for 1 hr under H_2) on Co–covered Si single crystals. An 80 nm thick Co–layer was magnetron sputtered onto a Si<100> substrate. ^{57}Co was then implanted with a total dose of 3×10^{14} atoms/cm^2 and an energy of 80 keV. This gives rise to an implantation range of about 24 nm, with a width of 16.5 nm [15]. The Mössbauer spectra are therefore representative for the behavior of the Co–layer.

After each annealing step of the Co–covered Si sample, we compared the obtained Mössbauer spectrum with the known [16] spectra of the various Co–silicides. Fig. 1 compiles the results of all our spectra. It shows that after annealing at temperatures up to 300°C, no silicides are formed yet, resulting only in a magnetically split spectrum of ^{57}Fe in Co, with a magnetic field of 318(3) kG. From the relative amplitudes of the lines, we calculated $\theta = 86°42'$ (θ is the angle between the magnetic field and the gamma ray direction, normal to the sample). Thus, there is a strong preferential magnetization along the plane, as can be expected for thin films. After annealing at higher temperatures (up to 900°C), Co_2Si, CoSi and $CoSi_2$ are formed subsequently. It is a remarkable fact that two different Co–silicides can grow simultaneously : between 350°C and 425°C Co_2Si and CoSi coexist, and between 475°C and 575°C CoSi and $CoSi_2$ are formed together. This phenomenon has only been observed for two transition metal silicides, namely for Co and Pt [2].

RBS measurements were performed simultaneously and confirmed fully our temperature dependent Mössbauer study. These RBS–data further show that the various silicide layers are formed in parallel layers on top of the Si–substrate. A more detailed study of these surface layer experiments is given in [17].

A surprising observation is that, although $CoSi_2$ has a cubic lattice (CaF_2) and thus is expected to give rise to a single line Mössbauer spectrum, we always find a broad side resonance next to the dominating single line (Fig. 2), the origin of which is not completely understood. Some possible interpretations will be discussed further (3.3).

3.2 Buried and surface $CoSi_2$ layers by ion implantation

3.2.1 The formation of buried and surface $CoSi_2$ layers by ion implantation

Si<111> samples are implanted with Co ions with implantation energies of 30, 50, 70 and 160 keV in a temperature range from 300°C to 400°C and a vacuum of 10^{-6} Torr. In order to form a buried and continuous $CoSi_2$ layer, the doses are chosen to be slightly larger than their critical values [6,7].1 To form the surface $CoSi_2$ layers, implantations are performed with doses about twice the values used for buried layers. For Mössbauer spectroscopy studies, the ion beam consisted of ^{59}Co, alternated with small fractions (< 1%) of ^{57}Co.

A two–step annealing process, normally 600°C for 1 hr followed by 1000°C for 1/2 hr is used to form the layers. Generally, we used conventional furnace annealing with an ambient of flowing N_2/H_2 (H_2 ~ 25%) gas. In some cases rapid thermal processing in a N_2 ambient is also used.

As an example, we consider a sample which was implanted with 0.9×10^{17} Co/cm^2 at 70 keV. The Mössbauer spectrum of the as–implanted sample at 350°C contains lines of various silicides, formed after implantation. The dominant single line at relative velocity v = –0.09(5) mm/s indicates that most of the Co has reacted with Si by forming precipitates with a local stoichiometry similar to $CoSi_2$. However, the resonances at positive velocity side show the existence of other local phases as well (Co_2Si, CoSi, interstitial Co,...).

The RBS spectrum shows a Co–implantation profile with a nearly Gaussian shape, and a Co–peak concentration of only 22%, which is not enough to form a continuous (macroscopic) $CoSi_2$ layer. The rather high minimum yield of the aligned spectrum (χ_{min} = 50%) indicates that the sample is still heavily damaged. Combining these channeling results with those of the Mössbauer measurements, we can conclude that after implantation, microscopic precipitates of various silicides are formed, the largest part being $CoSi_2$.

The spectrum taken after annealing at 600°C is the same as in the case of surface $CoSi_2$–layers on Si (Fig. 2), which means that we have the same kind of anomaly as in the former case. The RBS spectrum confirms the occurrence of a buried continuous $CoSi_2$ layer. However, when we measure in channeling geometry, it is found that the minimum yield χ_{min} = 25%, indicating that a good quality single crystal is not yet obtained. Only after annealing at 1000°C, the χ_{min} drops to 5% (Fig. 3). Considering this enormous improvement of the crystalline quality, it is a surprising fact that the Mössbauer spectrum doesn't change at all : the same type of spectrum as in Fig. 2 is obtained. This final improvement is therefore mainly long range reordering whereas the short range configuration, probed by Mössbauer spectroscopy, does not change any more.

Table 1 summarizes the results of various implantation energies. It shows that with lower energies, it is still possible to form continuous $CoSi_2$ layers, buried under a Si top layer (even with an implantation energy as low as 30 keV).

Recently, we succeeded for the first time in synthesizing double $CoSi_2$ heterostructures by IBS. Therefore, Co ions were implanted with two different energies in a Si<111> substrate. After annealing, a Si/$CoSi_2$/Si/$CoSi_2$/Si–substrate structure was formed [21,23].

A cross section TEM picture corroborates that finally, after the two step annealing, we obtain a monocrystalline and continuous buried layer with sharp interfaces (the interface roughness is estimated to be only about 15 Å).

The resistivity ρ of $CoSi_2$ at room temperature is 15–20 $\mu\Omega$cm which is much lower than that of Co_2Si and CoSi, so the resistivity measurements is also helpful to identify the phase of the buried layer. The ρ (330 K) value of 15.5 $\mu\Omega$cm and a residual resistance ratio ρ (300 K) / ρ(4.2 K) of 5.4 indicate that this buried layer is a $CoSi_2$ layer with good electrical transport properties (Fig. 4). In addition, the superconducting transition temperature T_c of this buried $CoSi_2$ layer is measured (full line in the insert of Fig. 4) and, for comparison, of another surface $CoSi_2$ layer (dotted line) prepared by MBE. They are 1.3 K and 1.25 K respectively and are comparable with the results from others [5,11].

3.2.2 Thermal stability of buried and surface $CoSi_2$ layers by ion implantation

Fig. 5 shows the RBS spectra of three Co–implanted Si samples (50 keV with a dose of 0.6×10^{17} Co/cm²). It shows that the sample after 1000°C has a buried layer with abrupt interfaces (Fig. 5a). Moreover, TEM (Fig. 5b) confirmed that a monocrystalline and continuous layer was formed. Fig. 5c shows the RBS spectrum of another buried $CoSi_2$ layer, subsequently annealed at 1100°C for 30 min. In comparison with fig. 5a, now the Co peak is lower and a tail is obvious, meaning that this annealing starts to change the Co profile of the sample. The corresponding TEM picture (Fig. 5d) shows that the interfaces are no longer as flat as after 1000°C annealing, and that voids are observed extending both above and below the buried silicide layer. The formation mechanism of the voids is still unclear. Finally, annealing at 1200°C for 5 min changes the Co profile significantly (Fig. 5e) and the TEM picture (Fig. 5f) shows that the continuous layer is broken up into islands distributed on top of the silicon substrate.

A similar study was done for surface $CoSi_2$ layers formed by ion implantation, and it was found that this kind of layer already breaks up into islands after 30 min annealing at 1100°C. Even more, the fact that the layer is broken up changes the resistivity as a function of temperature completely (Fig. 6). Not only ρ (300 K) has increased by about three orders of magnitude, but it also shows a rather complicated electrical behavior instead of a simple metallic one. This layer now has both semiconductor and metallic electrical properties : at temperatures higher than 200 K the metallic behavior is more pronounced, whereas at temperatures lower than 200 K the semiconductor electrical behavior is dominant. Measurements on a high temperature annealed MBE $CoSi_2$ sample [18] showed similar results as in the case of an ion implanted surface sample, confirming the conclusion that buried $CoSi_2$ layers have a better thermal stability than a surface $CoSi_2$ layer.

3.3 Comparison of ^{57}Fe in $CoSi_2$ after direct ion implantation and after ion implantation of a radioactive parent

Highly pure MBE–grown $CoSi_2$ was implanted with 3×10^{15} ^{57}Fe atoms/cm^2 at 40 keV with the target kept at 300°C. A target from the same MBE–grown sample was implanted under the same conditions with ^{57}Co with 1×10^{15} atoms/cm^2. We compared the emission Mössbauer spectrum of the ^{57}Co source with the absorption spectrum (using a ^{57}Co in Cu single line source) of the ^{57}Fe target, measured by conversion electron Mössbauer spectroscopy.

Fig. 2 and 7 show the Mössbauer spectra of the ^{57}Co and ^{57}Fe implanted MBE–grown samples, respectively. Both spectra can be fitted with the same set of parameters. The two observed resonances have isomer shifts of $\delta_1 = -0.09(5)$ mm/s and $\delta_2 = +0.41(5)$ mm/s respectively. The striking observation is that, whereas almost all ^{57}Fe atoms after ^{57}Fe implantation are found with isomer shift δ_2, almost all ^{57}Fe atoms after ^{57}Co implantation are found with isomer shift δ_1.

In the case of $^{57}Co(CoSi_2)$, the dominating Mössbauer resonance (Fig. 2) with isomer shift δ_1 is considered to be due to ^{57}Fe in the decay of ^{57}Co on regular Co lattice sites in cubic $CoSi_2$ [4–6]. The origin of the second resonance is much harder to account for. Its isomer shift is close to the value measured for CoSi, but careful X–ray analysis could not detect any crystalline trace of this phase. It cannot even be associated with the implantation process as it also appears after diffusion in bulk $CoSi_2$.

We can tentatively propose two interpretations for this anomalous resonance :

(1) It is associated with ^{57}Fe atoms in the decay of ^{57}Co atoms in a lattice configuration which is different from the normal substitutional Co atoms in

$CoSi_2$. As the electron density at the ^{57}Fe nucleus is close to the one observed in CoSi, a local deviation from the correct stoichiometry towards a more Co–rich one can be responsible for this.

(2) The ^{57}Co parent atoms are found in a correct $CoSi_2$ stoichiometry, but an anomalous charge state is stabilized at least during the 200 nanosecond lifetime of the exited Mössbauer state of ^{57}Fe.

When the ^{57}Fe isotope is implanted rather than ^{57}Co, it is a surprising observation to discover (Fig. 7) that the large majority of ^{57}Fe is found with the same hyperfine interaction parameters as the anomalous resonance of the ^{57}Co implantations. So far we do not have extensive data on different types of $CoSi_2$ implanted with ^{57}Fe, so that we cannot make a full comparison under different conditions. The present observations can somehow fit in both proposed interpretations for ^{57}Co :

(1) During the final relaxation phase of the collision cascade, it can not be excluded that an impurity atom like Fe preferentially ends up in a different local stoichiometric surrounding than (a Co atom in a) regular $CoSi_2$. What we then observe is that the local (Co–rich ?) configuration which is preferred by the majority of the Fe atoms is the same as the one preferred by a minority of the Co atoms.

(2) If the two isomer shifts are due to different electronic configurations in the same stoichiometric $CoSi_2$, we have to conclude that both are very stable, and that Fe atoms that have been implanted and came to rest long before the Mössbauer experiment prefer a different electronic configuration than the Fe atoms that came into existence only nanoseconds before the Mössbauer experiments.

3.4 Strain and orientation in epitaxial CoSi₂(111) layers formed by ion implantation

The disilicides $NiSi_2$ and $CoSi_2$ have the cubic CaF_2 structure with a lattice mismatch of –0.46% and –1.2% with respect to Si respectively. The misalignment (or kink angle) $\Delta\Psi$ between the relevant axes in the epilayer and in the substrate is a direct measure for the tetragonal distortion ϵ_T caused by the strain :

$$\epsilon_T = \epsilon_r^{\parallel} - \epsilon_r^{\perp} = \Delta\Psi \,/\, (\sin\Psi \cos\Psi) \tag{1}$$

where $\epsilon_r^{\parallel}$ and $\epsilon_r^{\perp}$ are the two components defined relative to d_{LF} (the interplanar spacing of the Layer material under Free lattice conditions) : $\epsilon_r^{\perp} = (d_L^{\perp} - d_{LF})/d_{LF}$ and $\epsilon_r^{\parallel} = (d_L^{\parallel} - d_{LF})/d_{LF}$. d_L is the spacing of the strained Layer. On the other hand, when defined relative to d_S, the substrate spacing, $\epsilon_X^{\perp} = (d_L^{\perp} - d_S)/d_S$ and $\epsilon_X^{\parallel} = (d_L^{\parallel} - d_S)/d_S$ are used. More detailed explanations concerning these components and how they can be measured by channeling spectroscopy or by X–ray rocking curve measurements, are given in [19–23]. According to the definition, it is clear that for a totally coherent or pseudomorphic $CoSi_2$ layer on a Si substrate, the parallel strain should have a value $\epsilon_r^{\parallel} = 1.2$ %. This is the case if the epitaxial layer has reached its critical thickness t_c. Concerning t_c, two theoretical models [24,25] have been developed. Although the predicted t_c values differ by two orders of magnitude, both models are claimed to be supported by experiments [26].

In the case of epitaxial $CoSi_2$(111) layers prepared by IBS, few strain studies have been performed and the results are more consistent than the data obtained on UHV $CoSi_2$ layers. For implanted layers with thicknesses in the order of 100 nm, a perpendicular strain $\epsilon_X^{\perp} = -1.68$ % was found, which is smaller than –3.5%, the value expected for a completely strained layer [27–28]. As we were able to form $CoSi_2$ layers as thin as 20 nm by ion implantation, we can study whether our thinner layers have a larger strain or tetragonal distortion or whether they are closer to pseudomorphic growth.

The orientation of epitaxial silicide layers is another interesting phenomenon to be studied. Two orientations exist on Si(111) : type A silicide has the same orientation as the substrate, while type B silicide is rotated 180° about the surface normal <111> axis. Unlike for $NiSi_2$, where the orientation can be controlled, for $CoSi_2$ layers prepared by UHV procedures on a Si(111) substrate, the type B orientation is always predominant [29]. On the contrary, $CoSi_2$ layers with a type A orientation can easily be grown by IBS. The questions whether the $CoSi_2$ layers formed by IBS have always the pure type A orientation, and whether the strain in the $CoSi_2$ layers will be related with its orientation, as is the case for $NiSi_2$ layers [30], will be studied.

3.4.1 Strain

The preparation conditions and some properties of the seven epitaxial $CoSi_2$ samples (1–7) are compiled in table 1. With the exceptions of sample 1, all samples are prepared by ion implantation followed by two step annealing. Samples 2 – 6 contain a single $CoSi_2$ layer, while sample 7 is a heterostructure containing two $CoSi_2$

layers (prepared by implanting Co atoms with two different energies). RBS/channeling and X–ray rocking curve measurements enable to determine the strain and orientation of the epilayers in these samples.

Fig. 8 and 9 show the angular scans for the different layers of two samples which contain a surface (sample 5) and a buried (sample 4) $CoSi_2$ layer respectively. The scans were made around the <114> axis (left) and the <1$\bar{1}$0> axis (right) in the (110) plane of the Si substrate. The angle between the <114> and <110> axes of the Si substrate is $2\Psi = 70.52^\circ$, indicating that the Si substrate is a perfect cubic single crystal free of strain. On the other hand, the angle between these two axes of the epilayer is $2\times(35.26 + \Delta\Psi)$ with $\Delta\Psi = 0.31^\circ \pm 0.03^\circ$ as shown in Figs. 8 and 9 and as listed in table 1. This result indicates that these epilayers are strained with a tensile strain parallel to the interface and a compressive strain perpendicular to the interface. Using eq.(1), a tetragonal distortion $\epsilon_T = (1.15\pm0.11)\%$ can be deduced from the measured $\Delta\Psi$. From X–ray rocking curve measurements, the perpendicular strain was found to be $\epsilon_r^{\perp} = -0.69\%$. With the assumption of volume conservation for the strained layer, a $\epsilon_r^{\parallel}$ value of +0.35% can be deduced. Finally, a value of $\epsilon_T = \epsilon_r^{\parallel} - \epsilon_r^{\perp} = (1.04\pm0.05)\%$ is obtained, which agrees well with the ϵ_T value calculated from the results of channeling measurements. As listed in table 1, and shown in figure 10, the six ion implanted samples have rather close perpendicular strain values with an average $\epsilon_X^{\perp}$ value of –1.9 %, which is larger than in the SPE grown sample, although our epitaxial layers are still far from a completely strained or totally coherent layer.

The strain measurements could eventually be influenced by the steering effect of the He^+ beam, because some of the channeled ions can be steered into misaligned channels after crossing the interface. In contrast to the results of Kohlhof [27] and Barbour [8], we have no indication that our measurements are influenced by this steering effect. This is discussed in [22].

3.4.2 Orientation

Figs. 8 and 9 show that when the angular scan is measured around the <114> axis of the Si substrate, the channeling dips from both the Si substrate and from the $CoSi_2$ surface and buried layer have the characteristics corresponding to the <114> axis, i.e. a higher χ_{min} value and a narrower width. On the other hand, when the scan is around the <110> axis of the Si substrate, both channeling dips show the same characteristics of the <110> axis. This confirms that these $CoSi_2$ layers which are prepared by IBS have the same orientation as their Si substrate, indicating that their orientations are pure type A or predominantly type A. However, when the $CoSi_2$ layers become rather thin, it was found to contain a mixed A+B type with a comparable amount of both orientations. Because the RBS/channeling technique can not determine the relative amount of type A and type B accurately in an epitaxial layer with mixed A+B type, we applied X–ray rocking curve measurements which have a high sensitivity and accuracy for such orientation determination. The results are listed in the last column of table 1 and Fig 11. It can be seen that even a very weak type B component of only 0.2% (sample 5) can still clearly be detected.

If the relative amount of type A is set out as a function of layer thickness, a strong correlation is found (Fig. 12). When the thickness decreases to 16 nm (sample 7), the $CoSi_2$ layer is almost completely type B oriented. Tung [31] pointed out that the dominance of type B orientation in the epitaxial silicide layers formed by UHV procedures can be attributed to its lower interface free energy so that the type B orientation in energetically favored. Unlike the UHV procedure, where the epitaxial $CoSi_2$ layer is grown layer by layer, the formation of $CoSi_2$ precipitates with different sizes in the Si substrate is the primary step in the IBS procedure. Bulle–Lieuwma [32] argued that the preferential growth of aligned precipitates causes the formation of continuous epitaxial layers with aligned (type A) orientation. Our results indicate that for thinner buried layers prepared by IBS, interface effects dominate over bulk effects, so that the preferential growth of type A precipitates is suppressed by the influence of the $CoSi_2$/Si interface, which seems to drive the system preferentially towards a type B orientation as shown in the layers grown by SPE (sample 1).

4. CONCLUSION

Mössbauer spectroscopy, complemented with RBS and channeling spectroscopy, has shown that after annealing of Co surface layers on Si, Co_2Si, CoSi and $CoSi_2$ are formed subsequently in parallel layers on the Si substrate. When Co is implanted in Si, buried and surface $CoSi_2$ layers can be formed after annealing at 600°C. The best crystalline quality is reached after 1000°C annealing, as confirmed also by TEM. Our study has further shown that different hyperfine interaction parameters are observed at Fe atoms that were directly implanted into $CoSi_2$, and Fe atoms that were formed due to the radioactive decay of implanted ^{57}Co atoms. Finally, a combination of RBS/channeling and X–ray rocking curve measurements showed that our $CoSi_2$ layers prepared by IBS are strained ($\epsilon_X^{\perp}$ = –1.9%) more than the thicker IBS–formed layers or MBE grown layers, but the values are still much lower than the values of $\epsilon_X^{\perp}$ = –3.5 % for a totally coherent layer. Epitaxial $CoSi_2$(111) layers with pure type A or predominant type A orientation can easily be formed by IBS. However, a component of type B appears if the thickness of the epilayer decreases to less than 36 nm, an reaches a relative amount of 100 % for a layer with a 16 nm thickness.

5. ACKNOWLEDGMENTS

This work was supported by the Belgian I.I.K.W. and G.O.A. (Government Concerted Action) projects. The authors would also like to thank J.M. Phillips for providing the MBE–samples.

REFERENCES

[1] S.S. Lau, J.W. Mayer and K.N. Tu, J. Appl. Phys. 49 (1978) 4005

[2] G.J. van Gurp and C. Langereis, J. Appl. Phys. 46 (1975) 4301.

[3] K.N. Tu, G. Ottoviani, R.D. Thompson and J.W. Mayer, J. Appl. Phys. 53 (1982) 4406.

[4] R.T. Tung, J.M. Poate, J.C. Bean, J.M. Gibson and D.C. Jacobson, Thin Solid Films 93 (1982) 77.

[5] A.E. White, K.T. Short, R.C. Dynes, J.P. Garno and J.M. Gibson, Appl. Phys. Lett. 50 (1987) 95.

[6] A.E. White, K.T. Short, R.C. Dynes, J.M. Gibson and R. Hull, MRS Symp. Proc. 100 (1988) 3.

[7] K. Kohlhof, S. Mantl, B. Stritzker and W. Jager, Nucl. Inst. Methods B39 (1989) 276.

[8] J.C. Barbour, S.T. Picraux and B.L. Doyle, MRS Symp. Proc. 107 (1988) 269.

[9] A.H. van Ommen, W.F.J. Slijkerman, K. Nakagawa, R.J. Smith, J. van der Veen and C.W.T. Bulle–Lieuwma, J. Appl. Phys. 64 (1988) 3005.

[10] S.P. Murarka, 'Silicides for VLSI applications' (Academic Press, New York, 1983).

[11] M.–A. Nicolet and S.S. Lau, in M.G. Einspruch and G.B. Larrabee (eds.), 'VLSI Electronics : Microstructure Science', Vol. 6 (Academic Press, New York, 1983, Chap. 6).

[12] J.C. Hensel, A.F.J. Levi, R.T. Tung and J.M. Gibson, Appl. Phys. Lett. 47 (1985) 151.

[13] J.C. Hensel, MRS Symp. Proc. 54 (1987) 499.

[14] E. Rosencher, F. Arnaud d'Avitaya, P.A. Badoz, C. D'Anterroches, G. Glastre, G. Vincent and J.C. Pfister, MRS Symp. Proc. 91 (1987) 415.

[15] J.F. Ziegler, J.P. Biersack and U. Littmark, in : 'The Stopping and Range of Ions in Matter', ed. J.F. Ziegler (Pergamon, New York, 1985) p. 202.

[16] I. Dézsi, H. Engelman, U. Gonser and G. Langouche, Hyp. Int. 33 (1987) 161.

[17] A. Vantomme, I Dézsi and G. Langouche, Hyp. Int. 41 (1988) 725.

[18] M.F. Wu, A. Vantomme, H. Pattyn, G. Langouche, K. Maex, J. Vanhellemont, J. Vanacken, H. Vloeberghs and Y. Bruynseraede, Nucl. Instr. Meth. B45 (1990) 658.

[19] H. Vanderstraeten, Y. Bruynseraede, M.F. Wu, A. Vantomme, G. Langouche and J.M. Phillips, Appl. Phys. Lett. 57 (1990) 135.

[20] H. Vanderstraeten, Y. Bruynseraede, M.F. Wu, A. Vantomme and G. Langouche, Accepted for publication in J. Phys. D.

[21] M.F. Wu, A. Vantomme, G. Langouche, K. Maex and J. Vanhellemont, to be publ. in the Proc. of the 1st Intern. Conf. on Epit. Cryst. Growth, 1–7 April 1990, Budapest.

[22] M.F. Wu, A. Vantomme, G. Langouche, J. Vanderstraeten and Y. Bruynseraede, Nucl. Instr. Meth. B54 (1991) 444.

[23] M.F. Wu, A. Vantomme, G. Langouche, K. Maex, H. Vanderstraeten and Y. Bruynseraede, Appl. Phys. Lett. 57 (1990) 1973.

[24] J.W. Matthews and A.E. Blakeslee, J. Cryst. Growth 27 (1974) 118.

[25] R. People and J.C. Bean, Appl. Phys. Lett. 47 (1985) 322; 49 (1986) 229.

[26] P.J. Orders and B.F. Usher, Appl. Phys. Lett. 50 (1987) 980.

[27] K. Kohlhof, S. Mantl and B. Stritzker, Appl. Surface Sci. 38 (1989) 207.

[28] A.H. van Ommen, J.J.M. Ottenheim, C.W.T. Bulle–Lieuwma, Appl. Surface Sci., 38 (1989) 197.

[29] R.T. Tung, J.M. Gibson, J.C. Bean, J.M. Poate and D.C. Jacobson, Appl. Phys. Lett. 40 (1982) 684.

[30] M. Okamoto, S. Hashimoto, B.D. Hunt and L.J. Schowalter, Mater. Res. Soc. Symp. Proc. 56 (1986) 157.

[31] R.T. Tung and J.M. Gibson, Mater. Res. Soc. Symp. Proc. 67 (1986) 211.

[32] C.W.T. Bulle–Lieuwma, A.H. van Ommen and L.J. van Ijzendoorn, Appl. Phys. Lett. 54 (1989) 244.

Figure captions

Fig. 1 : The normalized area of the various components in the formation of Co–silicide surface layers on Si as a function of annealing temperature, as deduced from the Mössbauer spectra.

Fig. 2 : ^{57}Fe Mössbauer spectrum of a source of ^{57}Co implanted in $CoSi_2$ versus a single–line ^{57}Fe–in–ferrocyanide absorber. For easy comparison with fig. 7, we have changed the sign of the velocity scale as if it were an absorber experiment.

Fig. 3 : RBS spectra of a buried $CoSi_2$ layer, after annealing at 1000°C. • random; ▵ aligned; — RBS simulation.

Fig. 4 : Temperature dependence of the resistivity of a buried $CoSi_2$ layer formed by implantation. Insert : T_c of the buried layer.

Fig. 5 : RBS spectra and cross–sectional TEM pictures. (a,b) The buried $CoSi_2$ layer after annealing at 1000°C. (c,d) The buried $CoSi_2$ layer after further annealing at 1100°C for 30 min. (e,f) The broken–up $CoSi_2$ layer after annealing at 1200°C for 5 min.

Fig. 6: Temperature dependence of resistivity of a surface layer formed by implantation before (curve 1) and after (curve 2) annealing at 1100°C for 30 min.

Fig. 7 : ^{57}Fe Mössbauer spectrum of an absorber of ^{57}Fe implanted at 300°C into MBE–grown $CoSi_2$, using a single–line ^{57}Co–in–Cu source.

Fig. 8 : Angular scan around the <114> (left) and <110> (right) axes in the $(1\bar{1}0)$ plane of sample 5.

Fig. 9 : Angular scan around the <114> (left) and <110> (right) axes in the $(1\bar{1}0)$ plane of sample 4.

Fig. 10: Strain in the $CoSi_2$ layers (samples 1 – 7 in Table 1) as a function of layer thickness, measured with RBS/Channeling and X–ray diffraction.

Fig. 11: X–ray rocking curves around the $<331>^{A}_{CoSi_2}$ axis (a1–a6) and the $<331>^{B}_{CoSi_2}$ axis (b1–b6) from the respective samples of Table I. The corresponding a and b curves have the same intensity scale. The splitting of P_{Si} and the asymmetry in $P^{A,B}_{CoSi_2}$ are due to the presence of $K_{\alpha 1}$ and $K_{\alpha 2}$ in the x–ray beam.

Fig. 12: Relative amount of type A $CoSi_2$ prepared by IBS and SPE as a function of the $CoSi_2$ epilayer thickness. The labels refer to the sample identifications in Table I. The line is a guide to the eye.

Nr.	Implantation	Annealing	Layer type	χ_{min} (%)	Thickn. (Å)	$\Delta\psi$ (°)	ϵ_T (RBS) (%)	$\epsilon_x^{\perp}$ (%)	ϵ_T (XRD) (%)	Type A (%)
1	SPE (650°C)	950°C, 30' + 1000°C, 30'	S	8	540	0.24	0.88	−1.66	0.683	1.6
2	30 keV, 0.47×10^{17} 300°C	600°C,60' + 1000°C, 30'	B	3	200	0.31	1.1	−1.90	1.050	30
3	40 keV, 0.58×10^{17} 300°C	600°C,60' + 1000°C, 30'	B	3	240	0.30	1.1	−1.87	1.002	87
4	70 keV, 0.86×10^{17} 310°C	600°C,60' + 1000°C, 30'	B	3	360	0.30	1.1	−1.82	0.927	99.6
5	50 keV, 1.4×10^{17} 300°C	600°C,60' + 1000°C, 30'	S	3	470	0.30	1.1	−1.87	1.002	99.8
6	30 keV, 0.94×10^{17} 300°C	600°C,60' + 1000°C, 30'	S	6	360	0.31	1.1	−1.90	1.050	100
7a	200 keV, 1.6×10^{17} 400°C	600°C,60' + 1000°C, 30'	B	3	600	0.30	1.1	−1.90		100
7b	30 keV, 0.42×10^{17} 320°C	850°C, 30'	B	21	160	0.30	1.1	−1.90		0

Table 1 : Identification, preparation parameters, strain and orientation of epitaxial $CoSi_2$ layers. The layer type indicates whether the layer is buried (B) or on the surface (S). The error on $\Delta\psi$ is ± 0.03°, on ϵ_x ± 0.03 % and on the orientation ratio $A/(A + B)$ ± 0.2 %.

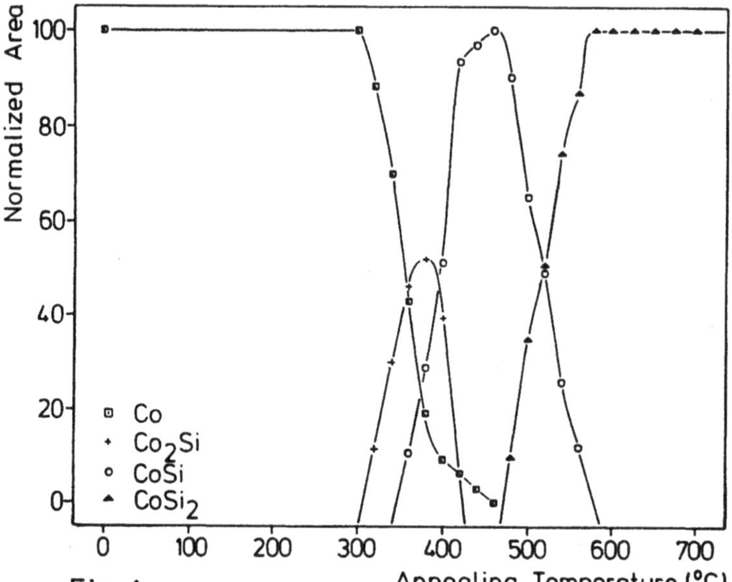
Normalized Area
100
80
60
40
20
0
Co
Co2Si
CoSi
CoSi2
0
100
200
300
400
500
600
700
Annealing Temperature (°C)

Fig. 1

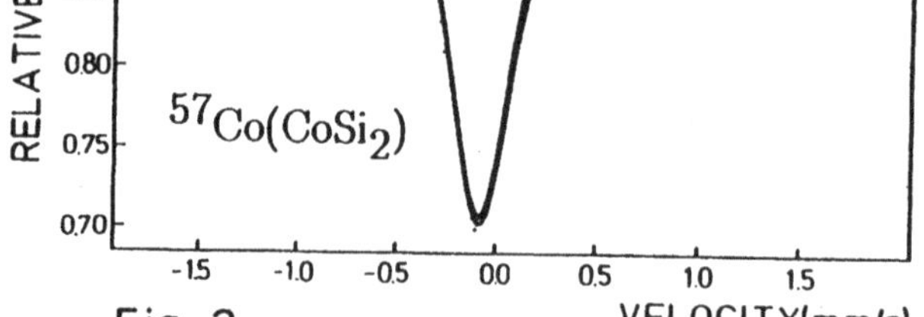
RELATIVE TRANSMISSION
1.00
0.95
0.90
0.85
0.80
0.75
0.70
57Co(CoSi2)
-1.5
-1.0
-0.5
0.0
0.5
1.0
1.5
VELOCITY(mm/s)

Fig. 2

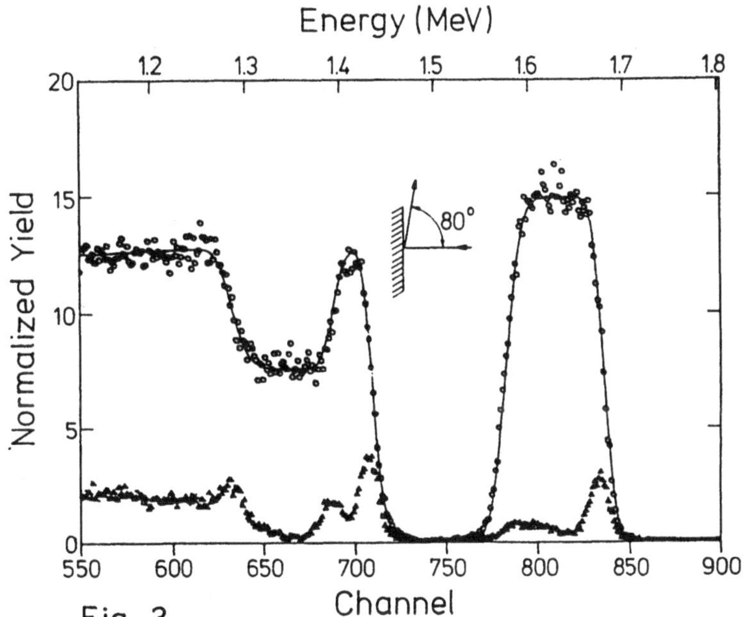
Energy (MeV)
1.2
1.3
1.4
1.5
1.6
1.7
1.8
20
15
10
5
0
Normalized Yield
80°
550
600
650
700
750
800
850
900
Channel

Fig. 3

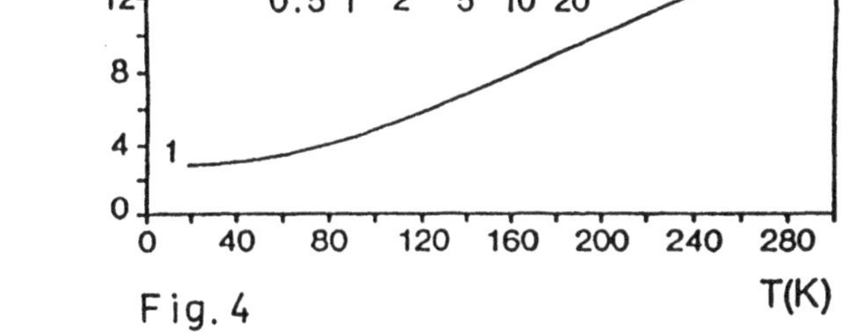
ρ (μΩ cm)
24
20
16
12
8
4
0
R(Ω)
0.5
0.4
0.3
0.2
0.1
0
0.5 1 2 5 10 20
T(K)
1
0
40
80
120
160
200
240
280
T(K)

Fig. 4

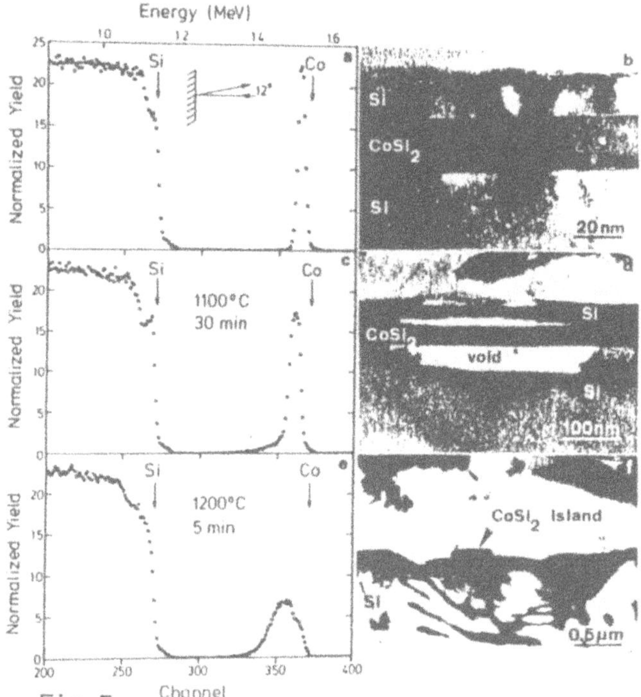
Energy (MeV)
Normalized Yield
Si
Co
12°
1100°C
30 min
1200°C
5 min
Channel
Si
CoSi2
Si
20 nm
Si
CoSi2
void
Si
100nm
CoSi2 island
Si
0.5μm

Fig. 5

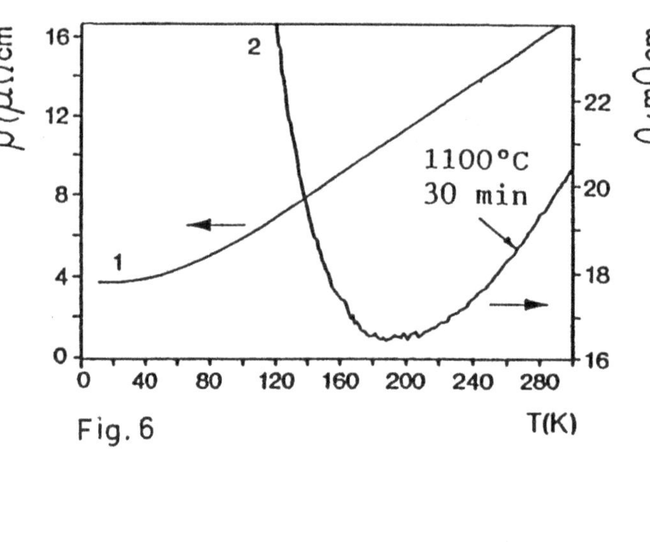
ρ (μΩcm)
ρ (mΩcm)
1
2
1100°C
30 min
T(K)

Fig. 6

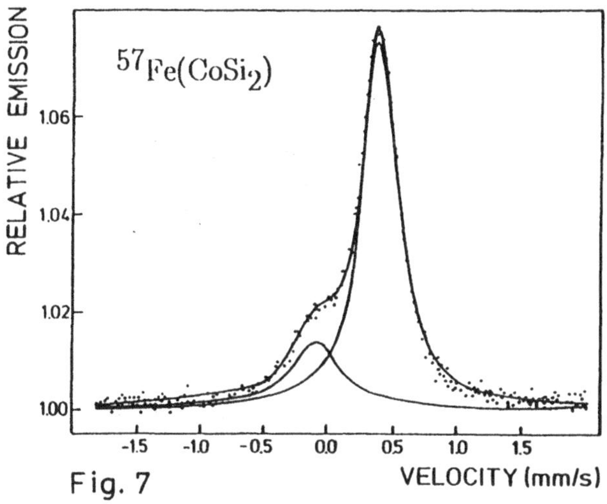
^{57}Fe(CoSi$_2$)
RELATIVE EMISSION
1.00
1.02
1.04
1.06
-1.5
-1.0
-0.5
0.0
0.5
1.0
1.5
VELOCITY (mm/s)

Fig. 7

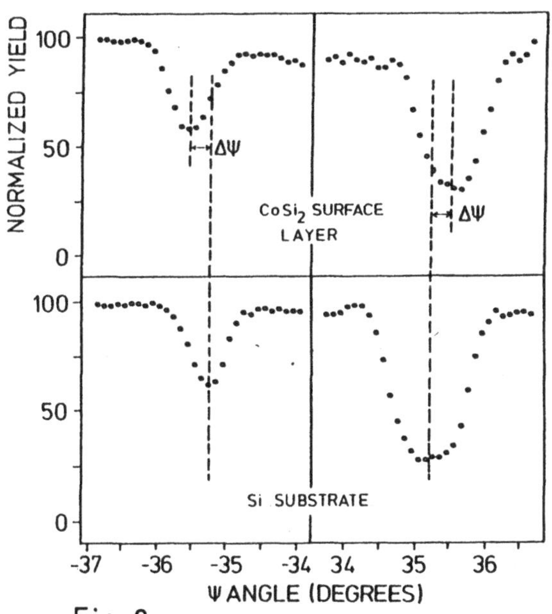
NORMALIZED YIELD
100
50
0
ΔΨ
CoSi$_2$ SURFACE LAYER
Si SUBSTRATE
-37
-36
-35
-34
34
35
36
Ψ ANGLE (DEGREES)

Fig. 8

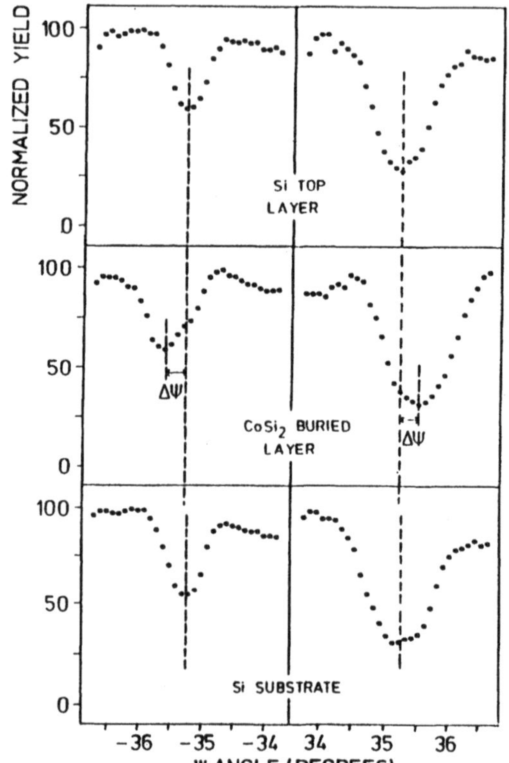

NORMALIZED YIELD
100
50
0
Si TOP LAYER
ΔΨ
CoSi2 BURIED LAYER
Si SUBSTRATE
-36 -35 -34 34 35 36
Ψ ANGLE (DEGREES)

Fig. 9

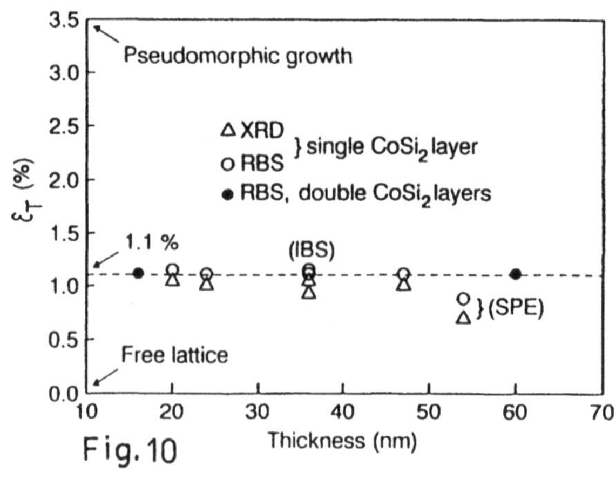

Pseudomorphic growth
Δ XRD
o RBS
} single CoSi2 layer
● RBS, double CoSi2 layers
1.1 %
(IBS)
} (SPE)
Free lattice
εT (%)
Thickness (nm)

Fig. 10

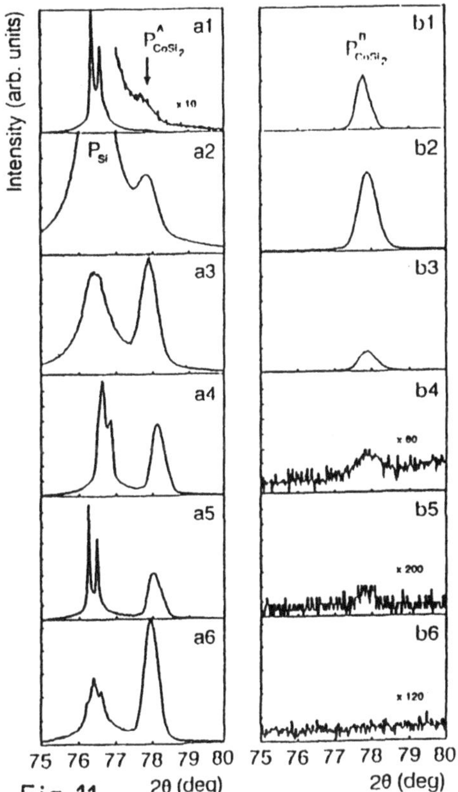
Intensity (arb. units)
a1
a2
a3
a4
a5
a6
b1
b2
b3
b4
b5
b6
P_{Si}
x 10
x 80
x 200
x 120
75 76 77 78 79 80
2θ (deg)

Fig. 11

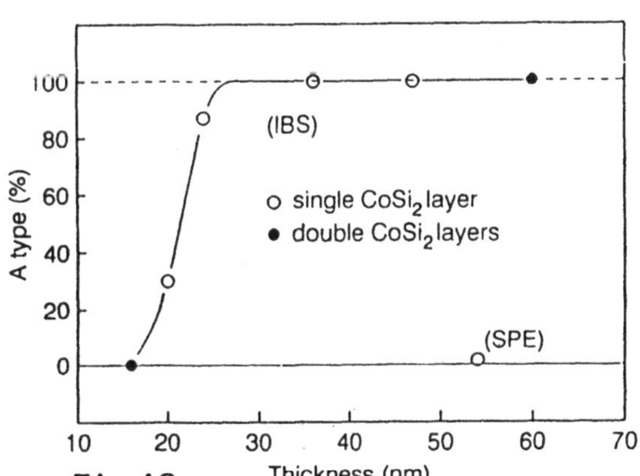
A type (%)
100
80
60
40
20
0
(IBS)
o single $CoSi_2$ layer
• double $CoSi_2$ layers
(SPE)
10 20 30 40 50 60 70
Thickness (nm)

Fig. 12

APPLICATION OF ^{119}Sn MÖßBAUER SPECTROSCOPY TO INVESTIGATIONS OF RET_2X_2-TYPE PHASES

Edward A.Görlich
Institute of Physics, Jagiellonian University, 30-059 Kraków, Poland

Abstract

Mößbauer spectroscopy of ^{119}Sn is an effective tool to study nuclear hyperfine interactions which in turn are important as a source of information complementary to that supplied by 'classical' methods. However, particularly when the effects are subtle, experimental conditions as well as data analysis should be handled with care. The attention is devoted primarily to the finite absorber thickness effects. As examples serve our recent results obtained from investigations of ternary tetragonal phases of RET_2X_2-type. In $EuPd_2Si_2$ mixed valent system ^{119}Sn was used as probe which detects the influence of Eu-valency change at a distant site of Si. Electrical resistivity measurements in $CeNi_2Sn_2$ may indicate a Kondo-type behaviour while Mößbauer effect leads to the conclusion in favour of a magnetically ordered state. The analysis of resonance absorption spectra of $CeAg_2Sn_2$ using a full transmission integral indicates the presence of a metallic β-tin in the sample, while data fitting within the thin absorber approximation has lead to the opposite conclusion. Neither in $CeNi_2Sn_2$ nor in $ErNi_2Sn_2$ the temperature dependence of the recoilfree factors alone cannot explain the observed broadening of spectra at low temperatures.

1 INTRODUCTION

Ternary tetragonal phases of a general chemical formula RET_2X_2 (where T stands for a d-transition element and X is most often silicon or germanium) attracted a great deal of interest and research efforts. Stability of crystallographic structure along with often occurring variety of magnetic structures [1] and less common electronic properties - made this family of compounds to a model system. We will focus our attention on $EuPd_2Sn_2$ into which a small amounts of ^{119}Sn isotope

[1]For more information see e.g. [1] and references therein

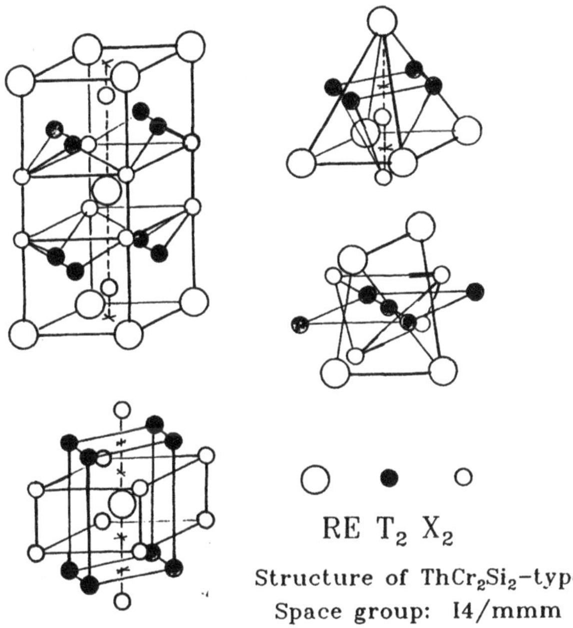

Figure 1: *$ThCr_2Si_2$-type of structure and the resulting atomic neighbourhoods*

were introduced to serve as a neutral witness of effects which are caused by the instability of the population of 4f-europium shell at the tin nucleus occupying the silicon site. Further, we have directed our interest toward systems with tin as an X-element. Information available in the literature until now is scarce. The presence of tin loosely bound $5p^2$-electrons leads to properties which are different from those of silicon or germanium compounds. In this article the results of the structural and electric resistivity investigations create a background for presenting the measurements of the resonance absorption of 21.8 keV gamma radiation of ^{119}Sn.

The goal and merit of this meeting allow to stress certain methodological aspects of the ^{119}Sn Mößbauer spectroscopy in some details.

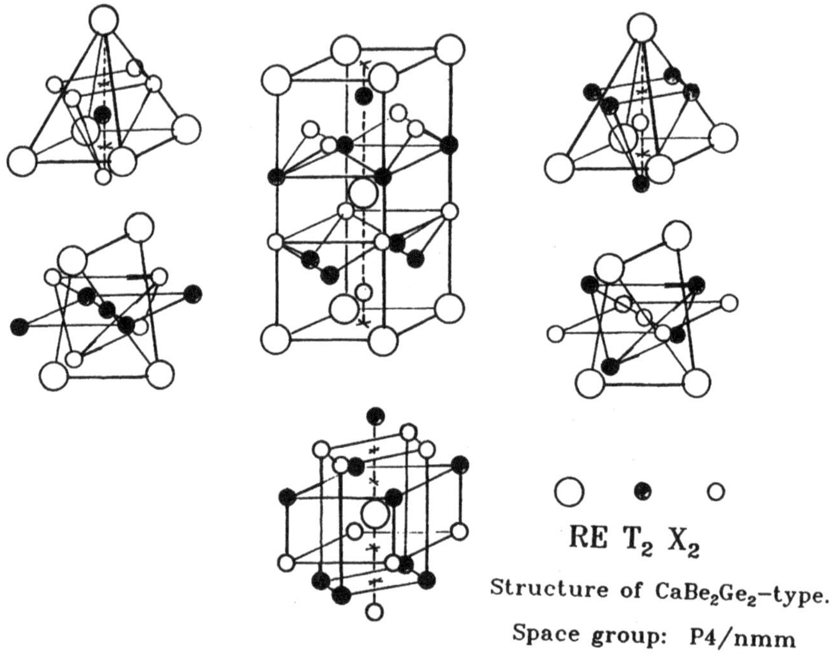

Figure 2: *$CaBe_2Ge_2$-type of structure and the resulting atomic neighbourhoods*

2 STRUCTURE AND SPECTRA COMPONENTS

2.1 Structure of RET_2X_2 Phases

The RET_2X_2-type phases were obtained as derivatives of a $BaAl_4$. In an original structure of $BaAl_4$ there exist two crystalographically nonequivalent Al sublattices. The most often observed atomic arrangement in RET_2X_2 compounds is of body-centered tetragonal $ThCr_2Si_2$-type (space group I4/mmm) which may be considered as resulting from a systematic replacement of an aluminum in each of those two lattice positions by T (3d- , 4d- or 5d-transition element) and X (p-element), respectively. A large majority of most widely studied compounds[2] with X = Ge or Si crystallize in this structure. Fig.1 shows the unit cell along with the near neighbourhood for each site. It is worth noticing that in such a lattice each kind of atoms has a unique type of coordination. That of transition element T may be considered as composed of two intersecting pyramids while the local surrounding of silicon or germanium consists of two tetrahedra and a planar square.

[2]See for example [1]

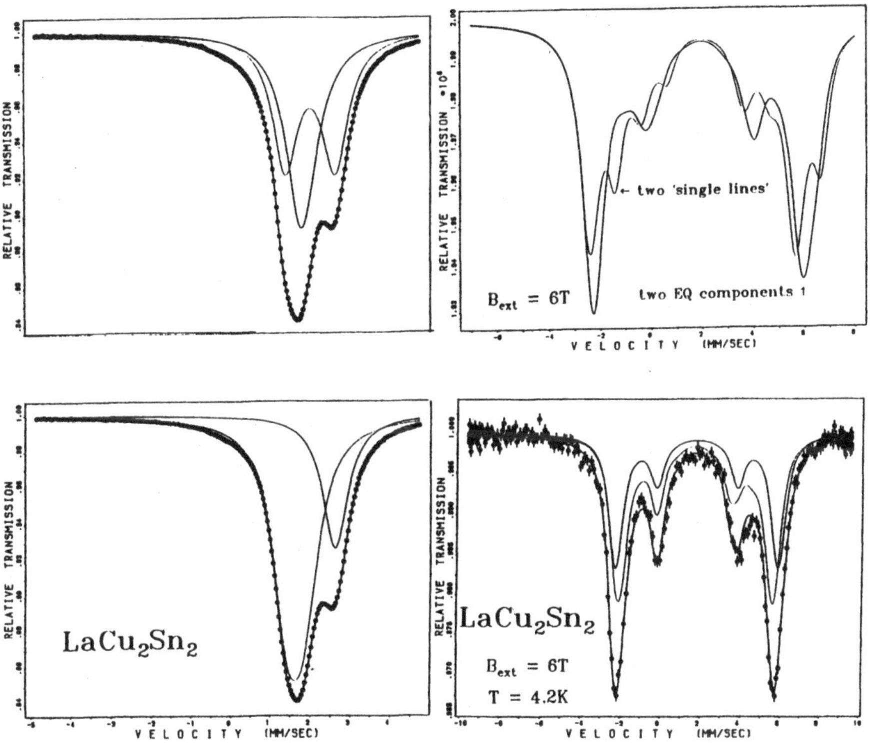

Figure 3: *Performing an experiment in the external magnetic field enables the decision between the two approaches possible*

However, another atomic arrangement preserving tetragonality of a unit cell is also observed for compounds of this stoichiometry. Interchanging places within every second pair of X- and T-atomic planes leads to a primitive tetragonal structure of a space group P4/nmm commonly referred to as $CaBe_2Ge_2$-type (Fig.2). In this structure both X- and T-atoms have two types of coordination each. The structure of ordered RET_2Sn_2 compounds ($RE = Ce, Sm, Er$ and Yb ; $T = Ni, Cu$)were determined in our room temperature X-ray measurements for a majority of cases as $CaBe_2Ge_2$-type [3], [9]. However, we have to admit that they show an affinity toward a crystallographical disorder connected most probably with exchanging places between Sn and T atoms. $ErNi_2Sn_2$ -phase, discussed in section **3.5**, shows this kind of jumble.

2.2 Decomposition of a Spectrum: $LaCu_2Sn_2$

From the structural considerations one expects for $LaCu_2Sn_2$ compound of the primitive tetragonal structure (Fig.2) the two nonequivalent Sn sites and therefore two components in the resonance gamma spectrum. But solely a χ^2-test gives no convincing decision between the two approaches, as shown on the left-hand side of Fig.3, of how to resolve the data into component subspectra:

(*a*) a single line and a quadrupole dublet with the close isomer shifts,
(*b*) two single lines of the markedly different positions.

The ambiguity of the Mößbauer spectrum decomposition into two components poses a serious challenge to the reliability of the physical information to be drawn out of the experiment. The problem was solved by an additional measurement carried out in the external magnetic field. The spectra to be expected when applying the magnetic field to the polycrystalline sample for the two possible situations are depicted in Fig.3 . The experimental data have been nicely fitted assuming the approach (*a*). This type of data analysis was further used in the case of crystallographically ordered RET_2Sn_2-compounds. The assignment of the spectral components to the lattice sites is a separate problem. We tentatively ascribe the component with the large quadrupole splitting to the pyramidally coordinated tin and a single-like line to the bitetrahedral site.

3 ABSORBER THICKNESS EFFECTS

3.1 Absorption Spectrum Shape

When a certain object is being observed by means of a method with inherent finite resolution the net result is expressed by the convolution integral. In the case of Mößbauer experiment in transmission geometry an incident γ-ray beam:

$$S(\varepsilon - v) = \frac{1}{2\pi}\frac{\Gamma_S}{(\varepsilon - E_S - v)^2 + \left(\frac{\Gamma_S}{2}\right)^2}$$

is being filtered by an energy dependent cross section of an absorber:

$$T(\varepsilon) = \left(1 - e^{\mu(\varepsilon)}\right) \qquad\qquad \mu(\varepsilon) = \frac{t_A\left(\frac{\Gamma_A}{2}\right)^2}{(\varepsilon - E_A)^2 + \left(\frac{\Gamma_A}{2}\right)^2}$$

to give the signal at the detector:

$$I(v) = I_o \int_{-\infty}^{-\infty} S(\varepsilon - v) T(\varepsilon) d\varepsilon \ .$$

The radiation passes through an absorbing medium characterized by the thickness parameter $t_A = f_A \sigma_o n$:
f_A - gives the probability of a recoilfree absorption,
σ_o - is the resonance cross section, and
n - is a number of Mößbauer nuclei per unit area in the absorber.
The increasing of t_A makes the observed effect larger but simultaneously the absorption line is broadened and looses its Lorentzian shape.
Finite absorber thickness effects often play an important role in ^{119}Sn Mößbauer

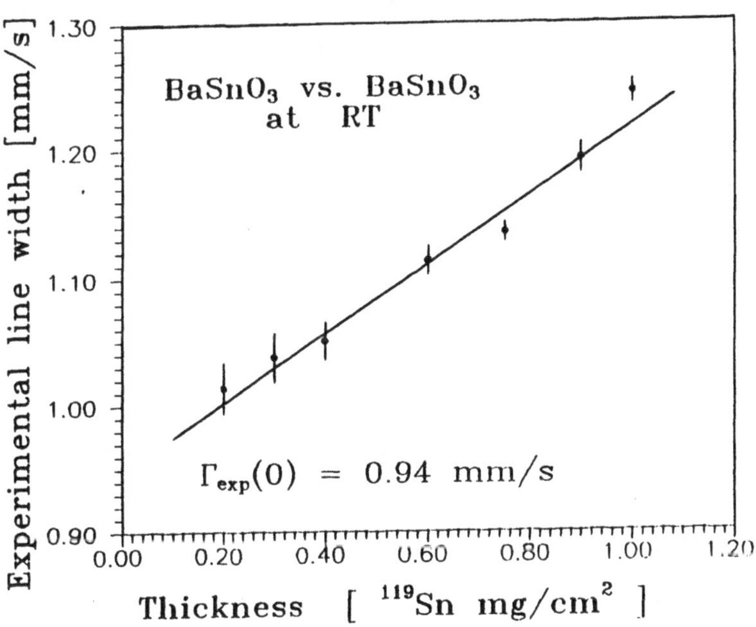

Figure 4: *Experimental line width plotted as a function of absorber thickness*

experiments due to the usually large recoilfree fraction. To visualize the extend of the influence of an absorber thickness on a line broadening Fig.4 shows experimentally determined line width at half maximum as a function of the ^{119}Sn concentration in the absorber. It is worthwhile to point out that commonly used absorber thickness optimizing procedures tend to optimize signal to error ratio rather than resolution power of an experiment. In the case of the mixed valent compound $EuPd_2Si_{1.992}Sn_{0.008}$, discussed below in paragraph **3.2**, we have used the absorber of 0.1 mg ^{119}Sn per cm^2 which is the value distinctly below the optimal one in an above quoted sense. This allowed to preserve a two dip structure of a tin spectrum (see Fig.5). Already slightly larger thickness would result in a spectrum of an apparently single line appearance and thus considerably decreasing the reliability of its decomposition. Further problem arises when a question is asked about the nature of unresolved spectrum broadening at low temperatures - whether it is of hyperfine origin or it is due to the absorber thickness increase connected to the temperature dependence of the recoilless fraction $f_A(T)$. In the framework of the Debye model the Lamb-Mößbauer factor reads:

$$f_A(T) = e^{-k^2\langle x^2\rangle} \longrightarrow exp\left(-\frac{6E_R}{k_B\Theta_D}\left[\frac{1}{4}+\left(\frac{T}{\Theta_D}\right)^2\int_0^{\frac{\Theta_D}{T}}\frac{x}{e^x-1}dx\right]\right)$$

where the recoil energy E_R of a nucleus due to the interaction with the γ-radiation of a wave vector k is equal to $\frac{\hbar^2k^2}{2M}$.
The issue of distinction between the possible sources of line broadening appeared to be of importance in the case of $CeNi_2Sn_2$ (cf. section **3.3**) and of $ErNi_2Sn_2$ (cf. section **3.5**). The correct accounting for the thickness effect contribution allows to evaluate hyperfine fields there.

3.2 Mixed Valency: $EuPd_2Si_2$

$EuPd_2Si_2$ crystallizes in a structure of the $ThCr_2Si_2$-type. Replacing 0.4% of the silicon atoms by ^{119}Sn-isotope the phase $EuPd_2Si_{1.992}Sn_{0.008}$ has been obtained. Mößbauer spectrum reveals in fact almost exclusively single component (Fig.5-bottom) as expected for a unique coordination of a silicon sites in this structure. The origin of a small line ($<$ 4% of ^{119}Sn spectrum area) is not clear but it founds its counterpart in ^{151}Eu spectrum (Fig.5-top). In fact none of the europium spectra for this compound published up to now are totally free from such a 'satellite' line. In contrary to the results obtained for the compound doped with larger amounts of tin [4] we do not see the effect of it on physical properties. Therefore we consider ^{119}Sn merely as an observer of phenomena relevant to $EuPd_2Si_2$. The temperature driven valence change of europium there [5], [6], [7], [8] causes a continuous shift of the ^{151}Eu Mößbauer line toward the position characteristic for 3+ europium state at low temperatures as shown in the inset of Fig.6. The purpose of our experiment [2] was to detect an influence of a valence transition between magnetic $4f^7$ state of Eu^{2+} ($J = \frac{7}{2}$) at higher temperatures and a $4f^6$ state (at lower temperatures) as it is observed at a distant site of silicon occupied by a ^{119}Sn probe. In fact that continuous change of europium 4f-state occupancy with temperature is reflected in the behaviour of the electric field gradient at the tin nucleus as well as the

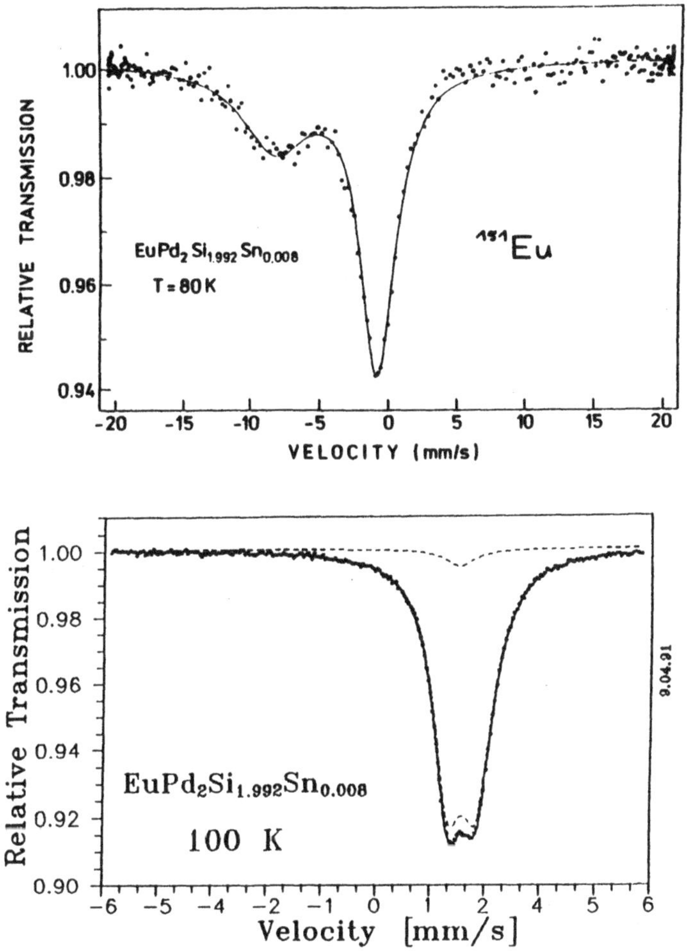

Figure 5: *Mößbauer spectra of* ^{151}Eu *(top) and* ^{119}Sn *(bottom) in* $EuPd_2Si_{1.992}Sn_{0.008}$

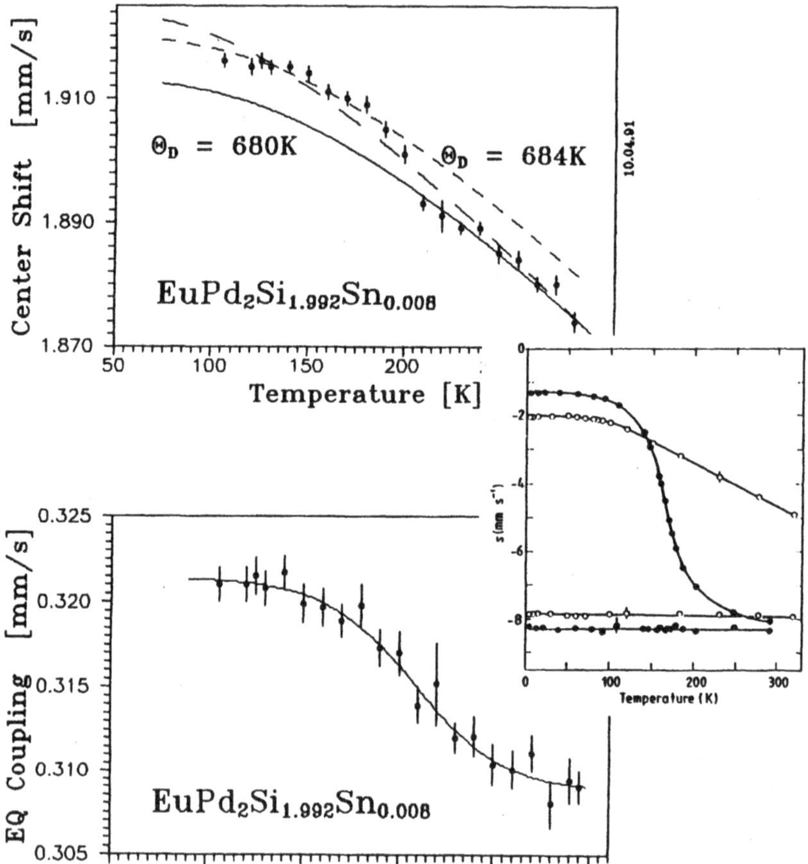

Figure 6: *Mößbauer spectrum center shift (relative to $BaSnO_3$ source at room temperature) and the electric quadrupole coupling of ^{119}Sn nuclei as a function of temperature. The inset comes from [8] and shows line positions vs. temperature for ^{151}Eu resonance in $EuPd_2Si_2$ (solid circles)*

tin isomer shift (Fig.6). The response of the above quantities to the temperature lowering is indicative of a depopulation of the localized 4f-state in favour of a delocalized conduction electron band state rather than increased occupation of the europium d-state. The results for a temperature dependence of an electric field gradient at ^{119}Sn may be nicely reproduced by a simple activation model [10] yielding a temperature dependent value of the excitation energy between $|4f^7\rangle$ and $|4f^6, c.e.\rangle$ states (Fig.6 - solid line). The observed change in the isomer shift (I.S.), that means electronic density at the nucleus, supplies the direct evidence of the substantial change in the spatial distribution of europium valence electrons: electronic density at the site of tin nucleus increases upon temperature lowering. It may, however, be noticed that in both temperature regions i.e. above and below valence transition the temperature variation of a line position C.S.(T) due to the second order Doppler shift is reproduced by the Debye model with practically equal values of Debye parameter Θ_D (Fig.6):

$$C.S.(T) = I.S. + \frac{\langle v^2 \rangle}{2c} \quad \xrightarrow{\textit{Debye model}} \quad I.S. + \frac{9k_B\Theta_D}{16Mc} + \frac{9k_B T^4}{2Mc\Theta_D^3}\int_0^{\frac{\Theta_D}{T}} \frac{x^3}{e^x - 1}dx$$

where $I.S. \propto (|\psi_A(0)|^2 - |\psi_S(0)|^2)\frac{\Delta\langle r^2\rangle}{\langle r^2\rangle}$.

It means that the dynamical properties of a tin atom in this lattice do not strongly change with the valence state of europium. It finds a support in an area vs. temperature dependence (Fig.7) - the data are well represented by a single $f_A(T)$ relation as expected on the basis of the Debye model (cf. paragraph **3.1**). The conclusion is also consistent with the observation that the linear thermal expansion [7] both below and above the transition region is approximately the same.

3.3 Is $CeNi_2Sn_2$ a Kondo System ?

Temperature dependence of the electrical resistivity of $CeNi_2Sn_2$ shows a metallic behaviour but with a clear deviation (over a broad temperature range) from the Grüneisen-Bloch linear (at higher temperatures) relation, which is not unusual for non-simple metals (Fig.8). Below $35K$ the resistivity strongly increases reaching a pronounced maximum at about $9K$. This indicates an enhanced interaction between a single 4f-electron of Ce^{3+}-ion with the conduction-band electrons. Contrary to the results of [12] and in agreement with [11] the X-ray diffraction analysis of the sample [9] indicated the primitive tetragonal structure of $CaBe_2Ge_2$-type. Magnetic susceptibility measurements reported in [11] show the Curie-Weiss behaviour and give the value of an effective magnetic moment $\mu_{eff} = 2.62\mu_B$ which

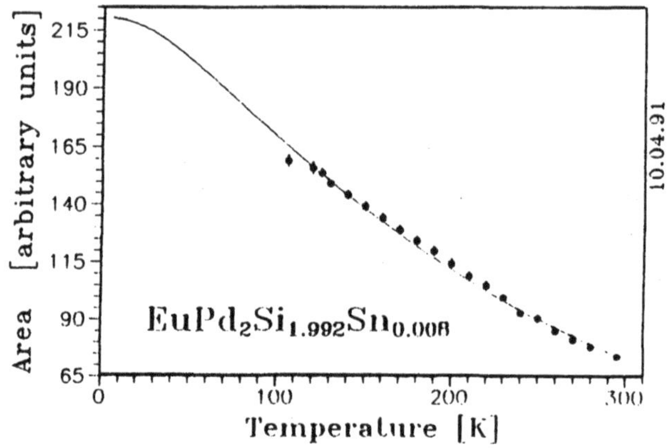

Figure 7: *Area of the ^{119}Sn Mößbauer absorption spectrum as a function of temperature. Solid line results from least square fit of the Debye model relation (see text)*

is indicative of a trivalent cerium state. The paramagnetic Curie temperature was found to be $\Theta_p = -5K$. Their heat capacity data yield a large value of the γ-coefficient of the linear term of $250\frac{mJ}{mole\ K^2}$. It places this compound close to the class of heavy fermion systems, as the large γ-term is equivalent to the strongly enhanced effective mass of the conduction electrons. Both the susceptibility (covering the region of 2-300K) and the heat capacity (from 2.5 to 20K) data of [11] show no indication of a magnetic ordering but only certain very initial stages of what might be collective electronic interaction at about $3K$.

In view of the above facts we may infer that $CeNi_2Sn_2$ creates a Kondo-lattice. The concept of the Kondo-lattice is applied to a strongly interacting electron or heavy fermion system. The negative temperature coefficient of $\varrho(T)$ in the paramagnetic region suggests the presence of the Kondo-scattering of the conduction electrons in this compound. Between 15K - 50K the resistivity is represented by the relation :

$$\varrho(T) = \varrho_o + \varrho_o^\infty + \varrho_{ph}(T) + c_K lnT$$

where

$\varrho_o + \varrho_o^\infty$ are the residual and the spin disorder, temperature independent terms, respectively,

$\varrho_{ph}(T)$ is the monotonic Grüneisen-Bloch phonon term,

$c_K lnT$ is a Kondo logarithmic term with c_K coefficient proportional to the third

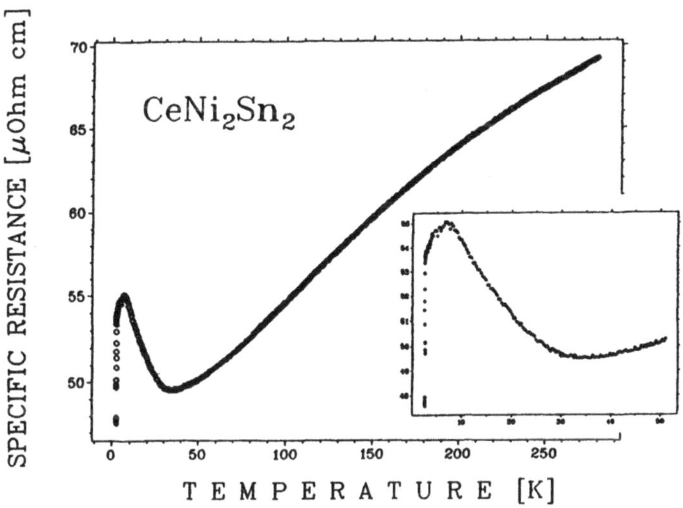

Figure 8: *Resistivity of $CeNi_2Sn_2$ as a function of temperature*

power of an exchange integral $J_{f\vec{k}}$ between conduction electrons and the $4f$-cerium electron. The interactions may lead to a formation of a nonmagnetic ground electronic state.

The purpose of applying the ^{119}Sn Mößbauer spectroscopy was to confirm the absence of a magnetic ordering (Fig.9). Surprisingly enough one observes a slight broadening of the resonance spectrum at temperatures below 10K. However, it may be the result of the recoilless factor increase which means the increase in the effective thickness in the absorber and thus line broadening. It appears that this effect is not enough strong to account for the observed spectrum width increment. In fact the f-factor saturates at low temperatures (cf. Fig.7) but the spectrum still gains in the width between $4.2K$ and $1.6K$. Therefore we may interpret the observed broadening as due to the small transferred hyperfine field at the tin nucleus of $3 \pm 0.5kOe$. That means that a certain type of a long-range order is set on below $10K$ in this compound.

3.4 Is $CeAg_2Sn_2$ Superconducting ?

Electrical resistivity measurements (Fig.10) supply the strong indication that a superconducting transition is present at $3.55(5)K$ (the middle point) in this sample. The preliminary magnetic susceptibility results [9] taken in the field of $450Oe$ in the temperature range between $5K$ and $250K$ obey the Curie-Weiss law yielding the effective magnetic moment of $1.82\mu_B$. That is distinctly below the value of $2.54\mu_B$ which is expected for Ce^{3+} $(4f^1)$ ion. The paramagnetic Curie temperature was determined equal to $-0.6(5)K$.

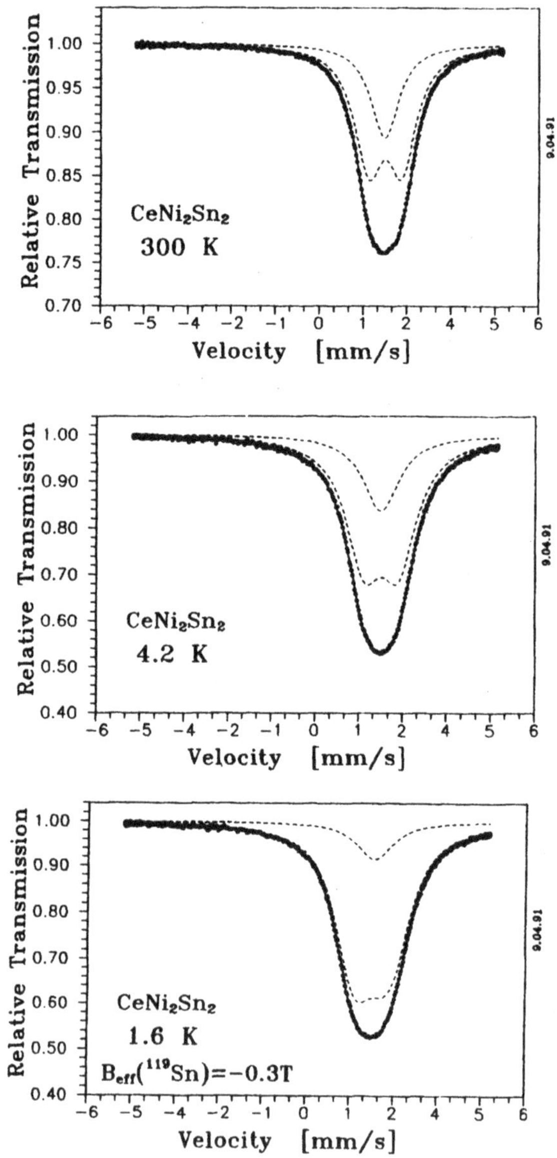

Figure 9: ^{119}Sn *Mößbauer absorption spectra* $CeNi_2Sn_2$

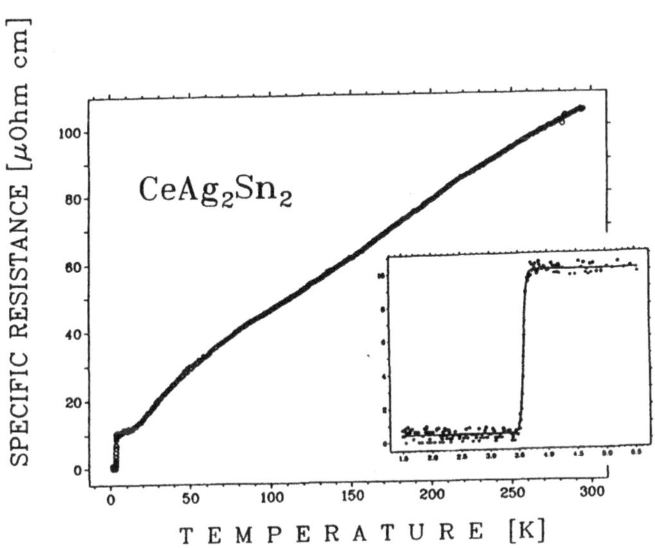

Figure 10: *Resistivity of $CeAg_2Sn_2$ as a function of temperature*

^{119}Sn Mössbauer effect appeared instrumental for establishing the quality of the sample. The metallic tetragonal β-tin (white) has the superconducting critical temperature of $3.72K$. Although no traces (i.e. less than $\sim 4\%$) of it was found in the X-ray diffraction analysis but the ME experiment may reveal smaller quantities of tin-rich phases. ^{119}Sn gamma resonance spectrum (Fig.11) clearly show more components than expected for a single phase of $CaBe_2Ge_2$ structural type. The room temperature isomer shift for $\beta - Sn$ relative to $BaSnO_3$-source is 2.56(1) [14]. The quadrupole dublet of the lower intensity and the room temperature isomer shift of $2.462 \pm 0.015\frac{mm}{s}$ ($\frac{1}{4}e^2qQ = 0.523 \pm 0.017\frac{mm}{s}$)cannot be ascribed to a bulk β-tin phase. To assure that the finite absorber thickness effect does not distort the results the spectra were fitted with the full expression for the convolution integral as given in the paragraph **3.1** (Fig.11). But then the center position of the quadrupole dublet discussed appeared to be $2.579 \pm 0.015\frac{mm}{s}$ and therefore allowing for the white tin as a possible contamination. Nevertheless, it is unlikely that there is enough metallic tin to form the percolations resulting in a network responsible for superconducting transition.

The sample satisfactory from the X-ray investigations point of view appeared to contain other phases when inspected with a method based on local probes. Summarizing the remarks above, one sees that neglecting thickness effects, although results in relatively small parameters' misestimation, may lead to the false or inadequate conclusions.

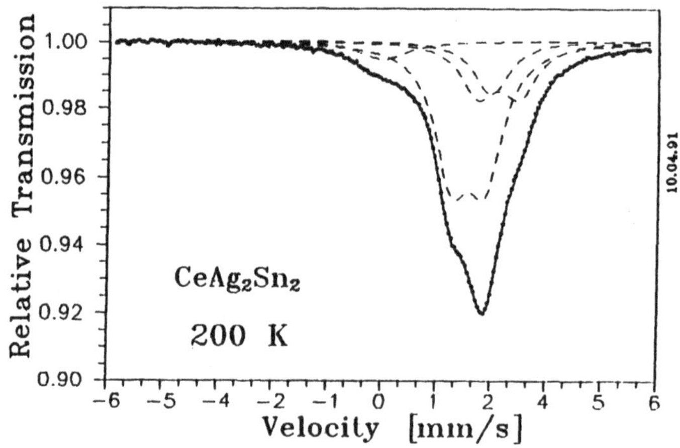

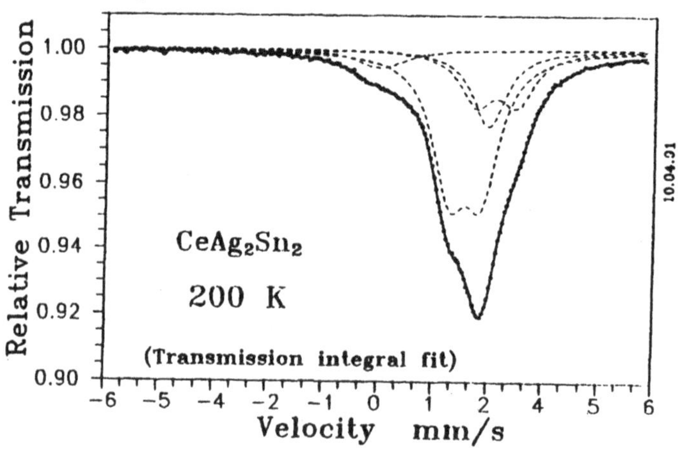

Figure 11: *^{119}Sn Mößbauer absorption spectra of $CeAg_2Sn_2$*

3.5 Weak Magnetism: $ErNi_2Sn_2$

Resistivity vs. temperature relation for $ErNi_2Sn_2$ most strongly, among the cases discussed, deviates over a wide temperature range from the simple metal behaviour (Fig.12). The small maximum at about 5K indicates an onset of a long-range

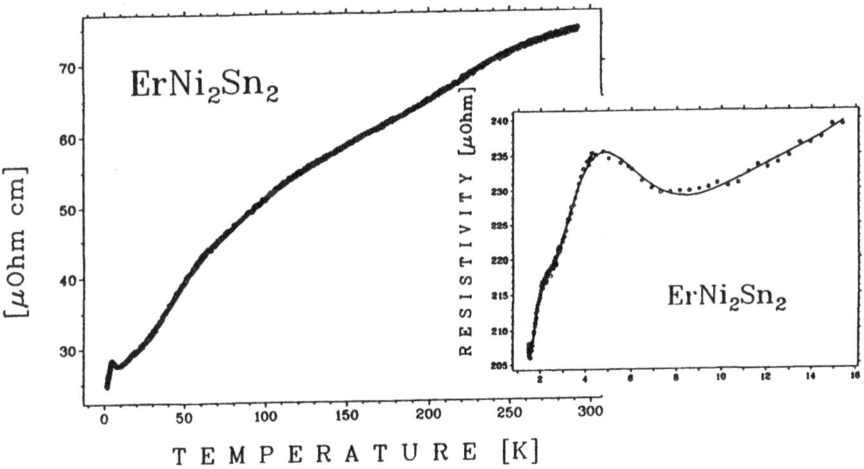

Figure 12: ***Resistivity of $ErNi_2Sn_2$ as a function of temperature***

magnetic ordering probably of an antiferromagnetic nature. The magnetic susceptibility data [9] show good agreement with the Curie-Weiss theory with an effective magnetic moment of $9.28(2)\mu_B$ and the paramagnetic Curie temperature $\Theta_p = -2.5K$. According to our X-ray diffraction results the phase has a tetragonal symmetry but there occurs a considerable Sn - Ni disorder. The statistical distribution of tin and nickel atoms over for them available sites results in an apparently single-component spectrum of ^{119}Sn resonance in $ErNi_2Sn_2$ (Fig.13). The clear asymmetry of lines in the $300K$ spectrum was accounted for assuming a presence of both the hyperfine field and the electric field gradient at a nuclear site. The reason for the small hyperfine field at $300K$ is the magnetic ordering of the nickel sublattice while the erbium sublattice orders only at about $5K$.
The increase of the recoilfree fraction with decreasing temperature is certainly not enough to explain the spectrum broadening as observed at $4.2K$. The transferred hyperfine field at the Sn site is thus equal to $11kOe$.

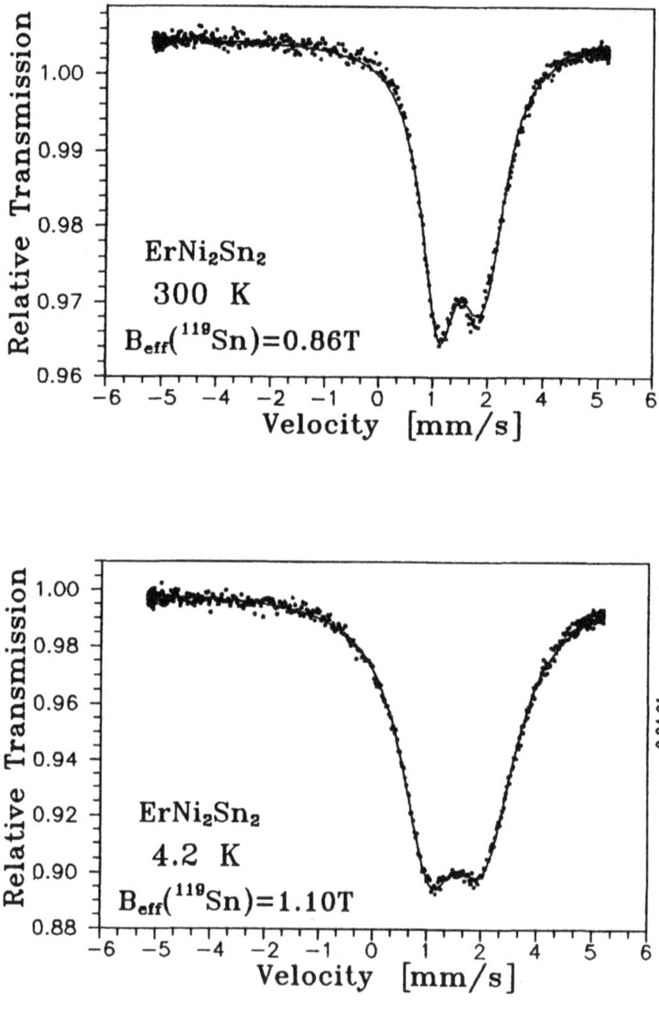

Figure 13: ^{119}Sn *Mößbauer absorption spectra of* $ErNi_2Sn_2$

4 CONCLUSIONS

RET_2Sn_2-compounds which have been only on a very limited scale investigated, appear to exhibit interesting physical phenomena. These generally metallic systems (in contrary to compounds with Si and Ge which most often show a semicon ductor-like behaviour) owe variety of their properties to the presence of $5s^25p^2$ valence electrons of tin which have joined rare earth f-electrons and transition element d-electrons in the interplay of collective phenomena. Tin has clearly contributed to the number and character of the conduction electrons but simultaneously it has retained the characteristic structural feature of *Si*- and *Ge*-compounds - the framework is created by the Sn - T bonds with rare earth occupying the free spaces.

On the methodological plane it is worth to conclude with the remark on the importance of accounting for finite absorber thickness effects in the ^{119}Sn Mößbauer experiment evaluation and thus, on the one hand, increasing the reliability of results, and on the other extending the possibilities of the method.

Acknowledgments

I am grateful to Professor G.M. Kalvius for making my stay in Munich (Physik Department E15, TUM) possible. The part of the experimental results were obtained there.

In Poland the project was financed in part by the MEN Fiz.22/91 grant.

References

[1] A. Szytuła and J. Leciejewicz in *Handbook on the Physics and Chemistry of Rare Earth , Vol. 12* eds. K.A.Gschneidner, Jr. and L.Eyring, Elsevier Science Publishers 1989, p.133

[2] E.A. Görlich and K. Łatka, *to be published*

[3] E.A. Görlich, K. Łatka and J. Moser *Hyp.Int.* **50**,723 (1989)

[4] S.N. Mishra and H.G. Devare *Hyp.Int.* **15/16**,633 (1983)

[5] E.V. Sampathkumaran, L.C. Gupta, R. Vijayaraghavan, K.V. Gopalakrishnan, R.G. Pillay and H.G. Devare , *J.Phys. C: Solid State Phys.* **14**,L237 (1981)

[6] G. Schmiester, B. Perscheid, G. Kaindl and J. Zukrowski in *Valence Instabilities* eds. P. Wachter and H. Boppart, North-Holland Publishing Company 1982, p.219

[7] B. Batlogg, A. Jayaraman, V. Murgai, L.C. Gupta, R.D. Parks and M. Croft in *Valence Instabilities* eds. P. Wachter and H. Boppart, North-Holland Publishing Company 1982, p.229

[8] M.M. Abd-Elmeguid, Ch. Sauer and W. Zinn *J.Phys. C: Solid State Phys.* **18**,345 (1985)

[9] E.A. Görlich, K. Łatka,A. Pacyna and A. Gleißner *to be published*

[10] I. Nowik in *Valence Instabilities and Related Narrow Band Phenomena* ed. R.D. Parks, Plenum Press 1977, p.261

[11] E.V. Sampathkumaran, S.K. Dhar, N. Nambudripad and R. Vijayaraghavan *Solid State Comm.* **67**(no.10),945 (1988)

[12] R.W. Skolozdra, W.M. Mandzik,Yu.K. Gorelyenko and W.D. Tkachuk *Fizika Metalov i Metalovedyeneye (in Russian)* **52**(no.5),966 (1981)

[13] R.W. Skolozdra and L.P. Komarovkaya *Fizika Metalov i Metalovedyeneye (in Russian)* **53**,1834 (1982)

[14] *Mößbauer Effect Data Index 1973* eds. J.G. Stevens and V.E. Stevens, IFI/Plenum Data Co. N.Y.1975, p.139

^{57}Co MÖSSBAUER EMISSION SPECTROSCOPY

H. Spiering, M. Alflen, C. Hennen, Zs. Kajcsos[+], F. Tuczek, P. Gütlich,
Institut für Anorganische Chemie und Analytische Chemie, Johannes Gutenberg-Universität, D-6500 Mainz
[+]Central Research Institute of Physics (KFKI), H-1525 Budapest

Dedicated to Professor Ekkehard Fluck on occasion of his 60th birthday.

ABSTRACT

After-effects following nuclear transformation of the electron capture decay ^{57}Co(EC)^{57}Fe have been studied in a variety of matrices both by Time Integral (TIMES) and Time Differential Mössbauer Emission Spectroscopy (TDMES). The after-effects comprise distortion of the environment, anomalous charge states different from the parent ^{57}Co, transient ligand field states of each charge state, and nonthermalized electronic multiplets split by spin orbit coupling and/or external magnetic fields. The emission spectra of ^{57}Co/MgO and ^{57}Co/$LiNbO_3$ are presented as examples for the different after-effects. The spectrometer for TDMES developed in the last years is briefly described and, in particular, the difficult problem of subtraction of accidental coincidences. The technique is applied in a study of a transient ligand field state in [^{57}Co/Co(phen)$_3$](ClO_4) and after-effects in ^{57}Co/Co(ClO_4)$_2$·6H_2O. A time integral measurement with γ-X-ray coincidence technique on ^{57}Co/Co(ClO_4)$_2$·6H_2O is presented and the importance of this technique for the study of after-effects is discussed.

1. INTRODUCTION

The nuclear decay of ^{57}Co leads to a high excitation of the nucleogenic daughter atom (^{57}Fe) and its environment. A K- or L- electron is captured by the ^{57}Co nucleus and a 0.7 MeV neutrino emitted feeding the 136.5 keV excited state of ^{57}Fe. The recoil energy of 4.6eV of the neutrino emission is far below the displacement threshold determined by the chemical bond energy of typically 25eV. Therefore, the physico-chemical consequences of the decay arise from the electronic deexcitation of the nucleogenic ^{57}Fe. In the following we consider a ^{57}Co atom in the divalent state so that after the electron capture a $^{57}Fe^{2+}$ with a hole in a deep shell is present.

The electron hole in the inner K or L shell moves outward via radiative (X-ray) or non-radiative Auger- processes. The different branches of this process and their relative probabilities are shown in fig. 1. The extremes are a Fe - ion in the Fe^{2+} excited electronic state (hole in a $M_{2,3}$ or $M_{4,5}$ state - 3.3 % of events), if a K_β-quantum (≈ 7keV) is emitted, or in a highly charged state up to Fe^{7+}, if Auger electrons of 5.4keV from the KLL Auger transition, of low energy 30-125 eV resulting from Coster-Kronig transitions, and of ≈ 650 eV from LMM Auger- transitions are emitted.

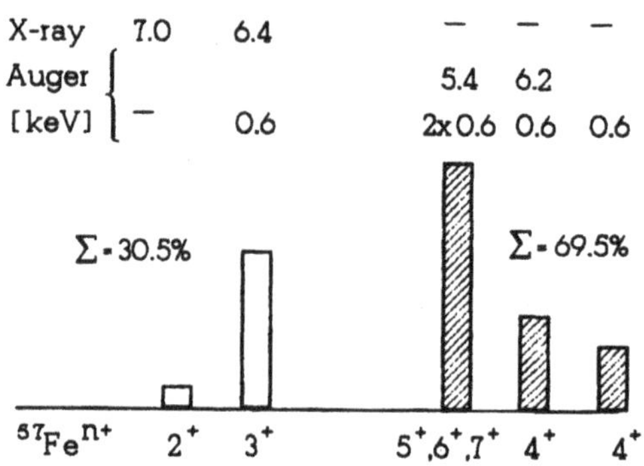

Figure 1: The deexcitation of a $^{57}Co^{2+}$ daughter atom of divalent $^{57}Fe^{2+}$ after EC and the predicted distribution of charge states. There are two LMM Auger electrons with energies between 560eV and 720eV (≈ 600eV) for KLL Auger transitions. The Coster-Kronig LLM transitions with M-electron energies ≤ 125eV are not plotted. The M-holes are counted for the final charge state which are indicated on the last line.

The emitted electrons, especially those of higher energy, themselves produce a lot of electron-hole pairs on their track which in turn may destroy molecules (radiolysis) or create other defects. The electrons will be recaptured by the holes and the Fe^{n+} ions. Highly charged Fe^{n+} ions ($n \geq 4$) may take electrons from the neighbouring ligand molecules which, due to the positive charge, loose their bond to the Fe ion. In the case that no X-rax energy (7.0 keVor 6.4 keV) escapes from the environment of the nucleogenic Fe ion the whole energy of 7.2keV is released to the lattice. A γ-X-ray coincidence experiment[1)] (counting only Mössbauer quanta (γ) with a preceeding K- or L-X-ray) selects the events where mainly (89%) one Auger electron of an energy of 560eV, 650eV, or 720eV is emitted and the Fe^{3+} ion is left in a state with two holes of excitation energies less than 100eV. 11% of the X-ray events lead to an Fe^{2+} state with an excitation energy of 53eV ($M_{2,3}$ hole) or only a few eV in the case of an $M_{4,5}$ hole which represents a $3d^6$ excited state.

These different processes produce a number of after-effects of the nuclear transformation, each leading to a subspectrum in the Mössbauer emission spectrum.

The subspectra are described by the following attributes with the corresponding sensitive Mössbauer parameter given in brackets:

- formal charge state n^+ (isomer shift),
- the ligand field ground and excited states (isomer shift and contribution of the valence electrons to the quadrupole splitting),
- the distorted environment (contribution of the lattice to the quadrupole splitting)
- Zeeman states in an external magnetic field (magnetic splitting).

All these different attributes may be of transient nature in the Mössbauer time window of $\tau = 141ns$ of the 14.4keV transition. In that case an average population of the states, or dependent on the transition rates, also average attributes determine the subspectra. The Mössbauer spectrum is taken to be understood if all subspectra are identified and their initial populations, the populations immediately ($\sim 10^{-3}\tau$) after the EC when the Mössbauer level is populated, are determined. If all transition rates are small as compared to $1/\tau$, then the intensities of the subspectra are proportional to the initial populations of the corresponding states. In case of equal f-factors the relative populations can be calculated. Each subspectrum can, of course, show all difficulties of an ordinary absorption spectrum. For example, it can be a (thermal equilibrium) relaxation spectrum of an electronic multiplet split by second order spin-orbit coupling or/and an external magnetic field. If the transition rates are comparable to or larger than $1/\tau$ the complete transition matrix must be known in order to obtain the initial populations from the measured subspectra, which in that case are not static spectra, but smeared out according to coherent relaxation effects[2]. The determination of both the initial population and the transition rate of each state can only be achieved by TDMES measurements.

The evaluation of an emission spectrum consisting of many superimposed spectra may be a difficult or even an unsolvable task. In most cases emission spectra have been evaluated only to some extent in order to obtain the fraction of charge states or the relative population of few Zeeman states. If this task is done an interpretation on the basis of a physical model for the formation of charge states and the initial population of transient electronic states can be developed.

In the first part of this contribution the initial population of Zeeman split ground states of the Fe^{2+} and Fe^{3+} charge state as obtained from TIMES measurements will be discussed. TDMES measurements are the subject of the second part, the conventional one which use the 122keV nuclear transition populating the Mössbauer level as a start, and the γ-X-ray coincidence technique, where the 6.4keV and the 7.0keV X-ray quanta serve as start signals. The spectrometer, which was developed in our laboratory, with a coincidence rate improved by an order of magnitude requires a sophisticated data reduction procedure which will be described briefly.

2 LOW ENERGY EXCITATIONS AS AFTER EFFECTS IN APPLIED MAGNETIC FIELD

Populations of split ground state multiplets out of thermal equilibrium have been observed several times by TIMES both for the Fe(II)[3] and Fe(III)[4] charge states and also for electronic states of daughter ions of other Mössbauer isotopes [5]. For a long time there was, however, no attempt to explain these phenomena. The after-effects seem to be too complicated that a prediction of the population of low energy excited states could be ever possible. The population were considered as a result of a relaxation of a highly excited atom in a hot environment with a large cooling rate so that transition rates and thermal relaxation times of the spin lattice interaction change rapidly and the integration over these processes gives populations which are frozen in at low temperatures. Nagy[6] developed this picture in some detail and proposed the "decoupling model" (by association with the decoupling processes in the standard "big bang" model). The electronic populations are assumed to be strongly coupled to the heat spike only at the very beginning after the nulear decay. When the local temperature has decayed to a "critical" temperature, decoupling occurs due to the slowing down of the electronic relaxation rates. Whereas the local temperature reaches the equilibrium temperature of the lattice , populations corresponding to the decoupling temperature then should be observed in MES. The measurements with ^{57}Co doped into $LiNbO_3$, however, gave a new turn to the discussion which did not any more consider the thermal spike and the time scale of its decay to be essential for the initial population in low energy excitations.

We will first look, in section 2.1, at the time integral emission spectra of ^{57}Co/MgO which demonstrates the difficulty to evaluate emission spectra. In section 2.2 the calculation of initial populations in ^{57}Co/$LiNbO_3$ is sketched and the unique situation met there is compared with the case in ^{57}Co/MgO.

2.1 ^{57}Co/MgO

The system ^{57}Co/MgO was first studied by Chappert, Frankel, Misetich and Blum in external magnetic fields $\leq$ 1.5T long time ago[7,8]. They found three charge states n=1+,2+,3+. The Fe^{3+} subspectra were not studied in detail because of their low intensity. In addition, several questions have been left open concerning nearby defects which seem to couple to spin states giving rise to anomalous internal fields. In view of interesting and rather well understood results obtained for the ^{57}Co/$LiNbO_3$[9] we decided to reinvestigate also the ^{57}Co/MgO system[10].

The after-effects in MgO lead to a large number of subspectra, almost all types described in the introduction, and is therefore a good example to illustrate the

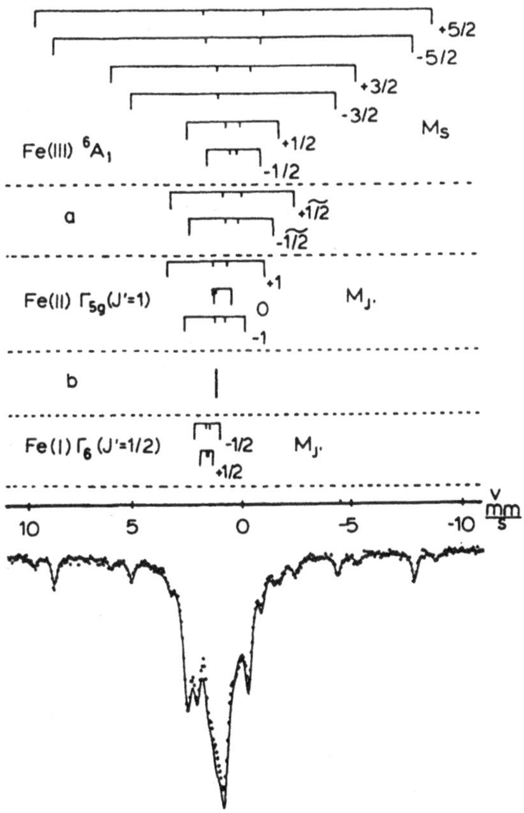

Figure 2: Mössbauer emission spectrum of $^{57}Co/MgO$ single crystal measured at 4.2K in an external magnetic field of 2.8T parallel to [111]. The stick diagram indicates the subspectra belonging to split electronic ground states of Fe^{3+}, Fe^{2+}, Fe^{1+} charge states, and nonground terms from a) Fe^{3+} and b) Fe^{2+}.

evaluation problems. The spectrum in figure 2 was fitted with 14 subspectra which are indicated by the stick diagram above. For two charge states two electronic ligand field states were identified. Since in MgO the ^{57}Co replaces Mg^{2+} in a perfectly octahedral environment the ligand field states are assigned according to the term symbols of the cubic group. If defects are not produced as after-effects in the lattice the cubic symmetry is maintained and therefore the ligand field does not give rise to a quradupole interaction. A brief comment on the spectra of each of the charge states is as follows:
Fe^{3+}: The $^6A_1$ ground state (originating from the ^6S free ion state) in an external

magnetic field B_{ex} gives rise to a superposition of six subspectra with effective hyperfine fields of $B_{eff} = B_{ex} + A*M_S$, $M_S = -5/2, -3/2, \ldots, +5/2$. The hyperfine coupling constant A takes for Fe^{3+} typically a value of 22T. The six four-line-spectra (γ-direction is parallel to the magnetic field) are symmetric according to the missing quadrupole interaction. The spectra could be taken as static in agreement with the large spin lattice relaxation times of the $^6A_1$ state at low temperatures (4.2K). The excited ligand field state is a Kramers doublet split by B_{ex}. The long lifetime of this state suggests that it belongs to a doublet term (originating from the ^2I free ion state) rather than a quartet term (^4G), so that the large spin change of $\Delta S = 2$ diminishes the transition rate to the ground state. The strong orbit lattice interaction of these states would produce a large quadrupole interaction, if nearby defects distort the cubic symmetry, which obviously is not the case.

Fe^{2+}: The nature of the magnetic field spectra of the Fe^{2+} ground state has been clarified by Ham[11] along with absorption experiments on $^{57}Fe/MgO$ by Leider and Pipkorn[12]. The triplet ground manifold with effective spin $J' = 1$ splits in a magnetic field into the $M_{J'} = -1, 0, +1$ states. In an applied magnetic field the symmetry is no longer cubic so that the $M_{J'}$ states exhibit a quadrupole interaction ($\eta = 0$) with $e^2QV_{ZZ}(\pm 1) = 0.32$mm/s and $e^2QV_{ZZ}(0) = 0.62$mm/s (for the unsplit state $\sum V_{ZZ}(M_{J'}) = 0$ as required for cubic symmetry). The $M_{J'}$ spin states are nonthermalized, and the decay rates $k(+1 \rightarrow -1)$, $k(+1 \rightarrow 0)$, and $k(0 \rightarrow -1)$ are of the order of the reciprocal lifetime τ and give rise to coherent relaxation effects. At liquid helium temperature the decay rates are determined by direct phonon processes and are proportional to the third power of the energy separation between the states, which is proportional to the applied field. The rates were calculated from the known orbit lattice coupling coefficients derived from static strain data of Watkins and Feher[13], so that the Mössbauer relaxation spectrum in the central part of fig. 2 has only the initial populations as parameters to be fitted. The exited ligand field state approximated by one broad Lorentzian line could not be assigned. Its magnetic splitting is covered because of its fast decay to the ground state.

Fe^{1+}: From the ground Kramers doublet of the cubic $^4T_1$ term (originating from the ^4F free ion ground term) two effective field spectra are observed in an applied magnetic field. The spin lattice relaxation within this doublet is again slow (the two states are related by time inversion operation whereas the orbit lattice interaction is invariant with respect to time inversion) and the two spectra show no quadrupole interaction. This means that this charge state is not stabilized by a nearby lattice defect.

From one spectrum like that of fig. 2 the information described above can of course not be extracted. The dependence of the spectrum on the external magnetic field up to $B_{ex} = 8T$ in two directions, [100] and [111], and on temperature from 4.2K to 20K were measured. Each of the subspectra has its own story with respect to the external parameters, and fitting everything together is the puzzle to be solved. The theoretical curve in fig. 2 still show small systematic deviations from the experimental points, but we are sure that the main features have been unravveled and the initial populations derived are reliable. To our knowledge this spectrum is the most complicated emission spectrum that have ever been decomposed into its subspectra.

The essential observation made with the $^{57}Co/MgO$ system are summerized as follows:

- Irregularities in the environment of the nucleogenic ^{57}Fe play no role.
- The initial populations $P(M_{J'})$ of the $Fe^{2+}(J'=1)$ ground state manifold in applied magnetic field show an alignment of the magnetically split $M_{J'}$ levels ($P(+1) = P(-1) = 0.23$ and $P(0) = 0.54$). The alignment does not depend on the direction and the absolute value of the applied field.
- The initial populations $P(M_S)$ of the $Fe^{3+}(S=5/2)$ ground state manifold in applied magnetic field are out of thermal equilibrium. The populations $P(M_S)$ and the alignment $P(M_S)+P(-M_S)$ depend on the absolute value of B_{ex} and on temperature $T \leq 20K$. For the [100] and [111] directions of the magnetic field the results are nearly the same, so that the angular dependence can be considered to be weak.

As already stressed, to understand after-effects means to understand the initial populations and their dependence on the external parameters. The qualitatively different behaviour of the ground states of different charge states implies no simple explanations. The property of Kramers states (Fe^{3+}) and non Kramers states (Fe^{2+}) will come into play and likely many details of the cascade, the energy levels and the coupling to the lattice determining the phonon transitions. The unique case of $^{57}Co/LiNbO_3$ uncovered the physical processes determining the initial populations as after-effects of a nuclear decay. This is described in the following section.

2.2 $^{57}Co/LiNbO_3$

Single crystals of $LiNbO_3$ were doped with ^{57}Co by diffusion. The Co ion replaces a Li^+ ion in trigonally distorted octahedral sites (as has been clarified by channeling experiments recently [14]). The ground state of the Fe^{3+} charge state is again the $^6A_1$ state. Only the initial population of this state will be considered here. In fig. 3 two emission spectra are shown which demonstrate the dependence of the population on the direction of the magnetic field ($B_{ex} = 4T$) with respect to the threefold symmetry axis of a single crystal of $LiNbO_3$. The population of the outer lines belonging to the $m_s = \pm 5/2$ Zeeman electronic states are much larger for B_{ex} orthonormal to the threefold axis than in the parallel case, although the energy splittings, which enter in any discussion about a thermal history, are in fact the same, since the zero field splitting of the A_1 state can be neglected as compared to its Zeeman splitting in a magnetic field of 4T. The subspectrum in the central region, which is only fitted in the upper spectrum,belongs to the Fe^{2+} charge state and will not be considered. A series of Mössbauer emission spectra were measured[15] in applied magnetic fields

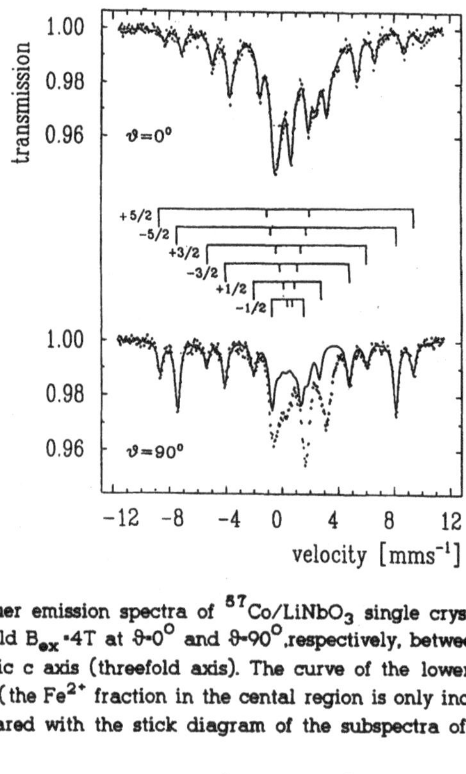

Figure 3: Mössbauer emission spectra of $^{57}Co/LiNbO_3$ single crystals at 4.2K in a longitudinal magnetic field $B_{ex}=4T$ at $\vartheta=0^0$ and $\vartheta=90^0$,respectively, between the field direction and the crystallographic c axis (threefold axis). The curve of the lower spectrum represents the Fe^{3+} fraction only (the Fe^{2+} fraction in the cental region is only included in the upper curve) and can be compared with the stick diagram of the subspectra of the six electronic states.

$2T \leq B_{ex} \leq 7T$, with field directions $0^0 \leq \vartheta \leq 90^0$ with respect to the c-axis of the crystal, and in the temperature range $4.2K \leq T \leq 100K$. The essential observations which are of importance for the understanding of the mechanism determining the initial populations of the Zeeman states are as follows:

– In the temperature range from 4.2K to 100K the Zeeman levels are populated out of thermal equilibrium. It is not possible to ascribe a "spin temperature" to the population. At low temperatures up to 20K, the population is independent of T. This fact rules out a relaxation of the population in the Mössbauer time window.

– The populations depend on the angle ϑ (see fig. 3) and the absulute value of the applied magnetic field.

— The alignment $P(M_S) + P(-M_S)$ depends on the direction, but not on the value of the applied field, whereas the polarization $\sim\sum P(M_S)$ depends on the direction and the value.

Comparing this behaviour with that of the $^6A_1$ sextet of Fe^{3+}/MgO allmost all dependencies are opposite. The independence of the alignment of the magnetic field value has been found for Fe^{2+}/MgO so that odd and even electron systems can also have similar properties, and to find a common physical basis seems to be even more difficult.

Three properties listed above point to the physical reasons of these anomalous populations. First, an angular dependence can only be due to non-S terms. If the crystal is rotated the $^6A_1$ wave functions are essentially fixed to the external field while all excited-term wave functions are more fixed to the crystal because of the coupling of the orbital momentum to the axial ligand field. Therefore, the transition matrix elements between excited terms and the $^6A_1$ term are expected to depend on ϑ for the axial ligand field in $LiNbO_3$. Second, the alignment $P(M_S) + P(-M_S)$ is the sum of the population of time conjugate states. When decaying to the $^6A_{1g}$ state each excited Kramers doublet will produce a characteristic alignment independent of the magnitude of the field as long as the states are the same and only the splitting changes and the two Kramers states preserve the property of time conjugation. This is the case if the direction of the field does not change and the Kramers doublets are well separated and do not mix by the magnetic interaction. Third, the quite different temperature dependences in the regions $T \leq 20K$ observed in $LiNbO_3$ and MgO suggest that thermal averaging of closely lying excited states takes place in MgO but not in $LiNbO_3$. Small energy separations of the excited states of Fe^{3+}/MgO are in accordance with the dependence of the alignment on the magnitude of the magnetic field. So we expect for $Fe^{3+}/LiNbO_3$ well separated ($\Delta E/k > 20K$) excited Kramers doublets. There are simple arguments that only from the lowest one the transition to the $^6A_1$ multiplet takes place.

The highly excited nucleogenic Fe^{3+} ion decays down the cascade of $3d^5$ electronic levels (cf. fig. 4). The ground state is the only spin sextet of the d^5 configuration. Thus, relaxation to the ground manifold is always spin forbidden ($\Delta S = 1,2$), whereas relaxation among the excited states may proceed either by internal conversion ($\Delta S = 0$) or intersystem crossing ($\Delta S = 1$). Concerning the latter process the relative decay rates depend on the strength of spin-orbit coupling, which in turn is higher between two energetically close lying excited levels than between any excited level and the ground state. Consequently, the lowest excited Kramers doublet (well separated by $200cm^{-1}$ from the next higher doublet) is populated prior to the final transition to the $^6A_1$ ground state. The initial population of the ground state sublevels is thus entirely determined by the relative transition probabilities from the excited state sublevels of this doublet weighted by their populations.

In optical spectroscopy the " spin memory " of radiationless transitions is well known in optical spin polarization experiments [16]. The polarization of a spin multiplet decaying over a large energy gap is transferred to another multiplet. Therefore, if a

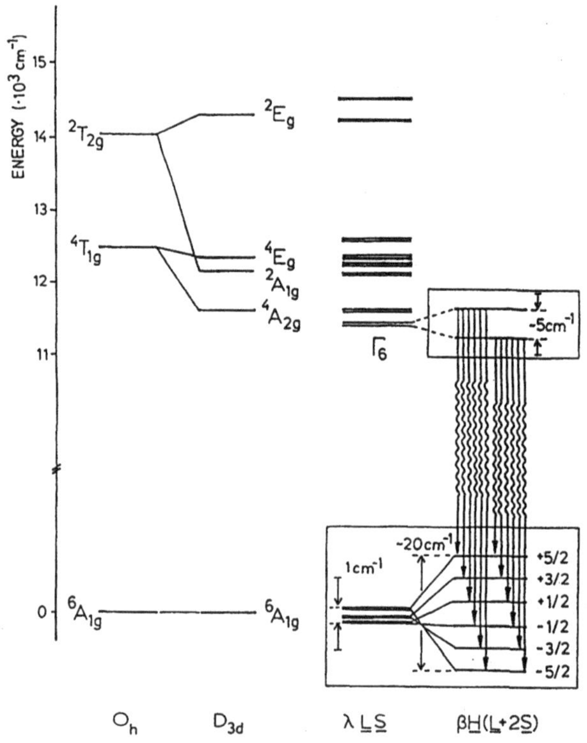

Figure 4: The low lying electronic levels of Fe^{3+} in octahedral ligand field, trigonal ligand field distortion, including spin-orbit coupling, and in applied magnetic field (4 T). The radiationless transitions of the last step in the cascade are indicated by wavy lines.

polarized doublet decays to a sextet, correlations between the six populations are just a result of the fact that they originate from the one polarization parameter of the doublet. We have calculated the spin orientation obtained by radiationless transitions [9]. The population ratio $R=P^+/P^-$ of the split Kramers doublet Γ_6 at $\vartheta=90^0$ (P^+ upper level) dependent on the magnetic field is shown in fig. 5.

The spin memory of radiationless transitions between electronic states with a large energy separation makes it possible to trace back the cascade to the first excited state. The question, about the origin of the polarization, however, is still not answered. On the contrary, if spin memory would be a general property of radiationless transitions, then the cascade should already start with polarized states. Any preexisting polarization of the ^{57}Co atom in applied magnetic field to be responsible has been ruled out by emission spectra of $^{57}Co^{3+}$ in the diamagnetic low spin state in ^{57}Co cobalt-acetylacetonate[17]. The mechanism of creation of polarization is very simple. First, it should be stressed that a relaxation within the Kramers doublet cannot account for the polarization, since such a relaxation would imply a magnetic field dependence by the process of spontaneous emission of acustical phonons (rates are proportional to the third power of the energy separation) and also a temperature dependence, which are not observed. Consequently, the polarization is necessarily built up in the cascade.

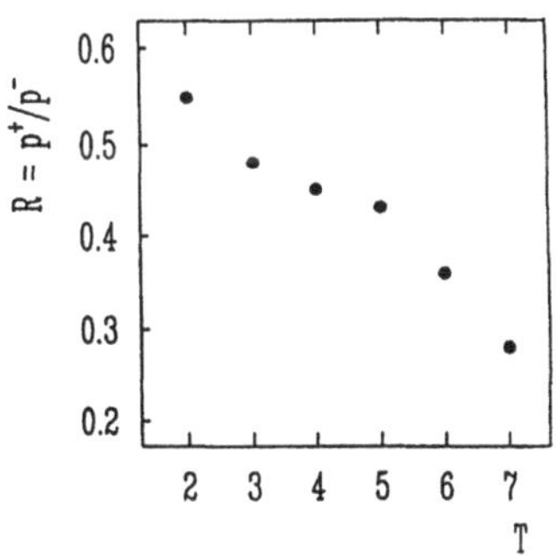

Figure 5: The ratio of the transient populations P^+ and P^- of the two levels of the first excited Kramers state Γ_6 (as calculated from the population of the ground state multiplet) as function of the applied magnetic field (in Tesla).

It is easily seen that radiationless transitions between electronic states with small energy separations, such that energy conservation requires creation of acoustic phonons, give rise to a spin momentum transfer to the lattice. . Let us consider one step from one Kramers doublet of energy E_k to the next lower one of energy E_{k-1}. In a magnetic field the doublets are split ($\delta_k = E_k^+ - E_k^-$) and four transitions, between like spin $E_k^\pm \rightarrow E_{k-1}^\pm$ and opposite spin $E_k^\pm \rightarrow E_{k-1}^\mp$, have to be considered. The energy separation between like spin states is similar, $(E_k^+ - E_{k-1}^+) - (E_k^- - E_{k-1}^-) = \delta_k - \delta_{k-1} \approx 0$, the separation between opposite spin states differs by the sum of splittings, $(E_k^+ - E_{k-1}^-) - (E_k^- - E_{k-1}^+) = \delta_k + \delta_{k-1}$. The probability of spontaneous emission of acoustical phonons is proportional to the third power of the energy separation according to the vibrational transition matrix element, while the electronic transition matrix elements of pairs of states related by time conjugation are the same. As a result, the transition rate of the spin flip transition $E_k^+ \rightarrow E_{k-1}^-$ is always larger than that of $E_k^- \rightarrow E_{k-1}^+$ which leads to a depopulation of the upper E^+-level in each step of the cascade. A more detailed description is found in [18].

The initial populations of the ground state of $^{57}Fe^{3+}$/ $LiNbO_3$ could be quantitatively reproduced by simulation of the cascade over 15 Kramers doublets[19]. This result shows that the observed populations do not contain any history concerning the after-effect of the nuclear transformation. Optical excitation of sufficiently high excited states would lead to the same population of the ground state levels .

The observations of $^{57}Fe^{3+}$ and $^{57}Fe^{2+}$in MgO as described above deserves a further brief comment. The diviating behaviour of the initial population of the sextet of Fe^{3+}/MgO has qualitatively been explained above by closely lying excited states Kramers doublets, the field dependence as well as the temperature dependence. The independence of the alignment of the triplet ground state of Fe^{2+} of the magnetic field points at a non magnetic singlet state feeding the ground state. The transition to the $M_{J'=0}$ is then most likely ($\Delta M_J = 0$), and those to the $M_{J'=\pm1}$ states are of the same probability (electronic transition matrix elements), if the energy separation is suffiently large as compared to the splitting of the triplet state. So in all cases of low energy excitation the cascade seem to be unaffected by the history of the thermal spike, and the initial populations reflect the properties of the electronic states and the phonon spectrum of the crystal.

3. TIME—DIFFERENTIAL MÖSSBAUER EMISSION SPECTROSKOPY

For studies of after-effects we have developed a new spectrometer for time-differential Mössbauer emission spectroscopy (TDMES) with high count rate capability and good time resolution[20]. The coincidence count rate (≥ kBq) was improved by an order of magnitude as compared to conventional spectrometers used so far. A time resolution of ≤ 3ns for the 14.4 γ-quanta achieved by the first photoelectron triggering technique (FPET) represents allmost the physical limit given by the number of photoelectrons of the NaI(Tl) detector [21]. The high count rate requires a sophisticated evaluation procedure of the accidental coincidences. This procedure will be briefly described in the next section. Application of the spectrometer to reinvestigate the transient electronic ligand field states in $[^{57}Co/Co(phen)_3](ClO_4)_2$, an old example in the literature studied by TDMES, is the content of the second section. The after-effects of $^{57}Co/\ Co(ClO_4)_2 \cdot 6H_2O$ as observed by TDMES is described afterwards, and in the last section the application of the spectrometer for γ-X-ray coincidence measurements on the same compound will be presented.

3.1 Evaluation of True Coincidences of TDME-Spectra

So far the evaluation of accidental coincidences have been published only for timing circuits based on a delayed coincidence method without a time-to-amplitude converter TAC[22]. In order to describe our evaluation procedure for accidental coincidences based on a TAC the principal features of our fast-fast coincidence system[20] are briefly outlined here. Scintillation detectors (plastic for the 122keV and NaI(Tl) for 14.4keV and 6.4keV, 7KeV X-rays) register the start and the stop quanta of the nuclear decay. Pulses from the NaI(Tl) detectors are fed into a fast energy-discrimination circuit which allows for simultaneous discrimination and timing to be made. The time measurement is carried out by a TAC. To achieve a coincidence rate of the order of kBq a source of ca. 3MBq produces about one MBq of starting (122keV, or γ-X-ray) and several kBq stoping quanta (14.4keV). The large difference in number of start and stop signals is due to the internal conversion factor of α=8.2 and the large solid angle (approaching 4π) for the start quanta and a relatively small solid angle for the 14.4keV stop quanta in order to limit the cosine smearing of the Mössbauer spectrum. To reduce count rate losses by the deadtime of the TAC, the physical start and stop sequence is reversed by a delay line.

Figure 6 illustrates the different situations a) - e) for the stop line of the TAC before and after the delay t_δ. On the start line of the TAC on the bottom of the figure a true 14.4keV Mössbauer pulse at two different times (before and after the delay on the stop line) is shown. Other start pulses which belong to nonresonant γ-quanta contribute to the background of the Mössbauer spectrum, whether they carry time information or not, and therefore need not be considered. The two situations a) and b) show the reverse (in time) decay time spectrum from t=0 to $t=t_\delta$ (see text of fig. 6) with the prompt event at t_δ. If the correlated physical start is not on the stop line c) the TAC will be stopped by an event preceding the 14.4keV and counted in a channel $t>t_\delta$. On a first sight the Mössbauer spectrum should have no time information and therefore should be the same as a time integral spectrum. But in d) it is shown that in channel $t>t_\delta$ also those events are counted which have a correlated true pulse on the stop line, which is earlier than the start (the 14.4keV is later than t_δ). This means the Mössbauer spectrum in channel $t>t_\delta$ is a superposition of a time integral spectrum (random coincidences) and a delayed coincidence spectrum in the time range $[t_\delta,\infty]$ indicated in fig. 6 by the time range 1, $-\infty \le t \le 0$ of the TAC. This type of uncorrelated events which are not really random since the spectrum carries time information will be called virtually random. They lead to a complicated subtraction procedure for the evaluation of the true coincidence Mössbauer spectrum of each time window. In e) a random event stops the TAC at time t'. Now there are three possibilities, i) there is no correlated pulse on the stop line, ii) the correlated pulse is of type 1 as in d), or iii) it is of type 2 preceding the 14.4keV pulse in times less than $t_\delta - t'$. These virtually random events give a Mössbauer spectrum which is the superposition of true coincidence spectra of time channels $[t',t_\delta]$ and $[-\infty,0]$. So in order to evaluate the true coincidences in a time window at t' true coincidences of the windows $t>t'$ and $t<0$ have already to be known, a task which can only be solved approximately. We start with the time window at t_δ. The spectrum of the virtually random coincidences in region 1: $[-\infty,0]$ are obtained from the spectrum of the window at $t>t_\delta$ after subtraction of a time integral spectrum. Having evaluated the true coincidence spectrum at t_δ the preceding time window can be evaluated

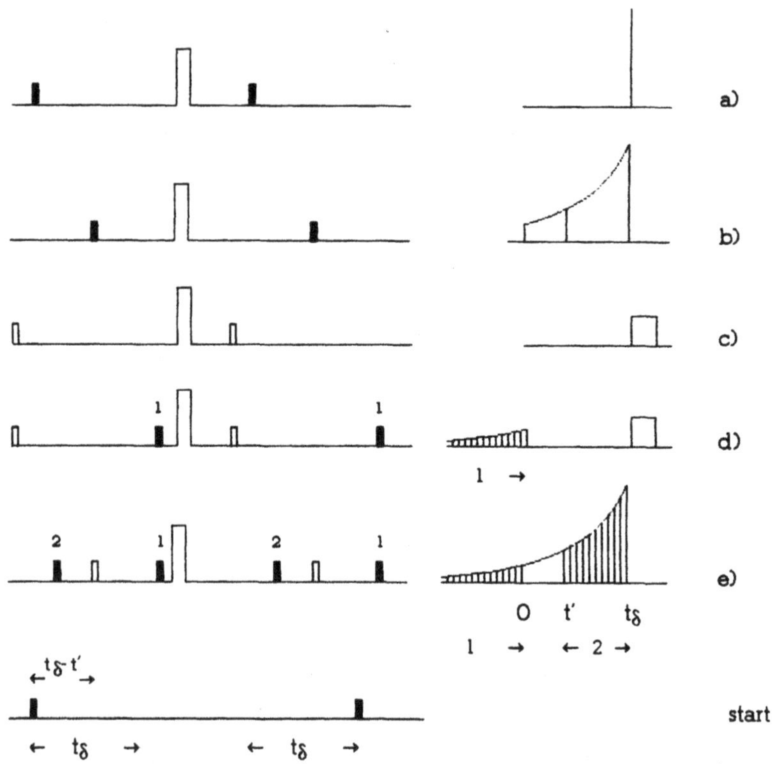

Figure 6: A delay of t_δ on the physical start line (122keV or X-ray quanta) reverses start and stop a-e) for the TAC. The electronical start line shows a pulse of a 14.4keV Mössbauer quantum at two different times, before and after the delay on the stop line. a) A prompt true pulse on the stop line (prompt decay of the 14.4keV level) is counted at t_δ. b) The 14.4keV pulse is behind the physical start by t_δ-t' and stops the TAC at t'. The decay spectrum is reversed and ends at a lifetime t_δ at time t=0. c) A pulse before the 14.4keV is necessarily uncorrelated (unfilled pulse). These pulses are counted later than $t=t_\delta$ in an extra single channel of the TAC. d) If the 14.4keV pulse is behind the physical start by a time larger than t_δ the true event (pulse 1) belonging to the time interval $[-\infty,0]$ of the TAC is counted in the window $t>t_\delta$. e) The 14.4keV pulse is stoped by an uncorrelated pulse at t'. The uncorrelated pulses of the channel at t' contain also correlated events belonging to the time interval $[-\infty,0]$ (pulse 1) and to $[t',t_\delta]$ (pulse 2).

and so forth. One can also start with the first time window at t=0. Then the virtually random coincidence spectrum is identical with the time integral spectrum. With the true spectrum at t=0 the spectrum of the virtually random coincidences of the next time window are obtained by subtraction of the true spectrum at t=0 from the time integral spectrum and so forth. It is obvious that both procedures are correct in the limit of zero width of the time windows.

The large number of time channels for 124 Mössbauer spectra (each having 1024 velocity channels) and the good time resolution of the spectrometer turned out to be a necessary supposition for the evaluation of accidental coincidences in case of high count rates, for which the spectrometer was designed.

For the subtraction procedure of the Mössbauer spectra of random and virtually random coincidences the fraction of such coincidences has to be calculated or obtained from Monte Carlo simulation for the actual experimental setup. We could derive closed formulae for the rates n^t := true coincidences/s, n^r := random coin./s, and n^v := virtually random coin./s which depend on following experimental parameters:

$n[s^{-1}]$ rate of initial nuclear decays (rate of 122keV quanta),
$\beta = \tau^{-1}[s^{-1}]$ inverse lifetime τ of the final nuclear state (14.4keV level),
$\alpha' = (\alpha+1)^{-1}$ internal conversion of the final nuclear decay,
d_{start}, d_{stop} detection efficiency in start and stop line, respectively,
n^b_{start}, n^b_{stop} $[s^{-1}]$ backgound rates in start and stop line, respectively.

The rate of the start line $n_{start} = n \cdot \alpha' \cdot d_{start} + n^b_{start}$ is subdivided into the rate of correlated n_{corr} and uncorrelated n_{uncorr} start-stop pairs such that $n_{start} = n_{corr} + n_{uncorr}$. These rates are obviously

$$n_{corr} = n \cdot \alpha' \cdot d_{start} \cdot d_{stop} \qquad \text{and} \qquad n_{uncorr} = n \cdot \alpha' \cdot d_{start} \cdot (1 - d_{stop}) + n^b_{start}$$

The rate on the stop line is $n_{stop} = n \cdot d_{stop} + n^b_{stop}$. With these definitions the coincidence rates in a time window $[t_1, t_2]$ can be expressed as a product of a rate and a probability.

$$n^i([t_1, t_2]) = n_{corr}\, P^i([t_1, t_2])\ ,\ i = t, v \qquad \text{and} \qquad n^r([t_1, t_2]) = n_{uncorr}\, P^r([t_1, t_2])$$

The probabilities are given by:

$$P^t([t_1, t_2 \le t_\delta]) = \beta/(\beta - n_{stop}) \cdot e^{-\beta t_\delta} \left(e^{(\beta - n_{stop})t_2} - e^{(\beta - n_{stop})t_1}\right)$$

$$P^v([t_1, t_2 \le t_\delta]) = n_{stop}/(\beta - n_{stop}) \cdot e^{-\beta t_\delta} \left(e^{(\beta - n_{stop})t_2} - e^{(\beta - n_{stop})t_1}\right) + $$
$$+ \left(1 + e^{-\beta t_\delta}\right) \cdot \left(e^{-n_{stop} t_1} - e^{-n_{stop} t_2}\right)$$

$$P^v([t_1 \ge t_\delta, t_2]) = e^{-\beta t_\delta} \left(e^{-n_{stop} t_1} - e^{-n_{stop} t_2}\right)$$

$$P^r([t_1, t_2]) = e^{-n_{stop} t_1} - e^{-n_{stop} t_2}$$

In figure 7 the relative portions of true, random, and virtually random coincidences are plotted for experimental parameters as typically used in a TDMES -

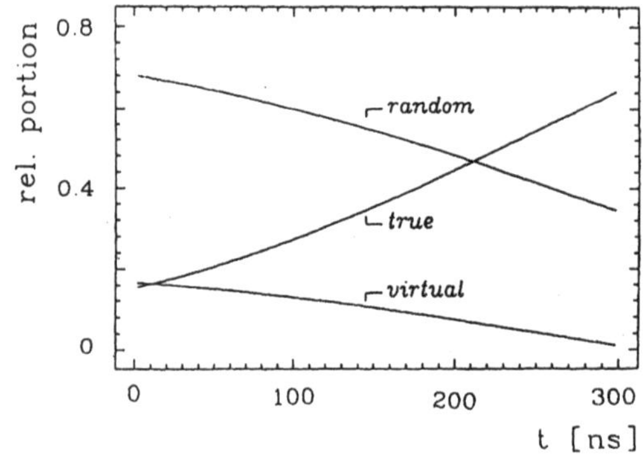

Figure 7: Calculated fractions of true, random, and virtually random coincidences. Mössbauer source strength n= 2MBq, detection efficiency d_{stop}= 0.35 (122keV) , d_{start}= 0.005 (14.4keV), background ratios n^b_{stop}/n = 0.15 and n^b_{start}/n_{start} = 0.25, delay time t_δ = 300ns, and internal conversion α= 8.2.

experiment. The number of random coincidences exceeds the number of true coincidences after about 100ns. The virtually random coincidences are almost negligible for short life times, but reaches the number of true coincidences at 2τ. In fig. 8 the

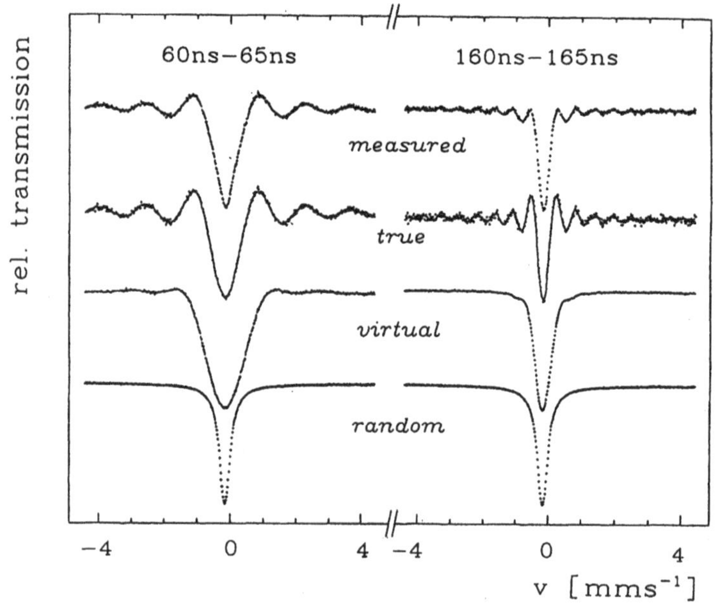

Figure 8 : TDMES-spectra of ^{57}Co/Rh single line source and $K_4[Fe(CN)_6]\cdot 3H_2O$ single line absorber. The measured spectra of two time windows are shown on the top. The time integral spectra (random) on the bottom are the same. The spectra of virtally random and true coincidences are evaluated by the subtraction procedure of accidendal coincidences. The spectra of true coincidences were fitted according to the time filtering theory[23].

measured spectra of two time windows of 5ns width and the measured time integral spectrum (random) are shown together with the evaluated spectra of true and virtually random coincidences of a single line source $^{57}Co/Rh$ and a single line absorber $K_4[Fe(CN)_6]\cdot 3H_2O$. The spectra of true coincidence have been fitted by the time filtering theory[23)] without any free parameter. The good agreement between theory and experiment gives confidence to the subtraction procedure of accidental coincidences.

3.2 Fe^{2+} Excited Ligand Field State (TDMES)

The observation of an excited ligand field state in MES study of the chelate compound $[^{57}Co/Co(phen)_3](ClO_4)_2\cdot 2H_2O$ (phen = 1,10-phenanthroline) was already reported in 1970[24)]. TDMES was applied in 1977[25)] and for the first time a decay of an excited state was unambigously observed by Mössbauer coincidence measurements. These measurements have been repeated with our new spectrometer, and the results obtained based on its higher resolution as well as the new insight in the electronic cascade by optical experiments has led to a more detailed physical picture of the deexcitation process of the nucleogenic Fe^{2+} complex.

In fig. 9 time integral spectra at different temperatures and in fig. 10 time differential spectra at T = 40K are shown. At room temperature two doublets are observed which can be assigned to a Fe^{3+} (12%) and a Fe^{2+} charge state. In the strong field of the bidentate phenanthroline ligands the electronic ground states of both charge states are low spin (LS) states, the $^2T_{1g}(O_h)$-Fe^{3+} and $^1A_{1g}(O_h)$ - Fe^{2+} states. The LS state was identified as the ground state by comparison with the parameters of the doublet in the analogous iron compound $[Fe(phen)_3](ClO_4)_2$ studied by Mössbauer absorption spectroscopy. The spectrum at 80K shows two further doublets which increase at the expense of the $^1A_{1g}$ LS doublet with decreasing temperature. These doublets were assigned to excited ligand field high spin (HS)- $^5T_2(O_h)$ states which decay to the LS ground state as unambigously shown by the TDMES in 1977. The Fe^{3+} fraction and also its lineshape is independent of temperature. Opposite to the situation in MgO the relaxation within the $^2T_{1g}$ Kramers doublets is fast due to the spin-spin coupling to the Co^{2+}-lattice. The time integral spectra show a time averaged fraction weighted by the lifetime of the Mössbauer level. The initial population and the lifetime Θ of the excited state are evaluated from the TDMES-measurement. In fig. 10 a series of time windows of a width of 40ns clearly show the decreasing HS fractions (relative intensity of the left absorption line decreases) with increasing time. The decay rates and the initial populations of both HS states (which could not be differentiated in earlier experiments[25)]) are collected in table 1.

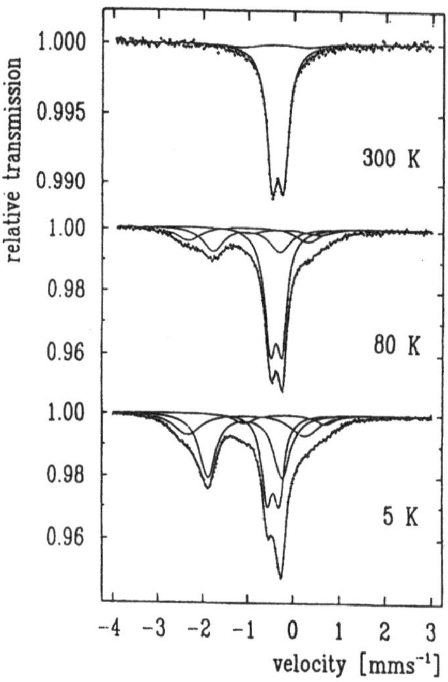

Figure 9: Mössbauer emission spectra of [^{57}Co/ Co(phen)$_3$](ClO$_4$)$_2$ as the source kept fix (reversed velocity scale!) at three temperatures. At RT the quadrupole doublets belong to the LS ground states $^2T_1$ of the Fe^{3+}(12%) and $^1A_1$ of the Fe^{2+} charge state. At lower temperatures two doublets of the excited ligand field HS state of Fe^{2+} increase at the expense of the $^1A_1$ LS state.

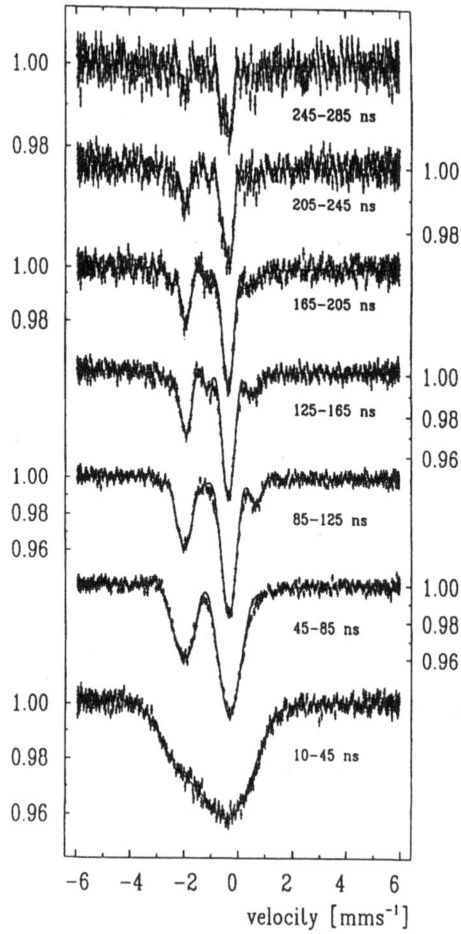

Figure 10: Time differential Mössbauer spectra of [^{57}Co/ Co(phen)$_3$](ClO$_4$)$_2$ at T= 40K. The time windows of 40ns width are choosen out of 124 windows of 5ns width. The decrease of the population of the excited HS state with increasing time is clearly visible by the intensity ratio of the two main absorption lines.

Other strong ligand field complexes were found to show a similar behaviour [26]. ME studies of Co complexes of intermediate ligand field strength made a direct identification of the excited state as the T_{2g} (O_h) HS state possible. Complexes of intermediate field strength still have a LS ground state at low temperatures, but the HS state can be thermally populated well below the decomposition temperature of the compound. The electronic state of the nucleogenic ^{57}Fe atom observed in MES of

	10K	40K	80K
n_{HS1}	0.55	0.57	0.49
n_{HS2}	0.17	0.17	0.17
n_{LS}	0.16	0.14	0.22
Θ_{HS1}/ns	645(63)	660(36)	260(16)
Θ_{HS2}/ns	$\sim 10^4$	≈2500	800(240)

Table 1: Data evaluated from TDMES spectra of $[^{57}Co/Co(phen)_3](ClO_4)$ measured at three temperatures. The initial populations of the two excited HS states and the LS ground state are given as fractions (error ≈0.01) of the total spectrum. The missing part of 0.12 to 1.00 is a LS-Fe^{3+} fraction which is independent of temperature.

these compounds doped with ^{57}Co is identical with the thermally populated HS state[27]. In all examples studied so far the Fe^{2+} fraction is completely in the HS state at all temperatures, so that the lifetime Θ_{HS} is definitely longer than τ and the initial population is close to $n_{HS}(t=0)=1$.

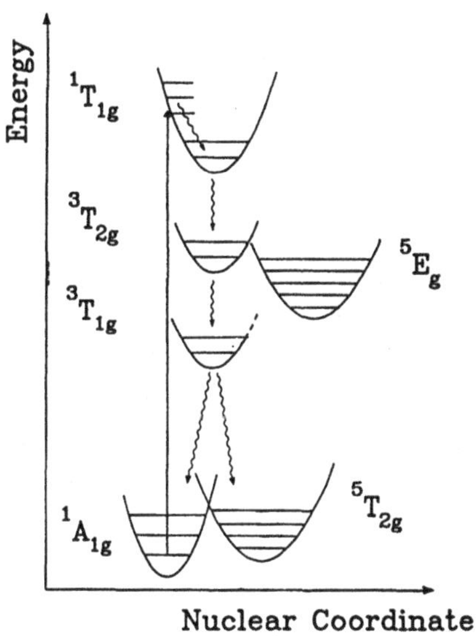

Figure 11: The adiabatic oscillator potentials of the the $3d^6$ electrons in a ligand field of intermediate strength in the direction of the A_{1g} nuclear coordinate. Optical spin allowed excitation from the $^1A_{1g}$ ground state are reresented by an arrow. Wavy arrows denote radiationless transitions.

The development in the spin-crossover research field, especially the phenomenon called "Light Induced Excited Spin State Trapping" (LIESST)[28] in Fe^{2+} spin-crossover compounds throw new light on the deexcitation of the excited ligand field HS state. The mechanism of LIESST can be explained on the basis of fig. 11, which shows a potential surface diagram in the direction of the nuclear A_{1g} coordinate of the lowest $3d^6$ states in a cubic ligand field of intermediate strength. The oscillator potential of the first exited $^5T_{2g}$ state is displaced by the difference in the metal ligand bond length of typically 0.2Å from the potential curve of the $^1A_{1g}$ ground state and a few $100cm^{-1}$ up to several $1000cm^{-1}$ higher in energy. Irradiating the cold sample induces spin-allowed transitions $^1A_{1g} \rightarrow {}^1T_{1g}$. The excited spin-singlet states are short-lived and decay back to the $^1A_{1g}$ state within ns or populate the triplet states $^3T_{2g}$ and $^3T_{1g}$ by intersystem crossing. These states again decay via intersystem crossing, either to the $^1A_{1g}$ ground state or to the $^5T_{2g}$ state. The radiationless decay of the quintet state to the ground state is the slowest step.. The temperature dependence of transition rates of several compounds were measured by pulsed laser excitation in our group[29]. We note here that there are large differences in lifetimes of more than 10 orders of magnitude for different compounds , with a tunnelling behaviour at low temperature and an Arrhenius type relaxation above ~100K as expected from the theory of radiationless transitions. The measured tunneling rates vary from $< 10^{-5}s$, the LIESST case of a trapped HS state, to 10^7s^{-1}. The rates correlate with the ligand field strength, i.e. the separation of the LS and HS state. This behavoiur was also observed by TDMES of the phenantroline compound in 1977. Moreover, the measured rates of the $[Fe(phen)_3]^{2+}$ complex diluted in a Nafion foil (Dupont type 117 ion-exchange membrane) in order to get sufficient light transmission are comparable with those observed by TDMES, so that there is no doubt about the nature of the exited state populated after the nuclear decay. The rates of all the spin-crossover compounds are much smaller than τ^{-1} in agreement with MES.

The new results of TDMES are qualitatively different from those[25] obtained before. The two excited HS states have quite different decay times and their initial populations are not 100% as evaluated from the old measurements. Our interpretation of the results follows the interpretation of the optical experiments. The nucleogenic Fe^{2+} ion reaches a highly excited electronic state, which decays via radiationless transition down the whole cascade of electronic states in the same way as depicted in fig. 11 for an optical excitation. However, the branching ratio is a property of the cascade and depends also on the excited state starting from, so that the same initial population in optical and in MES cannot be expected.

The measured temperature dependence of the initial population n_{HS1}, which also was not found earlier, seems to be a very interesting result. At 80K, n_{HS1} decreases by ≈ 10% as compared to the low temperature values. This behaviour must indeed be expected from optical experiments where a strong decrease of initial population with increasing temperature was observed[18]. So the decreased value of n_{HS1} at 80K is no problem, on the contrary the observation of the same values at 10K and 40K needs clearification. Very likely we look here into the history of the after-effect of the nulear decay. The high energy steps of the electronic cascade proceed in the subnanosecond region and can well be effected by the transient thermal spike. We believe that the transient temperature of the thermal spike determines the branching ratios at low temperatures. Only sufficiently high temperatures, higher than the transient temperature during the last steps of the cascade, change the branching ratios leading to a lower initial population of the HS state. There is in fact the possibility to prove this hypothesis. According to the number of Auger electrons in figure 1 the thermal spike should be smaller in case of a K-X-ray emission so that with a γ-X-ray experiment (TDMES) at low temperatures a higher initial population of the HS state should be observed.

Finally, a brief comment on the observation of two HS states will be given. The smaller quadrupole splitting of $\Delta E_Q(4.2K) = 1.65 mms^{-1}$ is believed to belong to the $^5T_{2g}(O_h)$ state, whereas the origin of the larger value of $\Delta E_Q(4.2K) = 2.80 mms^{-1}$ is unclear. The simplest explanation refers to a defect in the neighbourhood (radiolysis) which changes the ligand field and consequently the quadrupole splitting and the lifetime. In a γ-X ray coincidence experiment the fraction of this component is expected to be less or even absent. Another possibility can be seen in view of the long lifetime observed for the spin allowed $^5T_2 \rightarrow {}^5E$ transition in tetrahedral symmetry[30], which suggests an excited ^5E_g state above the $^5T_{2g}$ state as the long-lived component. In this case this species should survive in a γ-X-ray coincidence experiment.

3.3 Transient Lattice Distortions in $^{57}Co/Co(ClO_4)_2 \cdot 6H_2O$

Mulhem et al.[31] reported on time delayed coincidence Mössbauer spectra of $^{57}Co/Fe(ClO_4)_2 \cdot 6H_2O$ and $^{57}Co/Fe(BF_4)_2 \cdot 6H_2O$ at 200K. They compared time integral emission spectra and coincidence spectra in the time window [50ns, 400ns] and found different quadrupole splittings for both the Fe^{2+} and Fe^{3+} charge state and an increase of the Fe^{3+} fraction in the delayed coincidence measurement. Although the statistics of these measurements are very poor and all changes are in single or double statistical error the authors claim that the observed effects are reliable since four parameters $\Delta E_Q(Fe^{2+})$, $\Delta E_Q(Fe^{3+})$, the ratio of Fe^{2+} and Fe^{3+} fraction, and the line width showed the direction of change as expected from the autoradiolysis model (see next section). Since for the evaluation of the data only Lorentzian lines were used and any time filtering effects were neglected the fitted trends may also be the result of sytematic errors. Therefore, we reinvestigated one of these compounds, $^{57}Co/Fe(ClO_4)_2 \cdot 6H_2O$ at the same temperature of 200K.

In fig. 12 the time integral and TDMES-spectra with time windows of 50ns width are shown. The time integral spectrum has very broad lines, which can be fitted in different ways. Muhlem et al. fitted this spectrum by two doublets, one for each charge state. The good statistics compel us to introduce at least a second quadrupole doublet for the Fe^{2}-charge state and to allow for an asymmetry of the other Fe^{2+} doublet (texture). The large linewidths of the doublets are obtained by a Gaussian distribution of Lorentzians of natural width. The time differential spectra are a superposition of single Lorentzian time-filtering lines (see section 3.1). The width of the distribution and the intensity are free parameters. This fitting procedure is true if coherent relaxation effects are negligible. The result of such a fit does not confirm the observation of Mulhem et al. The quadrupole splittings and the ratio of the relative portions of the two charge states are independent of the time window. The width of the distribution, however, seems to be time dependent. The width (FWHM) of the absorption line as obtained by a superposition of Lorentzians is plotted in fig. 13 (this decrease of width must not be confused with the decrease by the increase of the lifetime of the selected Mössbauer events). The decreasing width of the lines of the Fe -charge state points at a healing of defects in the environment of this charge

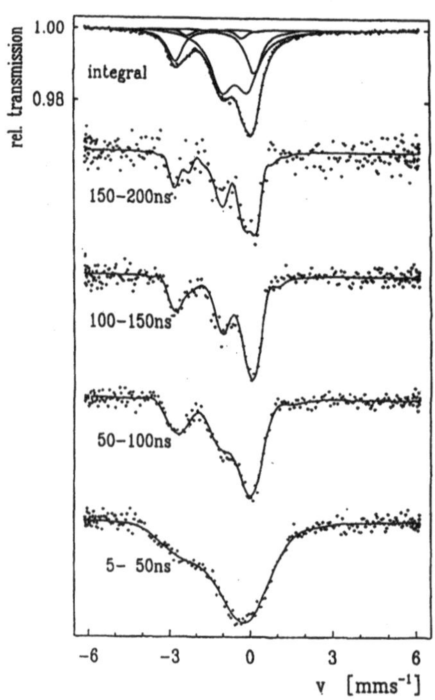

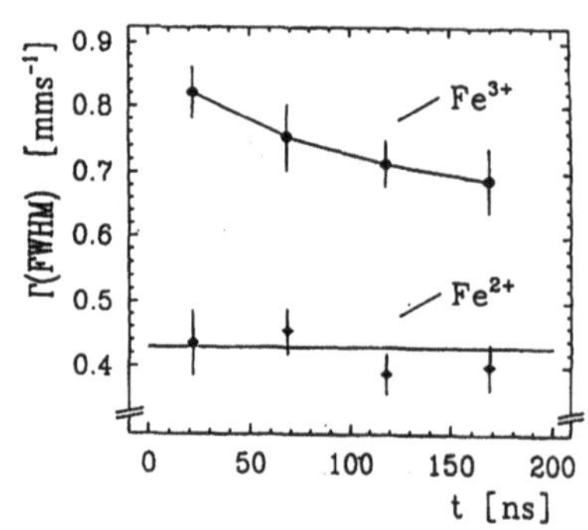

Figure 13: The width of the doublets of the two charge states are plotted versus the average lifetime t of the Mössbauer quanta in each time window. The width is (as a result of a distribution of lines of natural width) within the error independent of t for the large splitting of the Fe^{2+} charge state, but decreases significantly for the Fe^{3+} charge state.

Figure 12: Time integral and time differential Mössbauer emission spectra of $Co/Co(ClO_4)_2 \cdot 6H_2O$ at T=200K. The integral spectrum is fitted with two quadrupole doublets with an isomer shift of Fe^{2+} charge state and one doublet of smaller splitting assigned to a Fe^{3+} charge state. The time filtering spectra were fitted with fixed positions but free area and width of the distribution of doublets.

state. Though the results of Mulhem et al.[31] could not be confirmed, some changes in the Mössbauer time window remain. These result represents the first example of transient lattice defects in the Mössbauer time window.

3.4 γ - X-Ray Coincidence Mössbauer Emission Spectroscopy

The γ-X ray coincidence technique was introduced by Kobayashi and Fukumura[1] already in 1979 based on a conventional coincidence Mössbauer spectrometer. Instead of the 122keV γ-quanta the K-X-ray is used as start (electronical stop) signal for the 14.4keV detection. A time window of 400ns ~ 3τ was choosen to obtain an almost time integral spectrum which can be compared with the conventional time integral emission spectrum. The difficulty of this technique lies in the large electronic absorption of the 6.4keV and 7.0keV X-rays as compared to the 122keV γ-quanta. The sources have to be very thin and as much as possible uncovered in order to gain sufficient coincidence signals. Temperature dependent γ-X-ray coincidence measurements are therefore difficult to perform in the conventional manner with a source inside and the detectors outside the cryostat.

The first measurement on the hexahydrate $CoCl_2 \cdot 6H_2O$[1] was repeated in a later work of Kobayashi and Friedt[32] in 1986, where also the dihydrate was measured for comparison. The results were discussed in the frame of the autoradiolysis model, the development of which is briefly described in the following. Already in 1968, Gütlich et al.[33] observed the formation of Fe^{3+} species in Mössbauer absorption spectra of $FeSO_4 \cdot 7H_2O$ after external irradiation with ^{60}Co γ-rays. Later in 1976 such experiments were repeated by others[34] on the whole series of hydrates $FeSO_4 \cdot nH_2O$ (n=0,1,4,7) and on $FeSeO_4 \cdot nH_2O$ (n=1,4,7)[35]. They simply found that the formation probability of Fe^{3+} is proportional to the dose and, more important, to the number n of water molecules. The explanation already suggested by Gütlich et al. is conclusive. Fast electrons produced by the high energy γ-ray ionize the water molecules which subsequently form H atoms and OH radicals. The OH radical oxidizes the Fe^{2+} ion according to the reaction $Fe^{2+} + OH^{\cdot} \rightarrow Fe^{3+} + OH^-$, yielding the stable $[Fe^{3+}(H_2O)_5(OH^-)]^{2+}$ species. This interpretation was supported by the fact that the same MA spectra were recorded from separately prepared $[Fe^{3+}(H_2O)_5(OH)^-] \cdot SO_4$[32] At present the hypothesis is widely accepted that Fe^{3+} states in hydrates are formed in a secondary step through oxidation of Fe^{2+} by OH radicals, so that the picture of a complete recapture of electrons after the Auger cascade is used.

The γ-X-ray measurements of Kobayashi and Friedt on $CoCl_2 \cdot nH_2O$ (n=2,6) throw new light on the autoradiolysis model. The emission spectrum of anhydrous

$CoCl_2$ is a pure Fe^{2+} spectrum[37], so that any Fe^{3+} component in the hydrated salts is expected to be due to water molecules as in the case of the radiation experiment on the sulfates. In figure 14 the fractions of the subspectra of the dihydrate (coordination of four chlorine atoms and two water molecules) of the coincidence and anticoincidence spectrum are visualized by a bar diagram. The fractions of the anticoincidence spectrum are calculated from the time integral measurements by use of the γ-X-ray fraction of 30.5%. The striking result is that the Fe fractions of the coincidence and

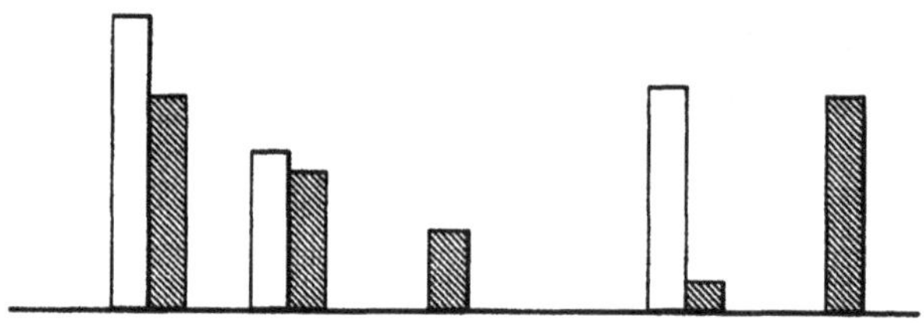

Figure 14: Relative intensities of the subspectra of the γ-X-ray coincidence spectrum (blank bars) and of the calculated anticoincidence spectrum (hatched bars) of $^{57}Co/\ CoCl_2{\cdot}2H_2O$ (data taken from ref. 32).

anticoincidence spectra of the dihydrate are the same so that the proportionality to the electron dose is completely lost. The number of radicals produced by the Auger electrons is about 2 times larger in the anticoincidence case (see figure 1) without any effect on the Fe^{3+} fraction. The simultaneous fit of the conventional and the anticoincidence spectrum reveal further interesting details of the after-effect. Two Fe^{3+} and three Fe^{2+} spectra were found. The $Fe^{2+}(1)$ is attributed to the undistorted lattice site, the parameters being the same as found for the Fe doped compounds[30]. A reference absorption spectrum of Fe^{3+} is available for the complex $[FeCl_2(OH_2)]^+$ at 90K[37]. The quadrupole splitting of $0.87 mms^{-1}$, however, is just in between the two values found in the emission spectrum, so that the undistorted case cannot be identified. The following fact is amazing:

The coincidence spectrum reveals two $Fe^{2+}(1,2)$ subspectra and one $Fe^{3+}(1)$ subspectrum, whereas in the anticoincidence spectrum the subspectrum $Fe^{3+}(1)$ is almost absent (4.3%) and instead a $Fe^{3+}(2)$ subspectrum with twice as large a quadrupole splitting appears. An unusual $Fe^{2+}(3)$ component (12%) with a very small quadrupole splitting of $0.45 mms^{-1}$ is also observed.

Some of these observations were discussed by Kobayashi and Friedt in the frame of the radiolysis model (the authors did not notice the equal Fe fractions in the coincidence and calculated anticoincidence spectrum). The mechanism for the appearance of quadrupole doublets other than Fe^{2+}(1) is solely considered as being due to the lattice contribution of radicals or H_2O molecules displaced from regular lattice sites outside the nearest H_2O coordination sphere, and Fe^{3+}(1,2) charge states may be regarded as the result of nearest neighbouring and further oxidizing radicals.

We cannot follow their arguments. Apart from the general considerations above, the Fe^{2+}(3) state, for example, needs an extremly large lattice contribution to the quadrupole splitting in order to reduce 2.42mms^{-1} to 0.45mms^{-1}. Such a large lattice contribution from distant lattice distortions is unrealistic. This Fe^{2+}(3) state, which is only found in the anticoincidence spectrum, rather suggests a severely distorted coordination sphere accounting for a large lattice contribution. Since the Auger electrons can hardly produce such a distortion, we very likely observe here a Coulomb explosion of the ligands as a result of the highly charged Fe^{n+} ($n \geq 4$) transient states, long ago proposed by Nath et al.[37]. The highly charged transient states may also be responsible for the appearance of the Fe^{3+}(2) fraction observed only in the anticoincidence spectrum with a twice as large quadrupole splitting as the Fe^{3+}(1) component.

When these difficulties about the well established autoradiolysis model, raised by the result of the γ-X-ray coincidence experiment on the $CoCl_2 \cdot nH_2O$, became aware, the way the experiments have been carried out had to be questioned by severe

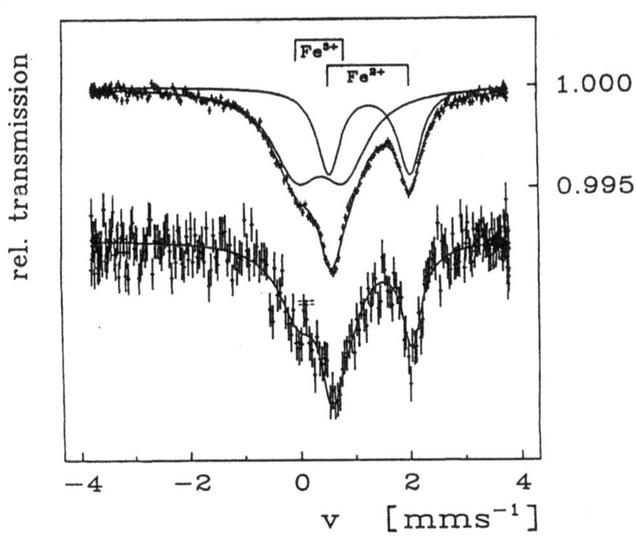

Figure 15: The conventional emission spectrum (above) and the time integral γ-X-ray coincidence spectrum of $[^{57}Co/Fe(H_2O)_6](ClO_4)_2$ at room temperature.

arguments. The hydrates were measured over several months with the conventional coincidence circuit to obtain sufficient statistics with a count rate of few Bq. The samples could not be enclosed in a cryostat and cooled as mentioned above, so that a decomposition of the hydrates exposed to air must have been difficult to avoid. Therefore, we measured the stable hydrate $^{57}Co/Co(ClO_4)_2 \cdot 6H_2O$ which, however, has the disadvantage of possessing a very small Debye-Waller f-factor. With two 1-inch NaI(Tl) detectors for the K-X-ray quanta a coincidence rate of 70Bq was achieved. With the evaluation procedure described above the complete time integral γ-X-ray coincidence spectrum was obtained. The emission spectrum and the γ-X-ray coincidence spectrum are shown in figure 15. The spectra were fitted by one quadrupole doublet each for the Fe^{2+}- and the Fe^{3+}-fraction where the positions are the same for both spectra. The linewidth is smaller by 10% for the coincidence spectrum as expected by the lower concentration of lattice distortions by the smaller number of Auger electrons. This effect has also been observed by Kobayashi et al. for all his measurements. However, the proportionality of the produced radicals to the electron dose cannot be confirmed. But an increase of the Fe^{3+}-fraction is indeed observed. Comparing the coincidence and the anticoincidence spectra the Fe^{3+}-fractions are 55% and 69%, respectively. Thus it appears that the formation of aliovalent charge states is a much more complicated process than is suggested by the simplifying models so far used to explain the observed charge state fractions.

4. Conclusion

After-effects of the nuclear transformation $^{57}Co(EC)^{57}Fe$ may be classified by three types of phenomena according to the information on the early stage after the transformation. Low energy excitations of the electronic ground state split by second order spin-orbit coupling or magnetic Zeeman interaction contain , generalizing the results from $^{57}Co/LiNbO_3$ and $^{57}Co/MgO$, no information about the early stage and the history of the deexcitation process. Studying the initial populations of excited ligand field states, which may be called high energy excitations, may open the window to the history of the thermal spike after the nuclear decay, as may be inferred from the anomalous temperature dependence of the initial populations of the HS-fraction in $[^{57}Co/Co(phen)_3](ClO_4)_2$. At low temperatures ($T < 100K$), these electronic excitations give rise to very complicated emission spectra which usually prevent a detailed evaluation of all the subspectra and their response to external parameters. At high temperatures, the different charge states are in the electronic ground state so that the number of subspectra decreases to the number of charge states of the nucleogenic iron, multiplied by the number of environments differing by the number, types, and arrangement of nearby defects. In many cases studied so far the charge states and the lattice defects survive the Mössbauer lifetime. The healing of defects in the Mössbauer time window has only been observed in the system $^{57}Co/Fe(ClO_4)_2 \cdot 6H_2O$, specifically the environment of the Fe^{3+}-charge state. The decay of Fe^{2+}- and Fe^{3+}-charge states has never been observed in the Mössbauer time window, although often such charge relaxations were claimed to be discovered. Only a $Fe^{1+} \rightarrow Fe^{2+}$ relaxation was found in $^{57}Co/ZnS$ and related semiconducting compounds[3)]. Therefore, the population of charge states and the distribution of lattice defects point back to an early stage of the nuclear transformation, and in view of hot atom chemistry they should attract more attention in the study of after-effects in the future. The new tool of γ-X-rax coincidence spectroscopy may have an impact in this research field.

Acknowledgement

This research project was financially supported by the Bundesministerium für Forschung und Technologie (Grant No. 03Gu2MAI) and the Fonds der Chemischen Industrie. We would like to thank the Mössbauer research groups in Erlangen and Budapest (KFKI) for the vivid and fruitful cooperation. We also owe thanks to the brilliant electronics shops of GSI (Darmstadt), of the Technische Hochschule Darmstadt, and of the research center Forschungsanlage Jülich for their help.

References

1. Kobayashi, T., Fukumura, K., Kitahara, T., Nucl. Instr. Methods 166, 257 (1979)
2. Kankeleit, E., Z. Physik A275, 119 (1975)
3. Bonville, P., Garcin, C., Gérard, A., Imbert, P. and Jéhanno, G., Phys. Rev. B23, 4293 (part I) and 4310 (partII) (1981)
4. Nagy, D.L., Doerfler, R., Ritter, G., Waigel, J., Zeman, N. and Molnár, B., Phys. Lett. 95A, 400 (1983)
5. Stöhr, J., Phys. Rev. B11, 3559 (1975)
6. Nagy, D.L., Second Seeheim Workshop : Trends in Mössbauer Spectroscopy, eds. P. Gütlich and G.M. Kalvius, Mainz, 241 (1983)
7. Chappert, J., Frankel, R.B., Misetich, A., Blum, N.A., Phys. Rev. 197, 578 (1969)
8. Chappert, J., Misetich, A., Frankel, R.B., Blum, N.A., Phys. Rev B1, 1929 (1970)
9. Doerfler, R., Gruber, W., Gütlich, P., Hasselbach, K.M., Leupold, O., Molnár, B.,

Nagy, D.L., Ritter, G., Spiering, H., Tuczek, F., Phys. Rev. Letters 57, 2849 (1986)
10. Tuczek, F., Spiering, H. and Gütlich, P., Phys. Rev. B41, 10933 (1990)
11. Ham, F.S., Phys. Rev. 160, 328 (1967)
12. Leider, H.R. and Pipkorn, D.N., Phys. Rev. 165, 494 (1968)
13. Watkins, G.D. and Feher, E., Bull. Am. Phys. Soc. 7, 29 (1962)
14. Rebouta, L., Da Silva, M.F., Soares, J.C., Hage-Ali, M., Stoquert, J.P., Siffert, P., Sanz-García, J.A., Dieguez, E. and Agullo-Lopez, F., Europhysics Lett. 14, 557 (1991)
15. Doerfler, R., Gruber, W., Horváth, D., Nagy, D.L. and Ritter, G., Hyperfine Interactions 29, 1229 (1986)
16. Imbusch, G.F. and Geschwind, S., Phys. Rev. Lett. 17, 2238 (1966)
17. Tuczek, F., Diplomarbeit, Johannes-Gutenberg Universität Mainz (1984)
18. Spiering, H., Alflen, M., Gütlich, P., Hauser, A., Hennen, C., Manthe, U. and Tuczek, F., Hyperfine Interactions 53, 113 (1990)
19. Manthe, U., Diplomarbeit, Johannes-Gutenberg Universität Mainz (1988)
20. Albrecht, R., Alflen, M., Gütlich, P., Kajcsos, Zs., Schulze, R., Spiering, H. and Tuczek, F., Nucl. Instr. Methods A257, 209 (1987)
21. Smith, D. and Muller, J.W., Rev. Sci. Instr. 60, 143 (1989)
22. Alflen, M., Kajcsos, Zs., Spiering, H., Gütlich, P., Albrecht, R., Schulze, R. and Vilhyalmsson, H., Nucl. Instr. Methods A265, 464 (1988)
23. Lynch, F.J., Holland, R.E. and Hamermesh, M., Phys. Rev. 120, 523 (1969)
24. Ensling, J., Fitzsimmons, B.W., Gütlich, P. and Hasselbach, K.M., Angew. Chem. 82, 638 (1970)
25. Grimm, R., Gütlich P., Kankeleit, E. and Link, R., J. Chem. Phys. 67, 5491 (1977)
26. Ensling, J., Fleisch, J., Gütlich, P.,Fitzsimmons, B.W., Chem. Phys. Lett. 45, 22 (1977)
27. Fleisch, J. and Gütlich, P., Chem. Phys. Lett. 45, 29 (1977)
28. Decurtins, S., Gütlich, P., Köhler, C.P., Spiering, H. and Hauser, A., Chem. Phys. Lett. 105, 1 (1984)
29. Gütlich, P. and Hauser, A., Coord. Chem. Rev. 97, 1 (1990)
30. Garcin, C., Imbert, P., Jéhanno, G., Régnard, J.R., Férey, G., Leblanc, M. and Gérard, A., J. Physique 47, 1977 (1986)
31. Mulhem, J.K., Balogh, J., Demeter, I., Horváth, D., Molnár, B., Nagy, D.L. and Szücs, I.S., Proc. Int. Conf. Appl. Mössbauer Effect, Jaipur 1981 (Indian National Sci. Acad., New Delhi, 1982), p.392
32. Kobayashi, J. and Friedt, J.M., Bull. Chem. Soc. Japan 59, 631 (1986)
33. Gütlich, P., Odar, S., Fitzsimmons B.W. and Erickson, N.E., Radiochimica Acta 10, 147 (1968)
34. Meyers, J., Ladrière, J., Chavee, M. and Apers, D., J. Physique 37, C6-905 (1976)
35. Ladrière, J., Krack, J.C. and Apers, D., J. Physique Colloq. 40, C2-434 (1979)
36. Friedt, J.M. and Danon, J., Radiochem. Radioanalyt. Lett. 3, 147 (1970)
37. Bordeleau, C. and Wiles, D.R., Inorganica Chimica Acta 41, 195 (1980)
38. Nath, A., Klein, M.P., Kündig, W. and Lichtenstein, D., in: Mössbauer Effect Methodology 5, 163, ed. I.J. Gruverman (Plenum Press, New York-London, 1970)

NUCLEI AS OBSERVERS ON SURFACES AND IN INTERFACES

G. Schatz

Faculty of Physics, University of Konstanz, D-7750 Konstanz, Germany

1. Introduction

Nuclear techniques play an important role in physics of solids and liquids. Recently, these techniques have been introduced in the field of surfaces and interfaces. One large group of nuclear methods is dealing with stable and radioactive nuclei as probes for solid state properties. Here, the hyperfine techniques like nuclear magnetic resonance (NMR) /1,2/, Mössbauer spectroscopy /3,4,5/ and perturbed angular correlations (PAC) /6,7,8,9/ have been successfully applied to study surface and interface phenomena. Especially PAC has proven to be rather powerful since surfaces and interfaces can be prepared with an extremely low probe concentration, i.e. surfaces and interfaces can be labeled with isolated radioactive probes. The observed hyperfine parameters are the electric-field-gradient tensor and the magnetic hyperfine field, acting at the probe nuclei. The electric field gradient , for example, can serve as a fingerprint to elucidate different probe sites at the surface and interface, impurity-probe configurations, diffusion of impurities and interface compound formation. The magnetic hyperfine field can be detected with monolayer resolution and can be used to study the magnetic behaviour of magnetic surfaces and thin films.

In this paper most recent examples of PAC surface and interface studies will be presented. The PAC method is a well-established tool for obtaining microscopic information on structural and electronic phenomena in solids. Several metal surfaces and interfaces have been investigated so far; in most of these studies radioactive ^{111}In probe atoms have been utilized.

2. The Perturbed Angular Correlation Method

The PAC method is based on the hyperfine interaction of radioactive probe nuclei with extranuclear magnetic fields or electric field gradients. The magnetic hyperfine field in magnetically ordered materials is mainly governed by the Fermi-contact interaction of polarized core and valence electrons. The electric field gradient, as the second spatial derivative of the electrostatic potential, is a tensor quantity which is produced by extranuclear charges. In metals the nearest-neighbour atomic environment with respect to the PAC probe atom determines this quantity and is strongly enhanced through conduction electron contributions.

For the present experiments we used the isomeric nuclear state ($I = 5/2$, $t_{1/2} = 84$ ns) intermediate between the 171 - 245-keV $\gamma\gamma$ cascade in ^{111}Cd, which is populated through electron capture of ^{111}In ($t_{1/2} = 2.8$ d). The detection of γ_1 selects a subgroup of states within the isomeric nuclear level, resulting in an anisotropic γ_2 emission in delayed coincidence with respect to the γ_1 detection. Interaction of the nuclear magnetic dipole moment ($\mu = -0.7656(25)\mu_N$) and the nuclear electric quadrupole moment ($Q = 0.83(13)$ barn) of the isomeric state with hyperfine fields splits this nuclear level, which causes a time dependency of the $\gamma\gamma$-angular correlation. This time dependence is characterized by frequencies ω_n, which are the transition frequences within the nuclear sublevels. In the presence of only a magnetic hyperfine field at ^{111}In(^{111}Cd) the Larmor frequency and its first harmonic appear; in the case of a pure electric field gradient three transition frequencies and for a combined interaction up to fifteen transition frequencies can occur /10/. With a five γ-detector setup, one records sixteen different coincidence spectra simultaneously from which eight counting-rate ratios R(t) are extracted /11/. These R(t) spectra can be expressed as a superposition of cosine modulations containing the transition frequencies ω_n. Fourier analyses of the R(t) spectra directly exhibit the transition frequencies; the set of hyperfine parameters, however, is extracted by a least-squares fit applied simultaneously to the full set of time-dependent counting-rate ratios. From this procedure the strength and direction of the magnetic hyperfine field as well as strength, symmetry, and orientation of the electric-field-gradient tensor are obtained.

3. PAC Investigations of Nonmagnetic Metallic Surfaces

3.1. Surface field gradients for different Cu surfaces: Cu(100), Cu(110) and Cu(111)

The dependence of the surface-field-gradient tensor on different surface cuts was first investigated for ^{111}In on various Cu surfaces /8,11,12/.

Copper single crystals, cut accordingly, have been cleaned and annealed in ultra-high vacuum and checked with Auger-electron spectroscopy (AES) and low-energy electron diffraction (LEED). Radioactive ^{111}In probes have then been deposited at the surface with a concentration of the order of 10^{-4} of a monolayer. A cross section of the used ultra-high vacuum apparatus, which has been specially designed for in situ PAC experiments, is shown in Fig. 1; further experimental details can be found elsewhere /11/.

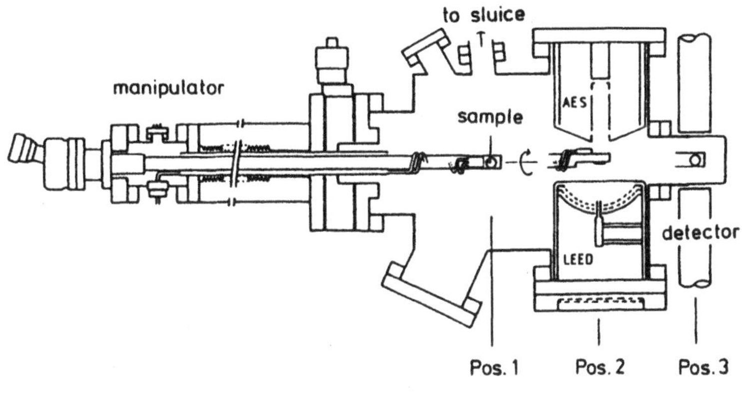

Fig.1: Cross-sectional view of ultra-high vacuum apparatus for surface PAC experiments.

In Fig.2 PAC time spectra are shown for the three main orientations of the face-centered cubic copper system, namely Cu(111) (top), Cu(100) (middle) and Cu(110) (bottom). The samples have been annealed to about 600 K for 20 minutes. In all cases the surface normal was lying in the γ-detector plane pointing under 45° between detectors. As one can see from the data most of the ^{111}In probes are exposed to well-defined electric surface field gradients, which is manifested by sharp transition frequencies seen in the Fourier transforms (right part). Due to the chosen geometrical arrangement, from the expected three transition frequencies (since the nuclear spin of the isomeric ^{111}Cd-level is $I = 5/2$) the second frequency is dominately seen. In the case of Cu(111) and Cu(100) the ratio between first and second frequency is found to be 1:2, an axial-symmetric electric field gradient with asymmetry parameter $\eta = 0$ is deduced. The following electric-field-gradient parameters measured at T = 295 K have been found:

Cu(111): $|V_{zz}| = 9.9(14)\cdot 10^{17}$ V/cm^2, $\eta = 0.05(5)$
Cu(100): $|V_{zz}| = 9.3(14)\cdot 10^{17}$ V/cm^2, $\eta = 0.00(5)$
Cu(110): $|V_{zz}| = 7.9(12)\cdot 10^{17}$ V/cm^2, $\eta = 0.74(1)$

The orientations of principal axes of the observed field gradient were obtained from a series of PAC-experiments where different surface-detector geometries have been used. From such experiments an unambiguous principal-axis orientation can be deduced /8,11/. For Cu(100) and Cu(111) the z-principal axis, which is the symmetry axis, is pointing perpendicular to the surface. The z-principal axis for Cu(110) is also perpendicular to the surface, whereas the y-principal axis is oriented along the densely-packed atomic rows (see atomic arrangements for each surface in Fig. 2, right side). In all three cases, symmetry and principal-axis orientation of the observed electric field gradients clearly reflect the configurational symmetry of the surface atoms.

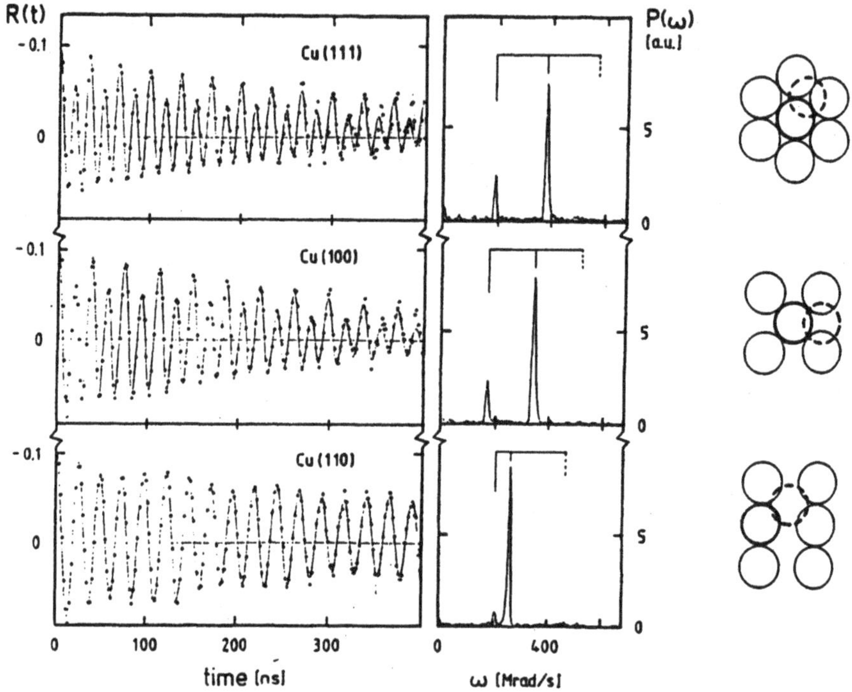

Fig.2: PAC time spectra together with their Fourier transforms for isolated ^{111}In probes on a Cu(111) (top), Cu(100) (middle) and Cu(110) (bottom) surface. The surface normals were lying in the detector plane under 45° with respect to two adjacent detectors. The time spectra are fitted with three frequencies corresponding to a unique electric field gradient. On the righthand side the atomic arrangement for the corresponding Cu surface is shown, the possible ^{111}In sites consistent with experiments are also indicated (bold circle: substitutional site, dashed circle: hollow site).

In the case of Cu(100) and Cu(111) only two sites of the ^{111}In probes can explain the experimental findings: firstly, a substitutional site, i.e. In is incorporated in the topmost layer or, secondly, a hollow site, i.e. In is occupying a regular Cu site in the next, still empty monolayer (see Fig. 2, right side). Cluster calculations by Lindgren /13,14/ and other pieces of information from diffusion studies on vicinal Cu(111) surfaces (see chapter 3.4) strongly support substitutional positions of ^{111}In on copper surfaces.

Recent experiments on Ag(100), Ag(111) and Ag(110) have revealed quite a similar behaviour as was found for copper surfaces. Again the measured surface field gradients exhibit symmetries and principal-axis orientations, which support the assumption of substitutional terrace sites for ^{111}In /15,16/. The somewhat lower field gradient values are consistent with the larger atomic distances in face-centered cubic silver.

3.2. Surface field gradient at ^{111}In probes deposited at different depth from the surface

As it was demonstrated above, ^{111}In probes which are occupying substitutional sites in the topmost atomic layer are exposed to rather strong unique electric field gradients. The electric field gradient acting on to probe nuclei should be governed mainly by the nearest-neighbouring atoms and the conduction electrons in the neutral Wigner-Seitz cell. Therefore once the ^{111}In probes are surrounded by the 12 next neighbours in face-centred cubic Cu a vanishing electric field gradient is expected. This situation is valid already for ^{111}In probes one layer below the surface layer in Cu(100) and Cu(111), since then a complete nearest-neighbour Cu shell is present. This effect has been shown experimentally in the case of In(111) /6/.

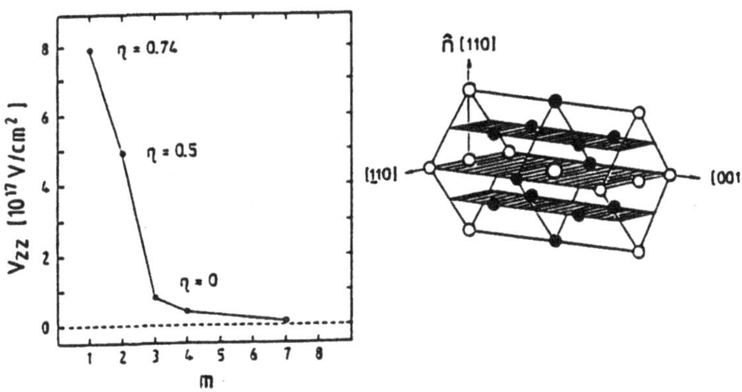

Fig.3: The largest component V_{zz} of the electric-field-gradient tensor measured at ^{111}In probes as a function of the depth m (in monolayers) from a Cu(110) surface. The asymmetry parameter η is also indicated. The inset illustrates the nearest-neighbour arrangement around the substitutional ^{111}In probe atoms.

For Cu(110) the situation is different, here two additional Cu monolayers are needed on top of the ^{111}In probes to restore the complete coordination (see Fig. 3, right side). This effect was investigated experimentally, by labeling Cu(110) surfaces with ^{111}In probes and successive coverage by additional copper. The result is shown in Fig. 3 where the electric field gradient is shown for ^{111}In probes situated at different depth from the surface. ^{111}In probes situated in the second layer (m=2) are still exposed to a rather strong electric field gradient, which stems from the fact that ^{111}In is only surrounded by 11 Cu atoms. It is interesting to note that this electric field gradient is quite similar to the value for a single vacancy trapped by ^{111}In in bulk copper /17/. Only after coverage by two layers, i.e. ^{111}In is now situated in the third layer, the electric field gradient tends towards zero, as it should be for a cubic surrounding. This result shows that the electric field gradient can be used to determine the position of ^{111}In probes with respect to the surface layer. This opens the possibility to follow migration of probes perpendicular to the surface on an atomic scale.

3.3. Indium Diffusion on Cu(100)

The labeling of copper surfaces by radioactive PAC probes offers the possibility to study diffusion processes. The basic concept applied is trapping and detrapping of a certain species under consideration at isolated ^{111}In probes. Trapping can be recognized by a detuning of the surface field gradient obtained for free ^{111}In probes.

Experiments have been performed to investigate indium diffusion on Cu(100) surfaces /18/. In a first step isolated ^{111}In probes have been deposited at Cu(100) and after annealing the PAC probes are exposed to a well-defined, axially symmetric electric field gradient (see Fig. 4a). As can be seen in the Fourier transform the amplitude for ω_2 is dominant, which is consistent with the principal z-axis system of the field gradient being oriented perpendicular to the surface (see chapter 3.1).

In a next step the Cu(100) surface, marked with isolated ^{111}In probes, was covered at 77 K with about 7 % of a monolayer of natural indium. The effect of this indium deposition on to the PAC spectrum is that a broadening of the electric field gradient distribution from 1 % to 3.5 % occurs, however, without

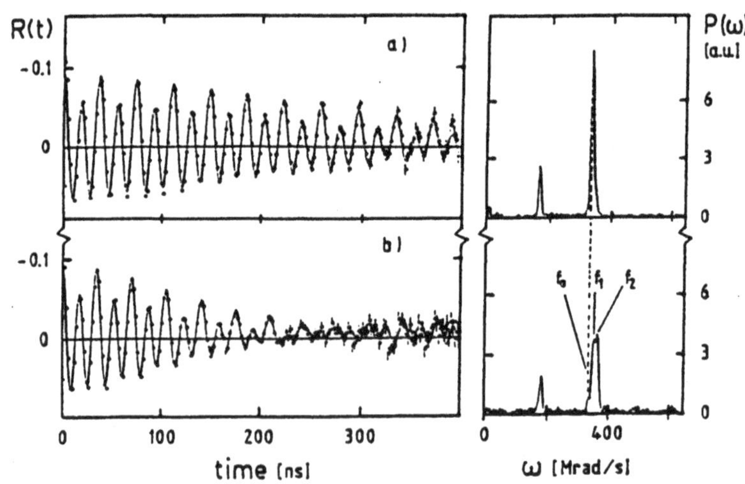

Fig.4: PAC spectra and their Fourier transforms for isolated ^{111}In on a Cu(100) surface (measurement temperature: 77 K) a) for a clean Cu(100) surface, b) for the Cu(100) surface covered with about 7 % natural indium and annealed up to 580 K.

any significant shift of the average frequency (f_0: ω_2 = 344(1) Mrad/s). Now the sample was subjected to an isochronal annealing program (annealing time: t_A = 20 min, measurement temperature: T_M = 77 K) to thermally activate diffusion of indium atoms.

At an annealing temperature T_A = 200 K a frequency shift of the electric field gradient of about 3 % to higher frequencies (f_1: ω_2 = 354(2) Mrad/s) is observed and at T_A = 240 K even another fraction with a higher frequency shift (f_2: ω_2 = 367(2) Mrad/s) appears. Over the entire annealing temperature range no other new frequencies occur. The PAC spectrum after annealing up to T_A = 580 K is shown in Fig. 4b. The Fourier transform clearly shows the splitting of ω_2 into three fractions, the corresponding splitting of ω_1 is not resolved.

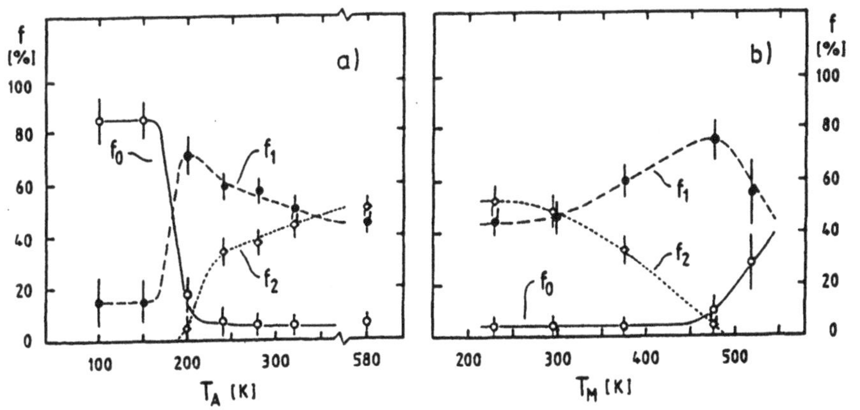

Fig.5: Fractions of ^{111}In probes, which are related to different In configurations, on a Cu(100) surface covered with 7% natural indium: a) as a function of annealing temperature (t=20 min) b) as a function of measurement temperature.

The annealing behaviour of the different fractions related to the three frequencies is depicted in Fig. 5a. The fraction f_0 is characterized by ω_2 = 344 Mrad/s, which is the same as found for the Cu(100) surface without natural indium deposition. Therefore we assume that this fraction of ^{111}In probes has no indium at nearest-neighbour sites on the surface, the observed increase of the field-gradient broadening is due to indium in the further surrounding. Heating the sample induces indium diffusion and with the choosen indium coverage already one or two jumps are sufficient to form In-In pairs. This effect occurs at temperatures between 150 K and 200 K, the fraction of free indium probes converts as indicated by a frequency shift of 3 % to f_1. At higher annealing temperatures this new fraction f_1 of In-In pairs partially converts into a fraction f_2, which might be identified as small indium clusters. The slow increase of f_2 at temperatures below 240 K can be explained by the fact that only few free In atoms are left to promote further cluster growth and therefore other clusters have to dissolve. From the transition temperature at 200 K an activation energy can be estimated. Underlying an Arrhenius-type behaviour and the assumption of the pre-exponential to be 10^{12} Hz an activation energy E_a=0.6 eV is obtained.

The ^{111}In fractions were also investigated as a function of PAC measurement temperature, which was kept for several hours typically. This experimental dependence is shown in Fig. 5b. Now with increasing temperature the fraction f_2 is reduced in favour of f_1; finally above 470 K also the fraction f_1 starts to diminish, i.e. in our model now In-In pairs break up. This behaviour was also studied theoretically by several groups /19,20,21/, who have performed lattice gas calculations which reproduce the main features found in the experiment.

3.4. Step sites for ^{111}In on Cu(111) surfaces

As discussed above ^{111}In probes occupy terrace sites at copper surfaces after thermal treatment. The question arises through what paths the ^{111}In probes find such terrace sites after deposition at 77 K and successive annealing. In the case of ^{111}In on Cu(111) it was tried to reconstruct the possible behaviour of probes until they settle down at terrace sites.

This question has been studied for ^{111}In on vicinal Cu(111) surfaces /22/. A Cu surface was prepared with the surface normal tilted by about 4° from [111] to $[\underline{11}2]$, resulting in a (667) plane. For this surface a LEED analysis showed parallel monoatomic steps, seperating (111) terraces (width: about 12 atomic rows). After cooling the sample down to 77 K the ^{111}In probe atoms were deposited on to the surface. Subsequently, PAC spectra were taken in an isochronal annealing program (t_A=10 min). After annealing up to 410 K a large fraction of probe atoms (here denoted by f_0) experience the well-known surface field gradient on Cu(111) as described above (see Fig. 6). However, at low annealing temperatures two other fractions f_{S1}, f_{S2} with different electric field gradients occur, corresponding to distinct probe atom sites on the surface. The field gradients for these two fractions are characterized by the following parameters /22/:

f_{S1} : $|V_{zz}| = 5.8\,(9)\cdot 10^{17}$ V/cm^2, $\eta = 0.37(2)$
f_{S2} : $|V_{zz}| = 7.8(12)\cdot 10^{17}$ V/cm^2, $\eta = 0.69(1)$

In order to relate these field-gradient parameters to certain surface sites of the probe atoms, the orientation of the principal-axis system of the field-gradient tensor was determined (for definition of angles see Fig. 7):

f_{S1} : $\delta = 31(2)^0$, $\varphi = 0^0$ (x-axis along steps)
f_{S2} : $\delta = 18(3)^0$, $\varphi = 90^0$ (y-axis along steps)

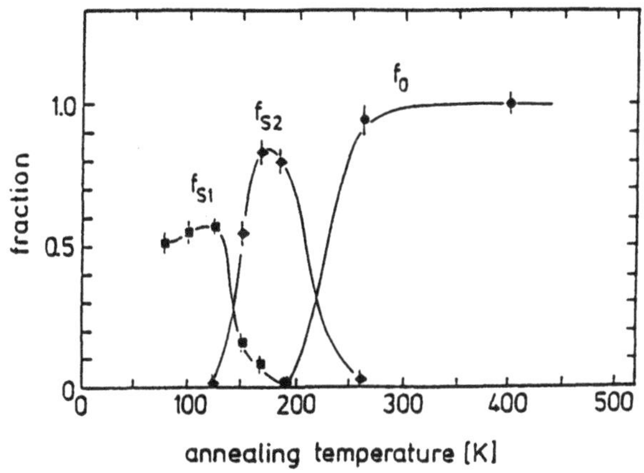

Fig.6: Fractions of ^{111}In probes on Cu(111), characterized by different electric field gradients, as a function of annealing temperature.

From this result one can see that in either case one of the principal axes is correlated to the step direction and therefore it can be concluded that the ^{111}In probe atoms, described by fraction f_{S1} and f_{S2}, occupy different step sites on the surface.

The behaviour of the three fractions after thermal annealing for 10 min is summarized in Fig. 6; clearly the conversion from step site f_{S1} to step site f_{S2} and further from f_{S2} to f_0 can be recognized.

From these different pieces of information a model was constructed, which describes the temperature behaviour and appearing electric-field-gradient orientations in a consistent way. The main features of the model are sketched in Fig. 7. It is assumed that the ^{111}In probes after deposition at 77 K are still mobile enough to reach nearby steps where they are trapped as adatoms (f_{S1}). The adatom step site converts around 140 K into a substitutional step site (f_{S2}) by trapping of a step vacancy. Then at annealing temperatures above 200 K the ^{111}In probe atoms diffuse into the topmost monolayer via vacancy mechanism in order to occupy substitutional terrace sites (f_0).

Quite a similar behaviour of thermal conversion of different surface sites has been observed for ^{111}In on vicinal Ag(100) surfaces /15/. Moreover, in this case a fraction of indium probe atoms associated with a rather weak electric field gradient was found, which within the above described model is interpreted as an immobile adatomic terrace site at 77 K.

From the observed transition temperatures for site conversion for Cu(111) one can easily obtain an estimate for activation energies determining the respective diffusion processes. Assuming an Arrhenius-type behaviour and adjusting the pre-exponential to be of the order of 10^{12} Hz the following activation energies are found /22/:

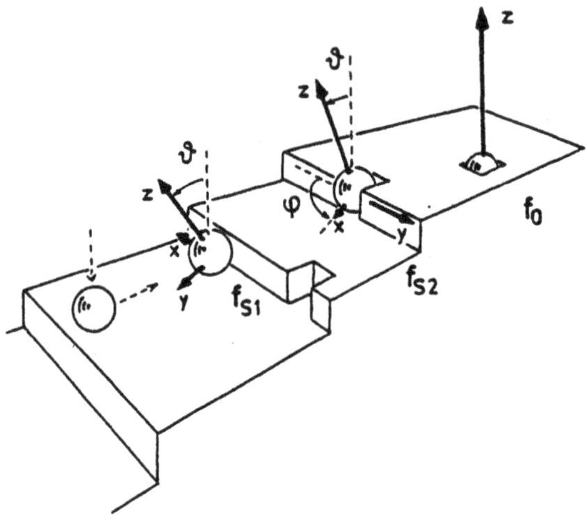

Fig.7: Model illustrating the observed electric-field-gradient properties and its association to different ^{111}In surface sites on vicinal Cu(111).

$$E_a(f_{S1} \rightarrow f_{S2}) = 0.41(3)\ \text{eV}$$
$$E_a(f_{S2} \rightarrow f_0) = 0.64(4)\ \text{eV}$$
$$E_a(f_0 \rightarrow f_{bulk}) = 2.2(3)\ \text{eV}$$

where $E_a(f_0 \rightarrow f_{bulk})$ describes the activation energy for the diffusion of ^{111}In from terrace sites into Cu bulk.

The experiments on vicinal Cu(111) surfaces provide direct evidence for a step correlated diffusion process of the ^{111}In probe atoms after deposition on to the surface. Consequently, they clearly confirm the assignment of substitutional terrace sites for the probe atoms at elevated annealing temperatures.

4. PAC investigations of magnetic surfaces

Magnetism of surfaces and thin films is the subject of continous experimental as well as theoretical efforts. The high local sensitivity makes the PAC method extremely favourable as a complementary approach to study surface magnetism. In the following an application of PAC to nickel surfaces is presented /23/.

Sample preparation and PAC experiments were carried out in an ultra-high-vacuum system, which was already introduced in chapter 3.1 (see Fig. 1). Disks cut from a [111]-oriented Ni single crystal were cleaned in ultrahigh vacuum by Ar^+ sputtering and subsequent annealing up to 1250 K. The crystallographic structure of the surface was monitored by LEED; the chemical purity was checked by AES. The resulting total impurity contamination was below 1 % of a monolayer. Finally, radioactive ^{111}In probe atoms were deposited on to the surface with a concentration of 10^{-4} of a monolayer, which, however, introduces Cl contaminations of about 0.5 % of a monolayer. The samples were then annealed to 850 K for 10 min and PAC spectra were recorded at different measurement temperatures.

First we consider results for measurement temperatures above the Curie temperature T_C =631 K. In this case the magnetic hyperfine field is switched off and only an electric field gradient should be present at the probe nuclei. In fact, the frequency pattern shown in Fig. 8a is characteristic of the electric field gradient at ^{111}In in substitutional terrace sites on face-centered cubic (111) surfaces, as was already established for ^{111}In on Cu(111) and Ag(111) (see chapter 3.1). The temperature dependence of this electric field gradient was measured above as well as below the Curie point, where in addition, a magnetic hyperfine field is present. This is shown in Fig. 9a, from where a strength of $|V_{zz}|$ = $11.4 \cdot 10^{17}$ V/cm^2 (at 77 K) can be read off and a weak temperature dependence can be seen.

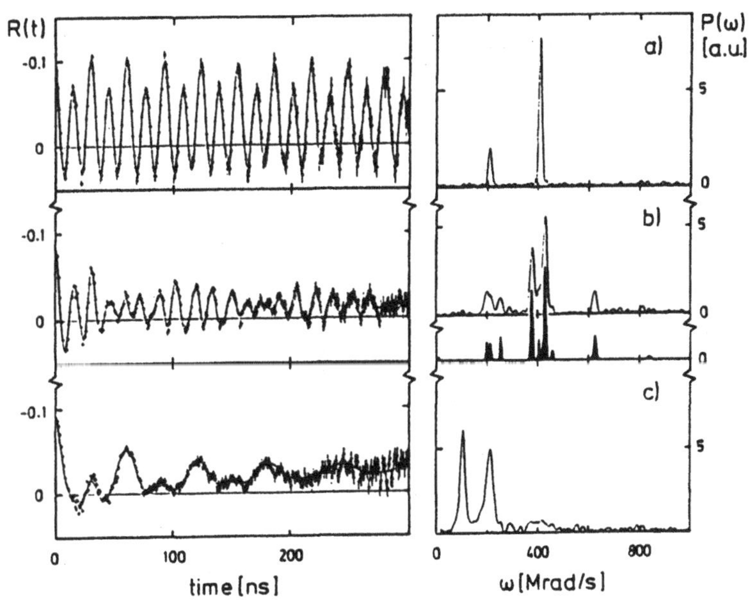

Fig. 8: PAC spectra and corresponding Fourier transforms for ^{111}In probes on Ni(111) at different measurement temperatures: a) above the Curie temperature at T_M=632 K, b) below the Curie temperature at T_M=300 K, and c) at 100 K and after coverage of 2.5 monolayers nickel. In b) the Fourier transform of the fitted function (shaded curve) is added for comparison.

Below the Curie point the ^{111}In probes are expected to be exposed to an electric field gradient as well as to a magnetic field. In Fig. 8b a PAC spectrum (left-hand side) recorded at 300 K is depicted. The Fourier transform of the data (right-hand side) reveals a number of transition frequencies, which can unambiguously be related to a single electromagnetic hyperfine field situation. In order to illustrate this agreement Fig. 8b comprises a Fourier transform of the respective fitted function in addition to the Fourier transforms of the data. The extracted electric field gradient has already been discussed above (see Fig. 9a). The strength of the magnetic hyperfine field B_{hf} measured as a function of temperature is shown in Fig. 9b. For comparison, experimental values of B_{hf} at ^{111}In on substitutional lattice sites in Ni bulk are enclosed.

The orientation of the surface magnetic hyperfine field B_{hf} is found to be temperature dependent. Above 300 K the direction of the field is lying within the (111) surface plane. Since no external magnetic field was applied, no preferred in-plane orientation of the hyperfine field can be expected. Below 300 K the magnetic field tends to be tilted out of plane; at 100 K a tilt angle of 3.5(5)° is observed.

In order to obtain the hyperfine field near the Ni surface, the ^{111}In-labeled Ni(111) surface was covered with an additional 2.5 nickel monolayers, which was controlled by a quartz-oscillator system. The result is shown in Fig. 8c. In this case the electric field gradient is expected to vanish due to the cubic symmetry of nearest- and next-nearest neighbours around the ^{111}In probes (see chapter 3.2). Therefore only the magnetic hyperfine field is present, which can be recognized in the Fourier transform as the Larmor frequency ω_L = 100(1) Mrad/s and $2\omega_L$. The temperature dependence of this magnetic hyperfine field is included in Fig. 9b. Irrespective of minor deviations this dependency mainly reflects bulk behaviour.

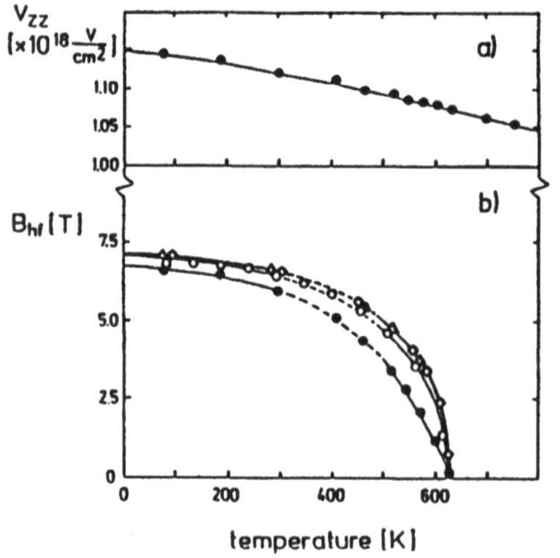

Fig. 9: Temperature dependence of the a) electric field gradient and b) the magnetic hyperfine field at ^{111}In in the topmost layer of Ni(111) surfaces (solid circles), in the 3.5th monolayer (open circles), and in nickel bulk (open sqares: present work, open triangles: Ref./40/). The solid lines up to 300 K are fits according to the expression B_{hf}=-$B_{hf}(T=0)(1 - bT^{3/2})$; the solid lines above 500 K up to the Curie point represent fits with the function $B_{hf} \alpha (T_C - T)^{\beta}$.

These PAC results for the magnetic hyperfine field at a Ni(111) surface reveal remarkable features. The surface hyperfine field is smaller compared to the bulk value over the entire temperature range up to the Curie point. In order to analyze the temperature dependence quantitatively, a $T^{3/2}$ law suggestive of spin-wave excitations was fitted to the data using $B_{hf} = B_{hf}(T=0)\cdot(1 - bT^{3/2})$. The agreement for both surface and bulk magnetic hyperfine fields up to temperatures of about 300 K is reasonable; however an exponent of 5/2 would improve the quality of the fits.

For the parameter b, a value of $b_{surf} = 2.4(2)\cdot 10^{-5}\ K^{-3/2}$ is obtained for the surface hyperfine field, which is about 50 % higher than the value $b_{bulk} = 1.6(2)\cdot 10^{-5}\ K^{-3/2}$ derived for ^{111}In in the 3.5th monolayer and in bulk. A similar effect was detected by Korecki and Gradmann in comparing the magnetic hyperfine field at ^{57}Fe for Ag-covered Fe(110) surfaces with Fe thin films /4/, which is discussed in terms of a loss of exchange bonds across the surface leading to standing spin waves /24/. On the other hand, for Ni(111) Rau and Kuffner /25/ detected a rather linear dependence of the electron polarization in the range from 300 K up to the Curie temperature as determined by electron-capture spectroscopy. Mean-field calculations by Kisker /26/ for the magnetization of the Ni(111) surface monolayer agree fairly well with the electron-capture-spectroscopy data. However, they do significantly deviate in the case of the bulk magnetization, which in its turn scales as B_{hf} found for ^{111}In in Ni bulk.

The surface magnetic hyperfine field, extrapolated to T = 0 K employing the $T^{3/2}$ dependence, turns out to be reduced by about 7 % [$B_{hf}(T=0) = 6.7(1)$ Tesla] as compared to the bulk value [$B_{hf}(T=0) = 7.2(1)$ Tesla]. This effect, actually a reduction of 15 %, is predicted by Freeman et al. /27/ for B_{hf} at nickel probes on Ni(100) surfaces. This is qualitatively understood in terms of a considerable change of the conduction-electron density at the surface, which leads to a diminished Fermi contact field at the probe nuclei. Experimentally, this was also noticed for Fe probes on Fe(110), where a reduction of about 5 % was identified /4/.

Finally, it is instructive to inspect the behaviour of B_{hf} close to the Curie point. Within experimental errors no indication for a shift of the Curie temperature for ^{111}In in the topmost layer is given. Analysis of the B_{hf} behaviour in the temperature interval from 500 K up to the Curie point exhibits an experimental dependence well described by the commonly used function $B_{hf} \propto (T_C - T)^{\beta}$, with β being the critical exponent for temperatures close to the Curie temperature. For ^{111}In probes in the surface we find $\beta_{surf} = 0.70(9)$; for bulk $\beta_{bulk} = 0.37(5)$ is extracted, which agrees with $\beta_{3.5ML} = 0.40(5)$ obtained for ^{111}In in the 3.5th monolayer. Although no β exponent for Ni(111) surfaces has yet been published, a comparison with previously obtained β values for Ni(100) and Ni(110) surfaces is quite informative. For Ni(100) surfaces, SPLEED experiments /28/ reveal $\beta_{surf} = 0.79(2)$; in case of Ni(110) the critical exponent is found to be $\beta_{surf} = 0.77(2)$ utilizing SPLEED /29/ and $\beta_{surf} = 0.70(10)$ measuring positronium decay /30/. On the other hand, for Ni bulk a value of $\beta_{bulk} = 0.358(3)$ was determined earlier /31,32/. Our results are in excellent agreement with all the results quoted above.

The nickel example demonstrates that with the PAC method the magnetic hyperfine field can be measured with high accuracy over the entire interesting temperature range. The location of the PAC probes is well determined through the measured properties of the electric-field-gradient tensor. The magnetic hyperfine field can be distinguished for probes dwelling at the surface as well as in near-surface layers and finally in bulk. This permits the study of the hyperfine parameters as a function of depth with monolayer resolution. The labeling of the system with only about 10^{-4} monolayers PAC atoms opens up prospects to apply PAC to many other magnetic surfaces as well as to ultrathin magnetic layers and interfaces /33/.

5. PAC investigations of Ni/In interfaces

The investigation of interdiffusion and compound formation at metal-metal and metal-semiconductor interfaces is an interesting topic and has been subject of numerous experimental studies, particularly because of its technical importance for example for the production of microelectronic devices. Interdiffusion may strongly influence the characteristics of these devices concerning mechanical as well as electrical properties. Especially when ohmic contacts are added to compound semiconductors, interdiffusion might change drastically the behaviour of the microcircuits. When Ni, for instance, is evaporated to InP, a phase separation into a Ni_2P layer near the substrate and into a NiIn layer near the surface is observed after annealing above 532 K, which weakens the stability of the contact /34/. On the other hand there are examples that interdiffusion of two materials might enhance thermal stability. This effect is observed for NiInW contacts on n-type GaAs, where a Ni_2GaAs layer is formed at the GaAs surface which suppresses the diffusion of In into the GaAs substrate. Thus donator compensation in the semiconductor is prevented and contacts with large fractional coverage x of $In_xGa_{(1-x)}As$ can be achieved, forming low resistance ohmic contacts. Also, excess In is found to form high melting point Ni_3In compounds which additionally improves thermal stability /35/.

The change of properties at interfaces due to diffusion processes in mind, we have studied the interdiffusion in the system Ni/In by utilizing the PAC technique, which permits to study interdiffusion as well as interface compound formation on a microscopic level via detection of electric field gradients at ^{111}In probes. Non-cubic, ordered compounds can be uniquely characterized and thus detected at a microscopic sensitivity by these electric field gradients. It should be mentioned that the electric field gradient is measured at ^{111}Cd, however, the actual position i.e. lattice site is determined by the parent probe atoms ^{111}In. Thus, for the process studied here, the probe atom is chemically identical with one constituent of the investigated material couples.

In a first experiment an In film 500 monolayers thick, homogeneously doped with the PAC probes ^{111}In, was evaporated on to a small sapphire plate at 100 K. Subsequently, at 77 K it was covered by a 500 monolayers thick Ni film. The atomic ratio Ni/In between the polycrystalline layers was about 2:1. Finally, the sample was subjected to an isochronal annealing program (t_A = 10 min) within a temperature range from 150 K to 900 K. PAC spectra where recorded well below the respective annealing temperatures.

As a result, as a function of annealing temperature the occurrence of six well-defined electric-field-gradient situations is observed, indicating that our ^{111}In probes occupy six well-defined substitutional lattice sites. The first fraction, that means percentage of probe nuclei exposed to a certain electric-field-gradient situation, of course, which occurs directly after the sample preparation, is related to ^{111}In probes in the pure In film. The other five fractions have to be related to stoichiometric Ni/In compounds. In order to clear up those different electric-field-gradient situations, that is to identify formed compounds, macroscopic samples were prepared with well-known Ni/In stoichiometries. After PAC measurement these samples were pulverized and their structure was determined by Debye-Scherrer X-ray analysis. Thus all but one compound (Ni_xIn_y) could be identified. We find the alloys Ni_2In_3, NiIn and Ni_3In, which have been investigated with PAC before /36/ and $Ni_{10}In_{27}$, all of them known from the bulk phase diagram.

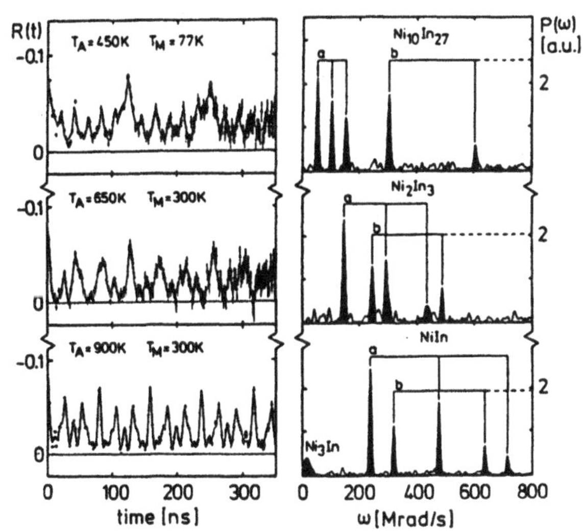

Fig. 10: PAC spectra and corresponding Fourier transforms for a (500 monolayers In / 500 monolayers Ni) thin film couple at different annealing temperatures T_A and measurement temperatures T_M. Clearly several Ni/In compounds are characterized. For $Ni_{10}In_{27}$, Ni_2In_3 and NiIn frequency triplets (a) and (b) are observed due to two different lattice sites for In.

Fig. 10 shows the PAC results for the thin film sample consistent of 500 monolayers In and 500 monolayers Ni at some annealing steps. Counting rate ratios R(t) of γγ-coincidence spectra for different γ-detectors are plotted on the left side and corresponding Fourier transforms are shown on the right hand side. The Fourier transforms clearly exhibit well-defined frequency triplets which correspond to certain electric field gradients at our ^{111}In probes. The fact that for some compounds two different electric field gradients are measured is easily explained by two possible non-equivalent lattice sites for In in these compounds.

PAC spectra and Fourier analysis of a 100 monolayers In / 50 monolayers Ni thin film couple is shown in Fig. 11, representing the compound Ni_xIn_y, whose stoichiometry could not be identified. Again we find two non-equivalent lattice sites for In in this phase. The new compound is characterized by the electric-field-gradient parameters $|V_{zz}{}^a| = 7.1(15)\cdot 10^{17}$ V/cm^2 and $|V_{zz}{}^b| = 11.0(15)\cdot 10^{17}$ V/cm^2 with asymmetry parameters $\eta^a = 0.83$ and $\eta^b = 0.57$, respectively. There is no hint from the Ni/In phase diagram for the stoichiometry of this alloy, and the fact that it was not possible to reproduce it as a macroscopic sample might indicate that it is stable only in thin film systems. Investigations by quantitative Auger analysis suggest, that the Ni content in this compound is below 27 atomic %.

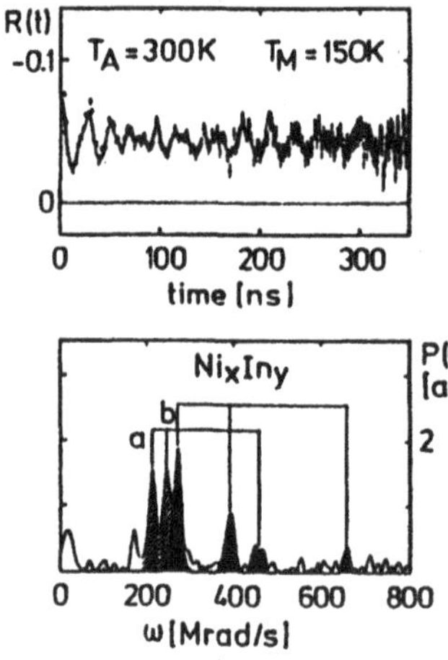

Fig. 11: PAC spectrum and Fourier transform of a (100 monolayers In / 50 monolayers Ni) thin film couple at annealing to T_A = 300 K and measuring at T_M = 150 K. The spectra show that the majority of ^{111}In probe atoms experience well-defined electric field gradients. This is due to a new compound Ni_xIn_y, with a composition that could be estimated to a Ni content lower than 27 atom %. Again there are two possible lattice sites for In in this compound.

After this first reference experiment investigations were done on In films of several thicknesses in the monolayer range using the low index Ni surfaces Ni(111), Ni(100) and Ni(110) as substrates. The samples were prepared by depositing ^{111}In probe atoms in the usual low concentration on the clean Ni surfaces at 77 K. Then the ^{111}In labeled crystals were covered by In films at the same temperature. The probe atoms are located right at the Ni/In interface now. Analysis by low-energy electron diffraction shows a pseudomorphic growth of In in [111] main direction on all investigated substrate orientations and independent on In film thickness above 1 up to 20 monolayers. These samples were subjected to an isochronal annealing program in the same temperature range as before.

For two representative Ni surfaces results are shown in Fig. 12, the Ni/In thin film reference experiment described in the beginning is also included. At 180 K about half of our probes at the interface become mobile and diffuse into the In layer, which can easily be detected by PAC. On further annealing Ni atoms become mobile and diffuse into the In layer, which leads to the formation of several interface compounds. From tracer diffusion data it is known that the activation energies for diffusion in In are generally much smaller than those for diffusion in Ni /37/. Therefore, we conclude that Ni is the diffusor responsible for compound formation.

We find a strong dependence of the compound formation on the film thickness. At thicker In layers all compounds known from the bulk phase diagram are observed, for thinner In films the In richer alloys are overrun. At the extreme coverage of 1 monolayer the Ni-richest compound Ni_3In is observed already above 440 K. There are slight differences between different orientations of the Ni single crystal substrates probably due to the different surface densities of the Ni-atoms and the slightly different (111) textural growth of In on these surfaces, which might influence the migration of Ni through the interface.

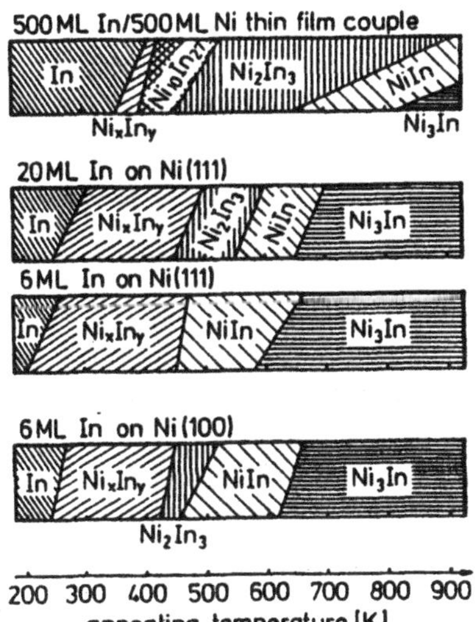

Fig. 12: Typical results for the annealing behaviour of the samples. As a function of In film thickness we find different courses of the compound formation.

As a main result for the Ni/In interface experiments we find, that the phase diagram for these metals has been run through from the In rich side to the Ni rich side. Interface compounds are formed by diffusion of Ni through the interface and incorporation in the In film or already formed compounds, respectively. Thus it is possible to observe a coexistence of different compounds at certain annealing temperatures, while the measured PAC signals for each alloy are unique.

These findings are in excellent agreement with the Walser-Bené rule /38/ as well as the "effective heat of formation" rule by Pretorius /39/ formulated for low temperature interdiffusion and phase formation at metal-metal thin film systems. Both rules predict that the In rich phases in the Ni/In system are formed first, because the lowest melting eutectic is on the In side of the phase diagram.

In addition we found a new Ni/In compound Ni_xIn_y with a Ni content lower than 27 atom %, which is not known from the phase diagram and therefore might not be stable as a bulk alloy.

Like in former experiments on different interface systems the PAC method has shown to be a powerful complementary tool to investigate interdiffusion and the appearance of stoichiometric compounds in Ni/In thin film systems. It is non-destructive and allows to study processes at interfaces without limitation in film thickness at a high depth resolution. These features are planned to be applied to the field of ohmic contacts on semiconductors in further experiments.

6. Acknowledgement

The continous support of the presented work by the Deutsche Forschungsgemeinschaft (SFB 306) is gratefully acknowledged.

7. References

/1/ B. Horn, E. Koch, and D. Fick; Phys. Rev. Lett. **53** (1984) 364

/2/ K. Wassmuth and D. Fick; Phys. Rev. Lett. **59** (1987) 3007

/3/ J. Korecki and U. Gradmann; Phys. Rev. Lett. **55** (1985) 2491

/4/ J. Korecki and U. Gradmann; Europhys. Lett. **2** (1986) 651

/5/ M. Przybylski and U. Gradmann; Phys. Rev. Lett. **59** (1987) 1152

/6/ W. Körner, W. Keppner, B. Lehndorff-Junges, and G. Schatz; Phys. Rev. Lett. **49** (1982) 1735

/7/ W. Keppner, T. Klas, W. Körner, R. Wesche, and G. Schatz; Phys. Rev. Lett. **54** (1985) 2371

/8/ T. Klas, J. Voigt, W. Keppner, R. Wesche, and G. Schatz; Phys. Rev. Lett. **57** (1986) 1068

/9/ E. Hunger and H. Haas; Surf. Sci. **234** (1990) 273

/10/ H. Frauenfelder and R.M. Steffen, in *Alpha-, Beta-, and Gamma-Ray Spectroscopy*, ed. K. Siegbahn (North-Holland, Amsterdam, 1965), p. 997

/11/ T. Klas, R. Fink, G. Krausch, R. Platzer, J. Voigt, R. Wesche, and G. Schatz; Surf. Sci. **216** (1989) 270

/12/ G. Schatz, T. Klas, R. Platzer, J. Voigt, and R. Wesche; Hyperfine Interactions **34** (1987) 555

/13/ B. Lindgren; Hyperfine Interactions **34** (1987) 217

/14/ B. Lindgren; Europhys. Lett. **11** (1990) 555

/15/ R. Wesche, R. Fink, T. Klas, G. Krausch, R. Platzer, J. Voigt, and G. Schatz; J. Phys.: Condens. Matter **1** (1989) 7407

/16/ R. Fink, R. Wesche, T. Klas, G. Krausch, R. Platzer, J. Voigt, U. Wöhrmann, and G. Schatz; Surf. Sci. **225** (1990) 331

/17/ Th. Wichert, M. Deicher, O. Echt, and E. Recknagel; Phys. Rev. Lett. **41** (1978) 1659

/18/ T. Klas, J. Voigt, W. Keppner, R. Platzer, R. Wesche, and G. Schatz; Hyperfine Interactions **34** (1987) 577

/19/ R. Blender, W. Dieterich, H.L. Frisch, and M. Yussouff; Z. Phys. **B71** (1988) 103

/20/ A. Natori, M. Baba, and N. Maruyama; Surf. Sci. 233 (1990) 392

/21/ K. Froböse; J. Phys.: Condens. Matter **2** (1990) 5933

/22/ T. Klas, R. Fink, G. Krausch, R. Platzer, J. Voigt, R. Wesche, and G. Schatz; Europhys. Lett. **7** (1988) 151

/23/ J. Voigt, R. Fink, G. Krausch, B. Luckscheiter, R. Platzer, U. Wöhrmann, X.L. Ding, and G. Schatz; Phys. Rev. Lett. 64 (1990) 2202

/24/ G.T. Rado and J.C. Walker; J. Appl. Phys. **53** (1982) 8055

/25/ C. Rau and H. Kuffner; J. Magn. Magn. Mater. **54 -57** (1986) 767

/26/ E. Kisker; J. Phys. Chem. **87** (1983) 3597

/27/ A.J. Freeman, C.L. Fu, M. Weinert, and S. Ohnishi; Hyperfine Interactions **33** (1987) 53

/28/ S. Alvarado, M. Campagna, and H. Hopster; Phys. Rev. Lett. **48** (1982) 51

/29/ S. Alvarado, M. Campagna, F. Ciccacci, and H. Hopster; J. Appl. Phys. **53** (1982) 7920

/30/ D.W. Gidley, A.R. Köymen, and T. Weston Capehart; Phys. Rev. Lett. 49 (1982) 1779

/31/ J.D. Cohen and T.R. Carver; Phys. Rev. **B15** (1977) 5350

/32/ H.C. Benski, R.C. Reno, C. Hohenemser, R. Lyons, and C. Abeledo; Phys. Rev. **B6** (1972) 4266

/33/ J. Voigt, X.L. Ding, R. Fink, G. Krausch, B. Luckscheiter, R. Platzer, U. Wöhrmann, and G. Schatz; Appl. Phys. **A51** (1990) 317

/34/ A. Appelbaum, P.M. Thomas, and P.A. Barnes; Proceedings of *Semiconductor-Based Heterostructures, Interfacial Structure and Stability*, Warrendale, PA, USA, (1986) p.406

/35/ Y.C. Shih, M. Muracami,and W.H. Price; J. Appl. Phys. **65** (1989) 3539

/36/ S.A. Lis, R.A. Naumann, and G. Schmidt; Hyperfine Interactions **5** (1978) 445

/37/ C. Weast; *Handbook of Chemistry and Physics*, (CRC Press Inc., Cleveland Ohio, 1990)

/38/ R.W. Bené; Appl. Phys. Lett. **28** (1982) 624

/39/ R. Pretorius, R. de Reus, A.M. Vredenberg, and F.W. Saris; Materials Lett. **9** (1990) 494

/40/ C. Hohenemser, A.R. Arends, H. De Waard, H.G. Devare, F. Pleiter, and S.A. Drentje; Hyperfine Interactions 3 (1977) 297

CEMS STUDIES OF SURFACE MAGNETISM

J.Korecki, M.Przybylski

Solid State Physics Department, Academy of Mining and Metallurgy
30-059 Kraków, Poland

ABSTRACT

Conversion Electron Mössbauer Spectroscopy is applied for ultra-thin films studies. Structural and magnetic properties of single crystalline samples, prepared and characterized *in situ*, could be analyzed with the monolayer resolution. In-plane surface anisotropy, hyperfine magnetic fields near Fe(110) free surface and thermal excitations in ultra-thin films, including monolayer, are discussed.

1. INTRODUCTION

There are two important conditions of the reliable experiment on magnetic surfaces, interfaces and thin films. First of all, real investigated objects have to reproduce the perfect crystal structure taken for theoretical calculation. It is a very difficult task and even clean surfaces, prepared under UHV conditions, suffer from inhomogeneities, pseudomorphic strains or island structure. Secondly, an applied experimental method must present high surface sensitivity and resolution. Only recently this two conditions have been fulfilled by application of different *in situ* techniques, like polarized electron spectroscopies[1)], refined classical methods (magnetometry)[2)] and nuclear methods reviewed in this volume[3)].

In this paper the Conversion Electron Mössbauer Spectroscopy (CEMS) studies of ultra-thin iron films are presented. The local character of CEMS and its selective possibilities (by a use of ^{57}Fe atoms as the probes of hyperfine fields) profit in analysis of the film structure and of the primary magnetic phenomena like magnetic surface anisotropy and magnetic surface and size effects.

2. CEMS

Mössbauer spectroscopy (MS) is the method of a local character. It measures the hyperfine interaction parameters intensively. The hyperfine interactions reflect the local electronic structure to a defined set of the nuclear energy levels which are revealed by the energy of the emitted γ-quanta. With a certain resolution given by the natural width of the nuclear levels (the life-time of the Mössbauer level) we can decompose the Mössbauer spectrum to discrete components. Each of them is characterized by isomer shift, quadrupole interaction and hyperfine magnetic field and corresponds to a specific atomic arrangement. For example for an ideal film consisting of ten atomic layers the spectrum should consist of ten components. It can be difficult to resolve all of them. Generally it can be expected that the electronic structure of surface atoms and that of the film interior differs enough to observe in the Mössbauer spectrum at least two components, if only the detection limit of the Mössbauer measurement is sufficient to get the signal from a single atomic layer.

An additional advantage of the MS results from the fact that this is an element and even isotope specific method. This allows to avoid a background signal from the substrate or coating. Among the experimental methods used to study surface and interface magnetism MS plays a special role. Frequency of its application can be compared only with various electron spectroscopies which are experimentally much more complicated and limited to the *in situ* investigations of uncovered surfaces. The isotope peculiarity mentioned above causes that the surface and interface application of MS is limited nearly exclusively to ^{57}Fe.

A special role of MS is due to the fact that it provides a direct measurement of the magnetic quantity such as the hyperfine magnetic field. This is the only experimental magnetic data, which can be compared directly with the recent theoretical calculations for surfaces and interfaces[4)]. It is rather difficult for magnetization measurements because of its integral character. Although it is possible to measure directly the value of the magnetic moment of the atoms in the monolayer[5)], but already the magnetization for the double layer results in the averaging of two possibly different values of the layer magnetic moments.

For surface and ultra-thin film studies the minimum desired sensitivity and resolution is one atomic layer. These conditions can be fullfiled by using CEMS. CEMS is a scattering technique. After the deexcitation of the Mössbauer nuclei different monoenergetic conversion and Auger electrons are emitted (7.3keV K-conversion electrons dominate in the case of ^{57}Fe). The energy at which these electrons emerge from the surface depends on their origin depth. The mean free path of electrons in solids ranges about $1.5K^{1/2}$nm, where K is the electron kinetic energy in keV. Because the energy loss per one collision is of the order of 20-30eV, this yields the electron escape depth of about 100nm. This number determines the information depth of the integral CEMS experiment in which all electrons emerging from the surface are detected.

The very low detection limit of CEMS for ^{57}Fe is determined by the ratio of the number of electrons reemitted in resonance way (conversion and Auger electrons) and non-resonance way (mainly 7.3keV photoelectrons). This corresponds to the ratio of the cross-section for the Mössbauer absorption and photoeffect: $\sigma_{Moss}/\sigma_{photo} = 3 * 10^2$. In the thin absorber approximation the resonant effect

$$\eta = \frac{N(0)-N(\infty)}{N(\infty)} \tag{2.1}$$

where N(0) and N(∞) are the counting rates at the resonance velocity and far-off resonance, respectively, can be estimated from simplified formulae (single line spectra):

$$\eta_{CEMS} = \frac{1}{2} f_s f_a \frac{\alpha}{\alpha+1} \frac{\sigma_{Moss}}{\sigma_{photo}} a \tag{2.2}$$

$$\frac{\eta_{CEMS}}{\eta_T} = \frac{\alpha}{\alpha+1} \frac{1}{\sigma_{photo}\, n} \tag{2.3}$$

Subscript "T" refers to the transmission geometry, f_s and f_a are the recoil-free factors for the source and absorber respectively, α is the internal conversion coefficient, a is the abundance of resonant isotope and n is the surface density of atoms in the absorber. The formulae (2.2) and (2.3) hold if only 7.3keV photoelectrons contribute to the nonresonant background. In practice of thin film investigations it is impossible to deal with unsupported samples. Consequently, the nonresonant background from a substrate reduces the signal-to-noise ratio. Nevertheless the formulae (2.2) and (2.3) clearly demonstrate the advantage of CEMS technique if applied for ultra-thin samples. Huge signal-to-noise ratios of 1000% for thin (400Å) ^{57}Fe absorbers and a detection limit of 10^{13} -10^{14} atoms/cm^2 (1/100 - 1/10 of a monolayer)[6)] can be achieved.

The integral CEMS measurements can easily be performed by embedding the sample in a gas-filled proportional counter[7)]. This method has the advantage of a high detection efficiency due to the large solid angle but it does not allow to study clean surfaces. The other important drawback, which is the limitation of the sample temperature around 300K due to the position of the sample inside the counting gas, has been partly removed by successful attempts to operate such a detector at low temperatures[8]. However, the better effects can be achieved by using channel electron multipliers i.e. channeltrons[9]. Although the detection efficiency is one order of magnitude worse, channeltrons operate in vacuum, what favours free-standing surface studies and the implementation of low temperatures[9].

The surface sensitivity of the integral CEMS can be considerably improved by performing the analysis of the conversion energy electrons, because their energy loss is related to the depth at which they are originated. In the Depth-Selective CEMS (DS-CEMS) electrons are detected by means of an electron spectrometer of several percent energy resolution. Mössbauer spectra are measured at various selected electron energies and possibly at various angle at which electrons are emitted from the sample. Usually DS-CEMS measurements are performed in the energy range between 6.6 - 7.3 keV which is the interval of almost exclusively K-conversion electrons. The partial contributions due to electrons generated at different depths result in the total Mössbauer spectrum measured at the given spectrometer energy setting. By the numerical analysis the total spectrum can be deconvoluted to contributions corresponding to the different distances from the surface, consequently giving certain depth resolution[10]. In comparison to the integral CEMS the depth resolution of DS-CEMS is by one order of magnitude better. It is still not enough for the probing of the surface electronic and magnetic properties for which the disturbance connected with the lack of translational symmetry ranges only few atomic layers.

The local Mössbauer analysis can be realized also in the experimentally simpler way: in a thin film, prepared of the non-Mössbauer ^{56}Fe isotope, at a certain distance from the surface, a probe layer (one or several atomic layers) of ^{57}Fe can be placed. By varying the position of the probe layer the hyperfine interaction parameters can be profiled.

An optimum solution of the experimental set-up for Mössbauer studies of well defined, single crystalline magnetic films, their surfaces and interfaces must fulfill the following requirements:

(i) ultra high vacuum performance (base pressure 10^{-8} Pa),

(ii) preparation of atomically clean, monocrystalline surfaces (films) characterized in situ by surface sensitive methods,

(iii) wide temperature range of the Mössbauer measurements and the monolayer detection sensitivity reached in a reasonable time (few hours) to avoid the contamination of the sample surface by adsorbed atoms of residual gases.

The monolayer sensitivity can be easily achieved by simple integral CEMS using a strong source (100mCi) and a channeltron as the detector if a typical non-resonant background comes from the substrate (sample holder). However, as will be described below, the tungsten substrate was used for the present studies. This increased dramatically the production of the non-resonant electron background due to photo, Auger and Compton electrons from tungsten. In this case a use of the electron analyser as an energy filter could bring a meaningful improvement in the signal-to-noise ratio and the reduction of the measuring time.

The ideas presented above are summarized in Fig.1. The corresponding UHV apparatus, combining surface technology (Molecular Beam Epitaxy - MBE) and surface sensitive techniques (Low Energy Electron Diffraction - LEED and Auger Electron Spectroscopy - AES) with CEMS becomes the standart tool in surface magnetism[11]. The methodology is described below for ultra-thin Fe(110) films on W(110), prepared with atomic cleaness under UHV conditions and characterized *in situ*.

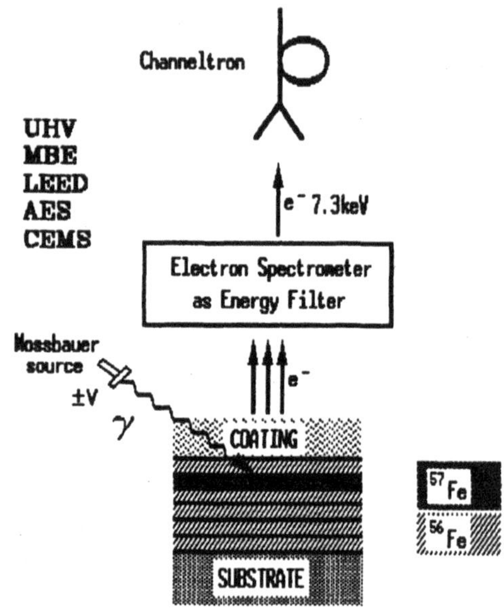

Figure 1. *The idea of the in situ ^{57}Fe monolayer probe CEMS experiment*

3. SAMPLES - PREPARATION AND STRUCTURE

Atomically flat monocrystalline ultra thin films of Fe(110) on W(110)[12,13] realize structure which is very close to ideal one, being used as a model for theoretical calculations. The epitaxial growth is possible due to the sufficiently small misfit of the Fe(110) and W(110) lattice spacing (f_{Fe-W} = -9.4%). Moreover, the wetting conditions of tungsten by iron are fulfilled (the surface free energy of the substrate $\gamma_W = 2.9$ J m^{-2} is larger than that of the growing iron films $\gamma_{Fe} = 2.0$ J m^{-2}). It means that the films can accommodate to the substrate and that they can grow layer-by-layer. Indeed, this type of the growth mode is observed experimentally. The growth mode depends strongly on temperature. For example a pseudomorphic double layer growth followed by the layer-by-layer growth of the iron films with periodic lattice distortion due to an interaction with the misfitting tungsten substrate is observed at 500K. The distortions fade out at about 10ML and the LEED patterns of thicker iron films correspond to the perfect Fe(110) face.

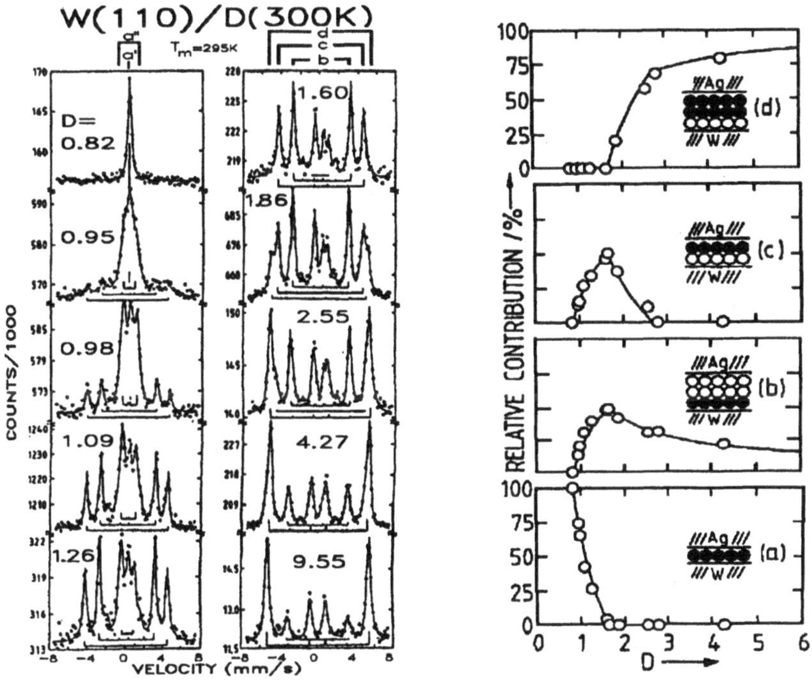

Figure 2. *(a) Mössbauer spectra of Fe(110) films consisting of D bulk atomic layers of pure ^{57}Fe, prepared on W(110). They were fitted by one single line (a') and four sextets [(a''), (b), (c) and (d)][13].*

(b) Relative contributions of components a = a' + a'', b, c, and d'. Theoretical predictions from layer-by-layer growth model, starting with two pseudomorfic monolayers, followed by further layers with bulk density, are given by solid lines[13].

In spite of extensive LEED and AES studies the detailed structure and the mode of growth for the few first atomic layers remained controversial. The key information came from CEMS measurements[13]. Selected spectra of pure ^{57}Fe films of different thickness, prepared on W(110) at T_{Fe} = 300 K, coated with Ag at T_{Ag} = 300 K and measured at T_m = 295 K, are presented in Fig.2a. Some of the measured spectra are rather complicated but all could be fitted with four structural components. These components can be attributed to structural environments assuming that the hyperfine interaction parameters differ for different atomic layers and are thus sensitive to the film structure.

The first structural component (a) considered in the spectra of the thinnest films, a single line and possible Zeeman sextet (B_{hf} ~ 4 Tesla), is assumed to be due to monolayer patches. The splitting into a non-magnetic single line (a') and a magnetic component (a'') is a special critical phenomenon of the Fe(110) monolayer on W(110), connected with the ferromagnetic coupling between atoms in the monolayer patches and in the double-layer patches. It is evident that a film thicker than one monolayer should contain monolayer and double-layer patches. The microscopic local character of CEMS allows us to see them separately. The room temperature Mössbauer spectra for D = 0.95, 0.98 and 1.09 consist of non-magnetic and magnetic components in the central part of the spectra (a' and a'', respectively) and of two additional magnetic components (b) and (c). Components b (B_{hf} ~ 19 Tesla) and c (B_{hf} ~ 28 Tesla) appear in films only slightly thicker than one monolayer and their intensity increase with increasing thickness. The relative contribution of both components is almost the same. Therefore, we propose a simple interpretation that these two components belong to the W and Ag interfaces of the double-layer patches, respectively. For thicker films (D ≥ 1.64) the relative contribution of component (c), (second atomic layer in contact with Ag) decreases, indicating an increasing number of atoms in the third atomic layer. A new component (d) in Mössbauer spectra appears. It represents all atoms beyond the first monolayer on W(110). Of course, the value of the magnetic hyperfine field of this component, which has almost bulk value (B_{hf} ~ 33 Tesla), represents an integral effect and neglects the structure of B_{hf} near the Fe/Ag interface. Sub-components could not be resolved by pure ^{57}Fe film analysis, because B_{hf} near the Ag interface varies from layer to layer only weakly[14)].

Justification for the structural interpretation of the spectra is given by the relative contributions p of all components a, b, c and d as the function of the film thickness D which can be followed quantitatively in Fig.2b. The experimental values agree perfectly with the relative contributions p derived from the theoretically predicted model of growth, in which the film starts by one pseudomorphic monolayer, followed by a second pseudomorphic monolayer and then further layers of the bulk density (full line in Fig.2b). Thus the layer-by layer growth mode at the given preparation condition is confirmed.

The film thickness of about 100Å was the upper limit of the investigated films. Films in this range are called ultra-thin and their Mössbauer measurements reveal features different from those found for bulk iron, as described below.

4. MAGNETIC PROPERTIES OF IRON SURFACE AND ULTRA-THIN FILMS

For really perfect samples the surface modification of magnetic properties penetrates surprisingly shallow. The ^{57}Fe films thicker than 100Å do not differ from bulk when measured using CEMS. In this case the films simulate quite well the semi-infinite solid. The pure surface effects (surface modification of the electronic and magnetic properties induced by the loss of the nearest neighbours) can be followed using ^{57}Fe probe layer situated near the surface of the ^{56}Fe film. When the film becomes thinner the quantum size effects become important and a transition from 3-dimensional to 2-dimensional behaviour is displayed.

4.1. In-plane Surface Anisotropy of Fe(110) Films

The geometry of the CEMS experiment is shown in Fig.3. The Fe(110) film is irradiated at a grazing angle of 15^0 in the $[\bar{1}10]$-azimuth. $\vec{J}_s$ is the spontaneous magnetization, which for the bulk iron lies in the easy [100]-type direction. The relative intensities I_i of the six line Zeeman spectrum for the purely magnetic interaction are connected with the angle α between $\vec{J}_s$ and the incident γ-ray by:

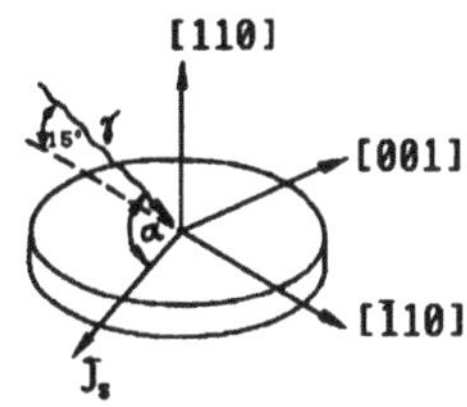

Figure 3. *Geometry of CEMS experiment in respect to crystal orientation*

$$
\begin{aligned}
I_1 &= I_6 = 3(1+\cos^2\alpha) \\
I_2 &= I_5 = 4\sin^2\alpha \\
I_3 &= I_4 = 1+\cos^2\alpha
\end{aligned}
\tag{4.1}
$$

The CEMS spectrum of the 105Å $^{57}Fe(110)$ film on W(110) (Fig.4) shows nearly bulk-like character as concerned both the hyperfine magnetic field B_{hf} and relative line intensities (3:4:1, $\alpha=\pi/2$) which reflect the [001] magnetization direction. As the films thickness decreases below 100Å the magnetization switches spontaneously to a [110] type direction as can be deduced from the CEMS spectrum showed in Fig.4 for a 17Å film. This phenomenon can be interpreted in terms of the surface magnetic anisotropy which is induced by the symmetry breaking at the surface or interface of the film. It is well established that for Fe(110) films grown on W(110) the uniaxial surface magnetic anisotropy forces the magnetization into the film plane (the film normal is the hard axis). The two-fold symmetry of Fe(110) surface gives rise to an additional anisotropy, the in-plane surface anisotropy. Since in the case of Fe(110) this anisotropy acts perpendicular to the bulk anisotropy, at the critical thickness d_c the Fe(110) film becomes spontaneously magnetized in the [110] direction. The critical thickness d_c relates the surface in-plane anisotropy constant $K_{s,p}$ and the bulk anisotropy constant K_1 of iron[15]:

$$d_c = \frac{8K_{s,p}}{K_1} \qquad (4.2)$$

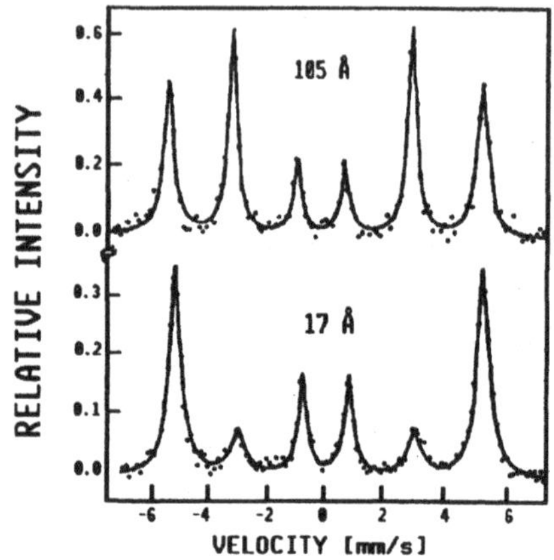

Figure 4. *Mössbauer sextets for Fe(110) films on W(110)*

The absolute determination of the $K_{s,p}$ constant is rather difficult, because the formula (4.2) assumes the symmetric conditions at both surfaces (interfaces) of the film. Nevertheless, the competition between in-plane surface and bulk magnetic anisotropies is straightforward for film thicknesses of the order of 80Å, depending on the surface cleanness, the film coating or the history of the thermal treatment of the film. This implies another factors (other than these of pure symmetry character) determining the effective films anisotropy, such as strains, roughnesses and obviously electronic interactions.

4.2 Magnetic Field Near The Surface

The symmetry of a surface atom is lower than in bulk, because the translational order in one direction is broken. Hence, the surface atoms have fewer neighbours than bulk atoms (lower coordination). The lower coordination has the obvious consequence for the electronic band structure. As usually, for a system of lower dimensionality the wave functions become more atomic-like. Also the *3d*-band, which in *3d*-metals responds for magnetic properties, becomes narrower. In other words the surface states are more localized than bulk ones. The question arises, how it influences the magnetic properties. In the frame of Stoner-Wolfarth itinerant model of ferromagnetism we expect a *3d*-metal to be spontaneously magnetized if the Stoner condition $U^*N(\epsilon_F) > 1$ is fulfilled i.e. when the density of states at the Fermi level $N(\epsilon_F)$ is high (U is the intra-atomic Coulomb interaction for two electrons with the opposite spins in the same orbitals). The magnitude of the magnetization results from details of band structure, such as a difference between the occupation of the splitted spin-up and spin-down energy bands. Let us consider a *3d*-metal with the half-filled *3d*-band (e.g. Fe). Because the Fermi level lies near the middle of the band its narrowing leads to a large increase of the density of states (DOS) close to the Fermi level. According to the Stoner condition this surface resonance favors a magnetic order and might cause the magnetic "live layers" as expected for paramagnetic vanadium[16].

Another surface effect, namely a surface modification of the ground state value of the magnetic moment has a much more complicated origin. Besides a band narrowing, another (more subtle) mechanism, like *d*-band shift and splitting, charge transfer between the *sp* and *d*-bands or the *sp-d* hybridization have to be considered. Figs.5 illustrate possible situations for a weak ferromagnet (e.g. iron) with the holes both in the majority and minority bands. A surface enhancement of the magnetic moment is expected for the weak ferromagnet simply due to the narrowing of the *d*-band and the corresponding redistribution of the *d*-electrons between the minority and majority bands, so that the total number of the *d*-electrons follows the neutrality condition.

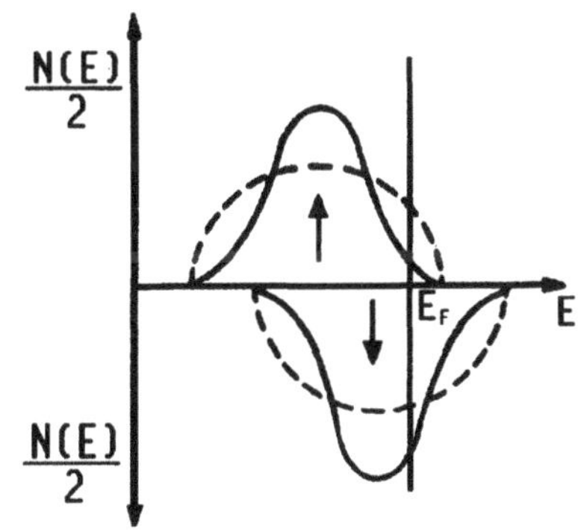

Figure 5. *Schematic bulk (dashed line) and surface (solid line) DOS for a weak ferromagnet*

To the similar conclusion lead also the detailed local spin-density function (LSDF) band calculations performed in slab geometry for iron[17,18]. The results for magnetic moments are coherent with a simple picture in which the magnetization is a monotonic function of the dimensionality. For the clean Fe(001) surface Ohnishi et al.[17,18] found the magnetic moment enhanced to 2.98μ_B/atom as compared with the value of 2.25μ_B/atom for the central layer in the 7-layer slab, whereas for the unsupported Fe(001) monolayer the magnetic moment reaches already 3.18μ_B/atom, not far from value of the free atom 4μ_B. The differences between (110) and (001) surfaces may be interpreted also in terms of different dimensionality measured by the number of the nearest neighbors, whose interaction dominates the *3d*-band width. The Fe-atom on the (001) surface has the least number of neighbors (four) as compared with the (110) surface (six n.n.) and atoms in the bulk bcc structure (eight n.n.) and consequently, the smallest bandwidth and the highest magnetization. The Fe atom on the (110) surface has an intermediate number of nearest neighbors and correspondingly, an intermediate moment of $2.65\mu_B$. The penetration of the surface effects into the bulk is very shallow - the screening length is of order of the interlayer spacing. 30% or 20% enhancement of the magnetic moment for the (001) and (110) surface respectively, occurs to the great extent in the outermost layer.

So far we have discussed only magnetic moments. Another quantity which refers to the local spin polarization is the hyperfine magnetic field. This is an effective magnetic field observed at the nucleus place as the effect of the coupling between the nuclear and electronic spins. For the bulk solids the hyperfine magnetic field B_{hf} measured in NMR or Mössbauer experiments is commonly taken as a measure of the local magnetization, being considered as proportional to the magnetic moment. This type of interpretation is often misleading owing to different contributions to the hyperfine magnetic field, other than proportional to the magnetic moment. Generally, B_{hf} is a sum of contributions:

$$B_{hf} = B_c + B_{orb} + B_{dip} \tag{4.3}$$

B_c, B_{orb} and B_{dip} are the Fermi contact term and the dipolar contributions from the orbital and spin magnetic moment, respectively. The orbital contribution B_{orb} arises from the dipolar interaction of a point magnetic dipole of the nucleus with a circulating charge of the angular momentum $<\hat{L}>$ and in the single electron approximation is given by:

$$B_{orb} = \frac{\mu_0}{2\pi}\mu_B\frac{\langle\hat{L}\rangle}{\langle r^3\rangle} \tag{4.4}$$

For bulk metallic iron and its cubic alloys the orbital contribution vanishes by the quenching of the orbital angular momentum. The magnetic dipole field B_{dip} at the given atom "j" can be written as:

$$B^j_{dip} = -\frac{\mu_0}{4\pi}\sum\left(\frac{\vec{\mu}_i}{r_i^3} - \frac{3(\vec{\mu}_i\cdot\vec{r}_i)\vec{r}_i}{r_i^5}\right) \tag{4.5}$$

In the conventional analysis the summation (4.5) can be broken down into on-site contribution and off-site contribution. The off-site contribution from all other atoms ($i \neq j$) placed in $\vec{r}_i$, each of them carrying the total magnetic moment μ_i, can be transformed just into the sum of demagnetizing field $B_D = -\mu_0 NM$ and the Lorenz field $B_L = -1/3\mu_0 M$ (for cubic symmetry) where N is the demagnetizing coefficient and M is the magnetization. For iron we get $\mu_0 M = 2.145$ T and so the off-site dipolar term, depending on the shape of the sample and its magnetization direction is of the order of 1 Tesla. The on-site contribution corresponds to the term with i = j in sum (4.5) and arises from the electron spin magnetic moment μ_j of the considered atom. In metallic iron this contribution is also small by spherical symmetry of states with L=0 (quenching of angular momentum).

The dominating contribution to B_{hf} seen by a *3d*-atom is the Fermi contact term B_c. It arises as the result of the interaction between the nucleus and an imbalance in the *s*-electron spin density at the nucleus and can be written as

$$B_c = \frac{2}{3}\mu_0\mu_B\,(\rho^{\uparrow}(0) - \rho^{\downarrow}(0)) \tag{4.6}$$

where $\rho^{\uparrow}(0)$ and $\rho^{\downarrow}(0)$ are the densities of electrons with the spin up and down at the nucleus. Only the *s*-electrons have a nonvanishing density at the nucleus and can contribute to B_c. For the interpretation it is useful to break down the B_c into a core electron contribution B_{core} (for iron from *1s, 2s* and *3s*-electrons) and conduction electron contribution B_{ce} (from *4s*-electrons resp.). The core polarization contribution B_{core} arises from the exchange interaction between the unpaired *3d*-electrons and *s*-electrons and so B_{core} scales usually with the magnetic moment. The same exchange effect polarizes also the conduction electrons but more important in this case is the effect of neighboring atoms. The conduction electron polarization term B_{ce} is then of non-local character and reflects the nearest and next-nearest neighbour influence.

The reliable calculation of the Fermi contact hyperfine field B_c became possible owing to local density functional formalism but still only very few papers [19] provide realistic data. The following results stand out for iron:

(i) the total B_c for bulk ranges about -37 T (B_c is opposite to the magnetization),

(ii) the core polarization contribution B_{core} scales with magnetic moment with a rate of about -13.7 T/μ_B,

(iii) the conduction electron term B_{ce} is also negative and amounts to about -5T.

The difference between the experimentally observed value B_{hf} = -34T and the theoretical one is usually ascribed to small orbital and dipolar contributions.

The calculations of hyperfine field only seldom go beyond pure bulk elements. Contrary to other physical quantities like conductivity or magnetization, which are mainly determined by conduction electrons, the core electron spin and charge densities contribute significantly to hyperfine fields. So, the full atomic wave functions have to be calculated. The resulting hyperfine field parameters are the effect of the very small imbalance of spin and charge densities. The verification of the calculated hyperfine fields at surfaces and their comparison with the experimental data is then an important test of the applied theoretical and experimental methods.

The calculated layer-by-layer B_c values and their decomposition into core and conduction electron contribution are summarized in Table I for the free-standing Fe(110) surface[18]. Similarly as for the bulk, the core contribution scales very precisely with the local moment, because the magnetic moment originates mostly from the localized d-electrons. The calculated value of the core contribution is about 13T/unpaired spin and is very close to the free-atom value. The magnitude of the negative core contribution increases at the free-standing surface, reflecting the surface enhancement of the magnetic moment.

Table I. *Theoretical layer-by-layer magnetic moment M (in μ_B) and contact hyperfine fields (in Tesla) decomposed into core and CE contribution*[11)]

	M	B_{core}	B_{CE}	B_{total}
S	2.65	-36.4	4.0	-32.4
S-1	2.37	-32.6	-6.1	-38.7
S-2	2.28	-31.3	-5.1	-36.5
S-3	2.25	-30.9	-5.3	-36.2
S-4 bulk	2.22	-30.5	-4.7	-35.0

As already told, the valence contribution in bulk ferromagnetic materials such as Fe and Ni is also negative. This negative polarization of the valence s-electrons in the bulk is due to the predominance of the indirect polarization (associated with the opposite spin conduction electrons) over the direct (following Hund's rule) coupling to the unpaired d-electrons. In a free atom, the direct polarization of the s-level dominates and produces a positive contribution to B_{hf}. A similar behaviour is observed for surface atoms being obviously more "atomic-like" than atoms in bulk. The total calculated surface value of B_{hf} is smaller in magnitude than it is observed in central layers.

Additionally, a spatial oscillations of B_{hf}, whose amplitude amounts to $\sim$6T for surfaces respectively, are predicted theoretically. The source of these oscillations is a competition between surface effects in core and CE terms. For the subsurface layer the magnetic moment and hence B_{core} are still enhanced, whereas a bulk-like value of B_{ce} is restored. It becomes clear that the hyperfine magnetic field at the surface cannot be related directly to the magnetic moment and that the standard interpretation of the ground state B_{hf} in terms of magnetization data is no more valid.

As yet only the hyperfine interaction measurements (CEMS, PAC) supply the direct experimental results which are able to verify theoretical calculations, although presented below CEMS measurements show some experimental limitation. This is the long acquisition time and consequently, the contamination of the sample surface from the residual gas atmosphere, specially at low temperatures. Therefore in the experiment[20)] the hyperfine interaction parameters measured near the Fe(110) surface using ^{57}Fe monolayer probe method have been determined by extrapolation of slowly varying values to the time immediately after the preparation. The near surface profile of B_{hf} obtained in this way at room temperature shows the characteristic oscillation: a minimum value at the surface, a maximum for the second layer and the fast saturation to the bulk value for the deeper layers.

The observed changes of B_{hf} result from a combination of the surface ground state modification and a temperature effect. It is impossible to eliminate the last one by the direct measurements at low temperatures because the Fe surface acts like getter. On the other hand, this is not difficult to anticipate the low temperature B_{hf} behaviour by simulating the temperature surface effect. It must be obviously a monotonous function of temperature and it causes the enhancement of the surface spin deviation by factor of about 2, as was measured for example for the Fe/Ag interface[14]. Exploring these findings it becomes clear, that any reasonable surface temperature effect cannot kill the oscillatory character of B_{hf} near the Fe(110) surface but it can only influence the amplitude of the oscillations.

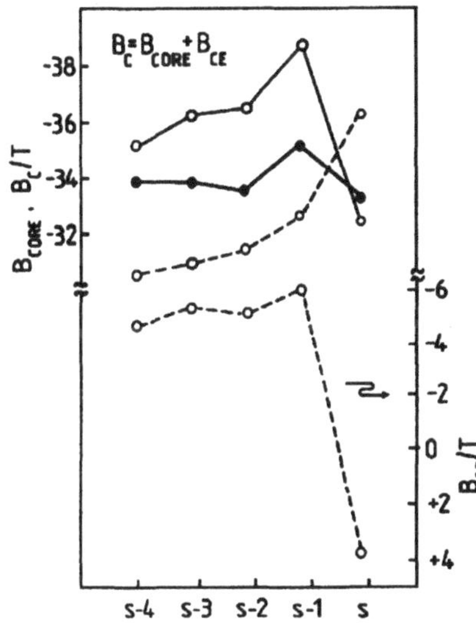

Figure 6. *Theoretical[18] (open points) and experimental[20] (full points) values of B_{hf} near Fe(110) surface*

The assumption that the temperature dependence of B_{hf} for a free surface is the same as for the Ag coated one results in a best confidence guess of the ground state B_{hf} changes near free-standing surface as shown by full points in Fig.6. The oscillation of B_{hf} is obvious: $B_{hf}(0)$ is reduced in the first atomic layer by about 0.75T with respect to the bulk value and enhanced in the second layer by about 1T. By this, the theoretical calculations[18], shown also in Fig.6 as the open points and the full line, has been qualitatively confirmed for the first time. However, the amplitude of the oscillation predicted theoretically is by the factor 4 larger than it is measured experimentally.

It must be emphasized that the oscillation of the contact hyperfine field B_c, often referred as a Friedel-type oscillation of conduction electron polarization, results in a similar extent from surface modifications of the core (B_{core}) and conduction (B_{ce}) electron contributions which are presented also in Fig.6 as the open points and the dashed lines. Surface modifications of the two terms can result in a peculiar spatial profile of B_c, whose absolute value is the main contribution to B_{hf}. It is also clear that any surface changes of B_{hf} cannot be related to the same type of changes in the magnetic moment. The behaviour of the negative core contribution B_{core} (the maximum of the absolute value at the surface) represents the behaviour of the magnetization and contrasts sharply with the total B_{hf}.

4.3. Magnetic Behaviour of Ultra-Thin Films

As it is seen from the previous paragraph, the surface effects concerning the local changes of the electronic structure range only a few atomic layers. Beside that, as the film thickness decreases, collective properties, like magnetic excitations or onset of magnetic order, can be modified. The fundamental question is the thickness dependence of the ordering temperature and the character of the temperature dependence of magnetization (or hyperfine magnetic field) going from 3-dimensional to 2-dimensional system. The actual situation is always complicated by the fact that we never deal with unsupported films and the interaction with interfacing materials must be considered. This effect clearly dominates for example in the case of Ni films on Cu as discussed by Schatz[3)] in other paper of this proceedings. On the contrary to Ni films, for which the B_{hf} reduces considerably as the film thickness decreases, for the case of Fe films on tungsten three atomic layer is enough to establish nearly bulk-like value of B_{hf} in two outermost layer. Only in the Fe layer closest to tungsten substrate the value of hyperfine magnetic field is very different (about 21T). Fig.7 summarizes the B_{hf} data as a function of temperature for the

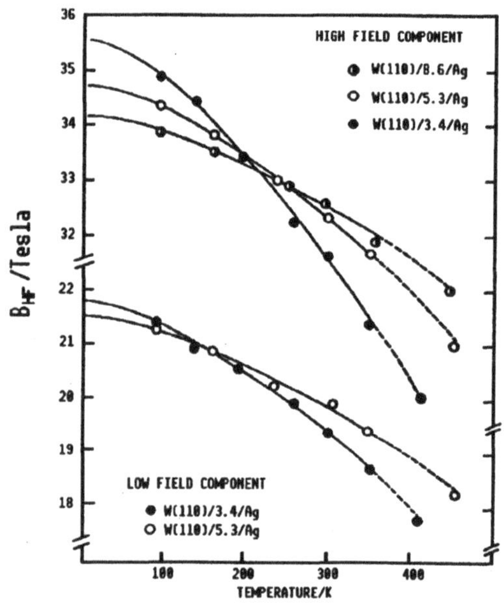

Figure 7. *Temperature dependence of B_{hf} in Fe(110) films of thickness D (in atomic layers) on W(110) coated by Ag. All data were fitted to $B_{hf}(T)=B_{hf}(0)*(1-bT^{3/2})$[21)]*

different iron films coated with Ag[21]. The low field component refers always to Fe layer interfacing with W, whereas the high field component reflects the average behaviour of the remaining layers in the films. They comprise also Fe/Ag interface at which B_{hf} increases with decreasing film thickness.

The temperature dependence of the hyperfine magnetic field can be fitted using $T^{3/2}$ Bloch law - a quasi-linear dependence which is postulated from the simple isotropic spin wave theory is definitely not observed.

The average film magnetization (or hyperfine magnetic field) is the result of local layer properties. The ^{57}Fe monolayer probe profiling of the hyperfine magnetic fields in 20.5 ML $^{56}Fe(110)$ film provides the full information on the spatial distribution of the reduced magnetization[14]. In the analysis using pure ^{57}Fe films[21] the information is integrated over atomic arrangements (atomic layers) whose hyperfine magnetic fields cannot be resolved in the Mössbauer experiment. However, the spatial variation of the magnetization can be deduced by assuming smooth magnetization variation and its film center symmetry. Fig.8 presents the reduced magnetic hyperfine fields at a constant temperature (300K), normalized to the ground state values. The full points are the experimental values integrated over the film region indicated by the horizontal segments and the straight horizontal lines represent the mean film values. The broken lines are the resulting spatial distribution profiles of the reduced local magnetization. Whereas the shape of the profile for the 20.5-layer film is quite straightforward, those of the 8.6- and 5.3-layer films are only a reliable guess matching the experimental observations. A consistent pattern of magnetization profiles in ultra thin films emerges: going to the thinnest films the magnetization across the film becomes homogeneous but simultaneously, a significant reduction of the film mean magnetization is to be noticed. This is illustrated by the inset

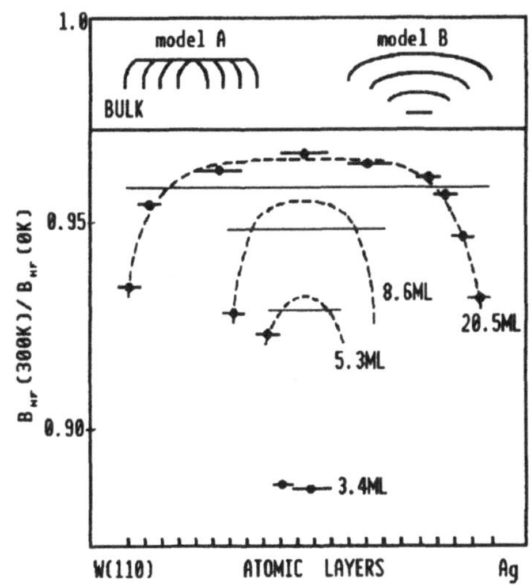

Figure 8. *Spatial distribution of the reduced magnetization at 300K in Fe(110) films on W(110)*[21]

in the top of Fig.8 as the "Model B", contrasted with the "Model A" presenting films in which the film center properties remain unchanged independently of the film thickness.

The Curie temperature of the films down to two monolayers exceeded the temperature at which the irreversible changes in the layered film structure took place (about 600K for the thinnest films). On the contrary, the monolayer film is paramagnetic at the room temperature (compare Fig.2a) and displays the onset of ferromagnetism below T=282K[22]. The temperature dependence of the monolayer B_{hf} is shown in Fig.9. The different extrapolations to 0K (linear or $T^{3/2}$) give the similar values of $B_{hf}(0)$ about 12T. Any interpretation of these experimental results must be done keeping in mind all peculiarities of the measured system. The tungsten substrate changes the electronic structure of the iron films by both direct interaction between W and Fe bands and by the influence on the atom spacing in the pseudomorphic Fe monolayer. These effects were taken into account in the theoretical paper by Hong et al.[23] leading to $B_{hf}(0)=10.7T$ in a remarkably good agreement with the experiment. This surprisingly low value of the hyperfine magnetic field, which is almost three times lower than in bulk iron, differs sharply from the behaviour of the magnetic moment. The results of the magnetometric measurements[24] are shown also in Fig.9. The experimental points were available only in the narrow temperature range 220K-300K and the extrapolation to 0K was made using the temperature dependence of B_{hf}. The ground state magnetic moment of $2.53\mu_B$ is slightly increased in comparison with $2.22\mu_B$ in the bulk. The strong reduction of B_{hf} sets an intuitively unexpected contrast with the enhancement of the magnetic moment and illustrates again the lack of proportionality between this two quantities for low dimensional systems.

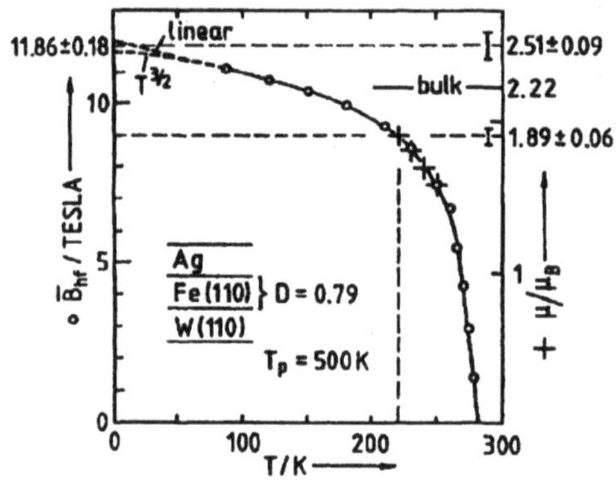

Figure 9. *Temperature dependence of B_{hf} for Fe(110) monolayer on W(110) compared with its magnetization (in μ_B/atom)*[24]

REFERENCES

[1]. *"Polarized Electrons in Surface Physics"*, ed.R.Feder, World Scientific, Singapore, 1985.

[2]. *"Magnetic Properties of Low-Domensional Systems"*, ed.Falicov and Moran-Lopez, Springer-Verlag, Berlin, 1986.

[3]. G.Schatz, in this Volume.

[4]. A.J.Freeman, C.L.Fu, S.Omishi and M.Weinert,in ref.[2], pp.3-66.

[5]. H.J.Elmers, G.Liu, U.Gradmann, Phys.Rev.Lett. 63(1989)566.

[6]. B.D.Sawicka, J.A.Sawicki, J.Stanek, H.Drwiega, Hyperfine.Interact. 5(1976)67.

[7]. K.R.Swanson, J.J.Spijkerman, J.Appl.Phys. 41(1970)3155.

[8]. J.Bayreuther, Hyperfine Interact. 47(1989)237.

[9]. J.A.Sawicki, B.D.Sawicka, Hyperfine Interact. 13(1983)199.

[10]. D.Liliequist and H.Ismail,Phys.Rev. B31(1985)4131.

[11]. J.Korecki, U.Gradmann, Hyperfine Interact. 28(1986)931.
Ch.Sauer, A.Holzwarth, Zs.Kajcsos, W.Zinn, Nucl.Instr.Math. B34(1988)377.

[12]. U.Gradman, G.Waller, Surf.Sci. 116(1982)539.

[13]. M.Przybylski, I.Kaufmann, U.Gradmann, Phys.Rev. 40(1989)8631.

[14]. J.Korecki, U Gradmann, Eur.Lett. 2(1986)651.

[15]. U.Gradmann, J.Korecki, G.Waller, Appl.Phys. A39(1986)101.

[16]. G.Allen, in ref.[2], pp.2-9.

[17]. S.Ohnishi, M.Weinert, A.J.Freeman, Phys.Rev. B30(1984)36.

[18]. C.L.Fu, A.J.Freeman, J.Magn.Magn.Mat. 69(1987)L1.

[19]. B.Lindsen, J.Sjöström, J.Phys.F 18(1988)1563.

[20]. J.Korecki, U.Gradmann, Phys.Rev.Lett. 55(1985)2491.

[21]. J.Korecki, M.Przybylski, U.Gradmann, J.Magn.Magn.Mat. 89(1990)325

[22]. M.Przybylski, U.Gradmann, Phys.Rev.Lett. 59(1987)1152.

[23]. S.C.Hong, A.J.Freeman, C.L.Fu, Phys.Rev. B381988)12156

[24]. U.Gradmann, M.Przybylski, H.J.Elmers, G.Liu, Appl.Phys. A49(1989)563.

INVESTIGATION OF THE ANTIFERROMAGNETIC ORDER IN THE 123 COMPOUNDS $REBa_2Cu_3O_y$ by NQR AND NMR

H. Lütgemeier and I. Heinmaa* IFF, Forschungszentrum Jülich, D–5170 Jülich

*Permanent address: Institut of Chemical Physics and Biophysics, Tallinn 200001, Estonia

ABSTRACT

The NMR and NQR spectra of Cu in the 123 compounds $YBa_2Cu_3O_y$ and related ones containing rare earth metals in the place of Y are discussed. Antiferromagnetic order in the Cu(2) sites exists in the Y and rare earth compounds at reduced oxygen concentration y but in the Pr compounds at any y. In both cases the Cu(1) sites do not carry a static magnetic moment. The EFG at the Cu(1) sites is not influenced by the replacement of Y by Pr, but reduced at the Cu(2) sites. At low oxygen concentration the hyperfine field at the Cu(2) sites of Pr123 reveals a field distribution which may be the indication of a modulated structure.

1. INTRODUCTION

The application of nuclear magnetic resonance, NMR, in the research of superconductors has a long tradition[1]. The NMR experiments performed with the classical superconductors have confirmed the BCS type pairing in simple metals like Al and the existance of a flux line lattice in type II superconductors. In these applications the interaction of the nuclear magnetic moment with the conduction electrons and the static local field allows to determine the change of the spin susceptibility at T_c from the change of the Knight shift, and the field distribution in the mixed state from the NMR line shape. Furthermore, the behavior of the spin–lattice relaxation rate at the transition from the normal to the supercondcuting state contains information about this state since the spin–paired electrons contribute no longer to the relaxation. The most important features of a BCS superconductor with an isotropic gap are the Hebel Slichter peak of the relaxation rate just below T_c due to the piling up of the density of states when the gap opens, and the Arrhenius type decrease at lower temperature.

In HT_c superconductors (HTSC), the situation is more complicated since most useful nuclei in these compounds have a spin I > 1/2 and carry a large electric quadrupole moment. Thus the quadrupole interaction is important in these systems and leads to complicated NMR spectra, at least for polycrystalline samples. From the quadrupole interaction structural information is obtained concerning the electric field gradient (EFG) at the different lattice sites. Furthermore the pure nuclear quadrupole resonance (NQR) spectra of most nuclei can be observed without an external magnetic field and allow an easy measurement of the spin–lattice relaxation.

A special feature of the HTSC is their relation to magnetic order of either Cu^{2+} or rare earth (RE) ions. Whereas the coexistence of superconductivity with antiferromagnetic (AF) order of the RE ions is well established, the coexistence with ordered Cu moments is still a matter of controversy. The influence of these magnetic moments and the type of the magnetic order can be studied by nuclear resonance, too.

In the following, some examples will be presented from our work at Jülich for the study of structural and magnetic properties of the HTSC and related compounds. We limit ourselves to the 123 compounds $YBa_2Cu_3O_y$ (Y123) and compounds with RE in the place of Y. A special emphasis will be given to some new results for the nonsuperconducting Pr123 compounds[2]. For all experiments polycrystalline powder samples have been used and measured by standard spin–echo technique mostly at 4.2 K.

2. Cu RESONANCE IN Y123

In $YBa_2Cu_3O_y$ all ions have been detected by NMR or NQR. Y is the only nucleus without a quadrupole moment since the spin is I = 1/2. Therefore this nucleus shows pure magnetic interaction and the narrow lines have allowed a detailed investigation of the Knight–shift and spin–lattice relaxation rate as function of the oxygen concentration and temperature[3].

The two Cu isotopes with I = 3/2 show strong quadrupole effects on both Cu sites (fig. 1). The small difference of the quadrupole moments of both Cu isotopes is the reason for the doublets in the Cu NQR spectra with the intensity ratio 2:1 and the frequency ratio 1:0.92. Since only one NQR transition 3/2, 1/2 exists for I = 3/2, the full EFG tensor with the largest component V_{zz} and the asymmetry parameter η $(0 \leq \eta \leq 1)$ can be determined only from the spectrum in an external magnetic field. The field–sweep spectra of grain oriented powder samples[4] and of small single crystals[5] with either $y < 6.3$ or $y > 6.8$ show an EFG of axial symmetry around the c–axis for the doublets at the higher frequency, and a nonsymmetric EFG for the low frequency doublets. Nevertheless the assignment of these lines to the Cu sites in the lattice is quite different.

At high oxygen concentration the line at 31.5 MHz for ^{63}Cu corresponds to the Cu(2) sites (CuO_2 planes) and that at 22 MHz with $\eta = 1$ to the Cu(1) sites (CuO chains). This assignment follows from the intensity ratio of the lines and from the fact that an EFG with symmetry axis in c direction can be expected only for the Cu(2) sites. This assignment follows also from the influence of the magnetic moments of rare earth ions replacing Y (see below).

At low oxygen concentration the situation is different. Here the 123 compounds show antiferromagnetic order with localized magnetic moments at the Cu(2) ions[6]. The zero–field spectrum of the Cu(2) sites is in this case an antiferromagnetic NMR spectrum in the region 70–110 MHz determined by an hyperfine field of 7.95 T [7] and an EFG with the symmetry axis perpendicular to the hyperfine field (fig. 2). The NQR doublets observed at lower frequency (fig. 1) correspond only to the Cu(1) sites: the line at 30.2 MHz with the symmetric EFG to those with twofold oxygen coordination, and that at 24 MHz with $\eta = 0.3$ [8] to those with three oxygen neighbours. The pure NQR spectra observed for the two–, three–, or fourfold coordinated Cu(1) sites prove that these do not carry a static magnetic moment even in the antiferromagnetic state. If magnetic moments exist at these sites, their correlation time must be shorter than 10^{-10}s. Otherwise either a static splitting or a loss of the signal by fast relaxation would appear.

3. MAGNETIC STRUCTURE OF $YBa_2Cu_3O_6$

In the antiferromagnetic state the order must be of a type that the dipolar field of the electronic moments localized at the Cu(2) sites cancel at the Cu(1) sites. This holds for the structure deduced from neutron diffraction[6] which is described by a wave vector (1/2,1/2,1) with no doubling of the unit cell in c–direction (fig.3a). In this structure the stacking sequence along the c–axis is + o – + o – where + and – describe the spin directions of the Cu(2) sites and o the zero spin at Cu(1). A different sequence is found in samples containing a small amount of Fe (about 1%) at the Cu(1) sites[9]. Here a Zeeman splitting of the NQR line by a local field of about 0.16 T is found (fig.4) which is equal to the dipolar field in a magnetic structure described by the wave vector (1/2,1/2,1/2) or the stacking + o + − o − (fig.3b) and magnetic moments of 0.67 μ_B **at the** Cu(2) sites.

At higher temperature the splitting decreases continuously (fig.5) indicating a continuous transition from structure B to A via a noncollinear canted structure (fig.6). The low–temperature structure depends clearly on the concentration x of the Fe impurities in $YBa_2(Cu_{1-x}Fe_x)_3O_6$ (fig.7). The splitting saturates at $x > 1\%$ indicating that this small amount is sufficient to change the structure completely. This is a consequence of the strong in–plane antiferromagnetic exchange interaction of the Cu(2) ions and the weak one between the planes. An extra ferromagnetic coupling between the planes induced by the impurities at Cu(1) sites will turn the staggered magnetization of the whole planes. At increasing oxygen concentration y the extra coupling is reduced. Thus a sample with $x = 0.2\%$ Fe showed the transition from A to B[10] at $y = 6.0$, but not at $y = 6.15$.

At first we had assumed that the reason for the extra coupling was the magnetic moment of the Fe ions at the Cu(1) sites which are in a frustrated situation in the high temperature structure. This is evident from ^{57}Fe Mössbauer studies which reveal the appearance of the hyperfine field for one component of the spectrum in the transition region from structure A to B [10], whereas the hyperfine field of the second component follows the staggered magnetization of the Cu(2) planes with $T_N = 420$ K. Further experiments, however, showed that doping with Al or Ga which mainly occupy the Cu(1) sites, results in the same effect as Fe, whereas doping with Ni (preferrently at the Cu(2) sites) has no influence on the NQR spectra. This means that the extra coupling is an effect of the different valency state of the 3+–ions at the Cu(1) site and not of the magnetic moment[11].

In contrast to our result that in the pure 123 compounds with Y, Gd or Nd only the structure A exists above 1.2 K, a change of the magnetic structure above 10 K has been found by neutron diffraction in different single crystals of Y123 and Nd123 [12,13]. Since many single crystals contain a large amount of Al from the flux, we assume that this transition is not a property of the pure 123 compounds but induced by Al. In contrast to the structure A, the symmetry of B allows a magnetic moment at the Cu(1) sites. The results of the different neutron scattering experiments is quite different ranging from below 0.1 [12] to 0.4 μ_B [13]. Our results, however, prove that the magnetic moment at the Cu(1) site is smaller than 0.03 μ_B, if the normal hyperfine coupling constant of Cu, 12 T/μ_B, is assumed.

4. INFLUENCE OF RE IONS

Besides the magnetic Cu ions, a second magnetic sublattice exists if Y is replaced by rare earth ions. For Nd and the heavier rare earth metals this happens without a distortion of the lattice structure and the superconducting properties. The magnetic moments of the RE ions show at low temperature a transition to antiferromagnetic order and a coexistence of antiferromagnetism and superconductivity. For Gd123 the Neel temperature is 2.23 K. The magnetic order of Gd does not influence the NQR frequencies 22.5 and 32.35 MHz of Cu(1) and Cu(2), respectively. This means that the dipolar field of the RE ions is completely canceled at the Cu sites by symmetry. The fluctuating components of the Gd spins, however, introduce strong longitudinal and transverse relaxation processes especially at the Cu(2) sites which are located nearest to the Gd sites (fig.8). Above T_N the transverse relaxation rate of the Cu(2) sites is increased by an order of magnitude compared with the values of Cu(1) and reveals a maximum at T_N. Below T_N the rate decreases since the Gd spins are fixed. At the same temperature the relaxation behavior changes from Lorentzian above T_N to more Gaussian like below T_N. But even at the lowest temperature of our experiment, 1.2 K, the longitudinal and transverse relaxation rates are increased at the Cu(2) sites compared with the Cu(1) sites and prove strong fluctuations in the Gd spin system.

At low oxygen concentration the two magnetic systems of the RE ions and the Cu(2) ions coexist without a strong interaction. The situation at the Cu(1) sites is not changed by the replacement of Y by Gd or Nd and no magnetic moment is found at these sites. For the Cu(2) sites, the zero field NMR spectra found below the Neel temperature of the Gd reveal an hyperfine field of 7.68 T which is reduced by 0.3 T compared with Y123.

5. Cu RESONANCE IN Pr123 COMPOUNDS

The lightest RE metal which can build up the 123 structure is Pr. In contrast to the other RE ions, Pr suppresses the metallic behaviour and the transition to superconductivity already at full oxygen loading, if more than 50% of the Y is replaced. At the same Pr content AF order of Cu has been observed by Mössbauer effect of Fe at Cu sites[14] and by muon spin relaxation[15]. The reason for the suppression of the superconductivity by Pr in contrast to the heavier RE ions is an open question. The larger ionic radius can lead to pair breaking by the stronger overlap of the 4f states with the electron states in the CuO_2 planes. Another reason may be that Pr enters the lattice as Pr^{4+} and fills up the holes responsible for the metallic behaviour. This model is supported by the similarity of the phase diagrams of the systems $Y_{1-x}Pr_xBa_2Cu_3O_7$ and $YBa_2Cu_3O_{7-x}$.

We have investigated the Cu resonance in a series of Pr123 samples of oxygen concentration between x = 6.0 and 7.0 prepared by the ceramic technique in different laboratories. All these samples reveal at 4.2 K NQR lines in the region between 18 and 32 MHz which are much broader than the lines in Y123 and do not allow to separate the lines of the two Cu isotopes. This means a broad distribution of the EFG values and indicates a much lower crystal perfection in this system. Nevertheless the comparison of the NMR spectra at field–sweep (fig.9) and the NQR spectra (fig.10) allows conclusions about the EFG. At high oxygen concentration (x = 6.9) there is only one NQR line at 22 MHz (for ^{63}Cu) with a nonsymmetric EFG ($\eta = 1$) as seen from the NMR spectrum. At lower oxygen concentration two other lines at 24 MHz with nonsymmetric EFG and at 30 MHz with symmetric EFG appear from which only the 30 MHz line remains at x = 6.0. These are just the frequencies found in Y123 for the Cu(1) sites with four, three and two oxygen neighbours. Thus the EFG at the Cu(1) sites is not influenced by replacing of Y by Pr.

No signals corresponding to the Cu(2) sites could be detected at 4.2 K. At 1.2 K, however, broad zero–field lines are found for each oxygen concentration in the frequency region above 60 MHz (fig.10), which must be ascribed to NMR lines of Cu in a hyperfine field and proves that the Cu(2) sites carry a stable magnetic moment as in the case of Cu(2) in $YBa_2Cu_3O_6$. Reyes et al.[16] present the NMR spectra of $PrBa_2Cu_3O_7$ for temperatures between 285 and 300 K with the line of the Cu(2) sites. This line disappears at 285 K indicating a well defined Neel temperature.

The shape of the zero–field spectra of Cu(2) for high oxygen concentration differs for the samples measured up to now [2] and must be connected with the crystal perfection. Similar to our spectrum for the sample with the narrowest line (fig.10), the zero field spectrum by Reyes et al. shows no resolved quadrupole splitting but is shifted to lower frequency by 5 MHz. Thus it is impossible to obtain a reliable fit of these spectra and to determine the quadrupole coupling. In any case, however, the EFG of the Cu(2) sites in Pr123 with high oxygen concentration is smaller than in YBCO for any oxygen concentration, and the hyperfine field is definitely smaller, 6.7(3) T, compared with 7.95 T in antiferromagnetic YBCO.

At lower oxygen concentration, x = 6.0 and 6.3, the Cu(2) site spectra reveal an unexpected shape with two well defined maxima at 80 and 95 MHz (Fig.10) which coincide with the values of the hyperfine field in Pr123 for x = 7 and Y123 for x = 6, respectively. The distance of the maxima is too large to be explained as lines of the two Cu isotopes, since the nuclear magnetic dipole and electric quadrupole moments of both isotopes differ at most by about 8%. Therefore the line shape reflects a distribution of the hyperfine fields which is an indication of a modulated magnetic structure. As in the case of x = 7 the spectra for low x do not show any quadrupole splitting proving again a smaller EFG at the Cu(2) sites than in Y123.

Thus we must conclude that the EFG at the Cu(2) sites is strongly reduced if Y is replaced by Pr, whereas it stays constant at the Cu(1) sites. This result is different from our earlier conclusion [2] where due to the broader linewidth in the first available samples a similar EFG also for the Cu(2) sites was assumed in YBCO and PrBCO. The reduced EFG means that the charge distribution in the Cu(2) planes is changed by the substitution and supports the model of different valency states of Y and Pr as the reason for the suppression of the superconductivity in Pr123. The magnetic structures of Pr123 and Y123 are similar in the fact that the Cu(1) sites do not carry a stable magnetic moment. The broad distribution of the hyperfine fields at Cu(2), however, indicates different types of magnetic order in these planes.

A last conclusion can be made on the magnetic order of the Pr ions in these compounds. Specific heat measurements [14] and neutron diffraction [17] reveal antiferromagnetic order of the Pr sublattice at about 17 K for x = 7. Moessbauer effect of Gd in Pr123 doped with Gd shows the same Neel temperature for x = 7, but only 7.5 K for x = 6 [18]. The intensity of the zero–field NMR spectra of the Cu(2) sites decreases very fast at increasing temperature above 2 K and disappears at about 3 K by fast transverse relaxation, whereas the Cu(1) site intensity reveals only the normal decrease corresponding to the Curie law. This means that even in the ordered state the magnetic fluctuations of the Pr sublattice are very strong, a feature found also in the Moessbauer spectra of Gd in Pr123 [18], where the hyperfine field increases down to the lowest temperature.

ACKNOWLEDGEMENT

We are indebted to many colleagues who supplied the samples used for this research. Special thanks are due to B. Maple for the new Pr123 samples with oxygen concentration 6.9 and 6.0, to R. Dupree for Y123 samples doped with Al, Gd and V and to H. Buchgeister and F.M. Hosseini for the Gd123 samples. I. Heinmaa thanks Prof. Zinn for the hospitality he received at the IFF of the KFA Jülich, and the DAAD for financial support.

REFERENCES

1. MacLaughlin, D.E., Solid State Physics 31, edited by F. Seitz and D. Turnbull (Academic Press, New York 1977).
2. Lütgemeier, H., JMMM 90&91, 633 (1990).
3. Alloul, H., et al., Phys. Rev. Lett. 63, 1700 (1989).
4. Lütgemeier, H., Physica C153–155, 95 (1988); Lütgemeier, H. et al., Physica C153–155, 731 (1988).
5. Pennington, C.H. et al., Phys. Rev. B37, 7944 (1988).
6. Burlet, P. et al., Physica C153–155, 1115 (1988).
7. Mendels, P. and Alloul, H., Physica C156, 355 (1988).
8. Heinmaa, I. and Lütgemeier, H., to be published.
9. Lütgemeier, H. et al., Physica C162–164, 1367 (1989).
10. Brand, R. et al., Hyp. Inter. 22, 1229 (1990).
11. Bottyan, L. and Lütgemeier, H., to be published.
12. Kadowaki, H. et al., Phys. Rev. B37, 7932 (1988).
13. Li, W.H., Lynn, J.W., and Fisk, Z., Phys. Rev. B41, 4098 (1990).
14. Felner, I. et al., Phys. Rev. B40, 6739 (1989).
15. Cooke, D.W. et al., Phys. Rev. B41, 4801 (1990).
16. Reyes, A.P. et al., Phys. Rev. B42, 2688 (1990).
17. Li, W.H. et al., Phys. Rev. B40, 5300 (1989).
18. Wortmann, G. and Felner, I., Solid State Comm. 75, 981 (1990).

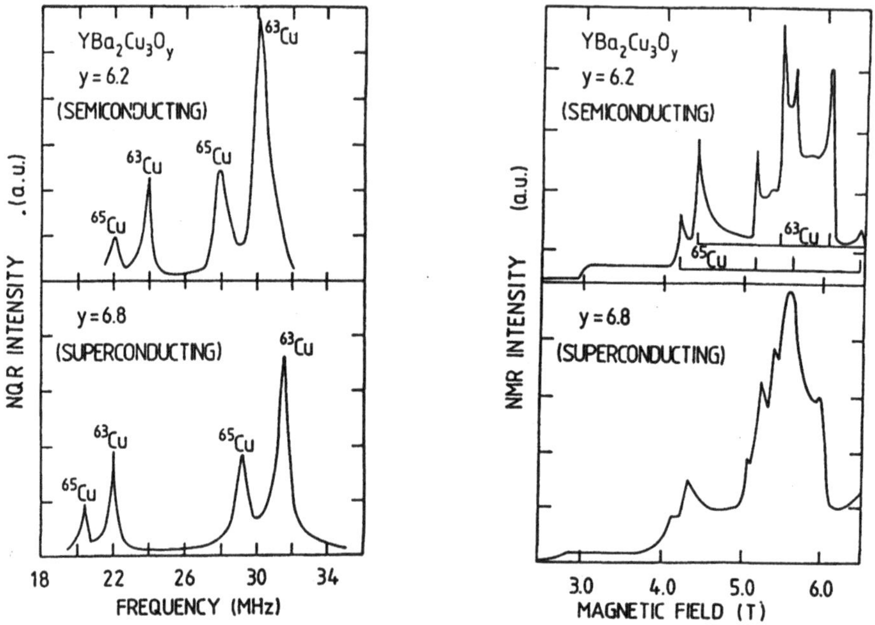

Fig. 1: NQR spectra (left) and NMR spectra at 64 MHz (right) of antiferromagnetic and of superconducting $YBa_2Cu_3O_y$ at 4.2 K.

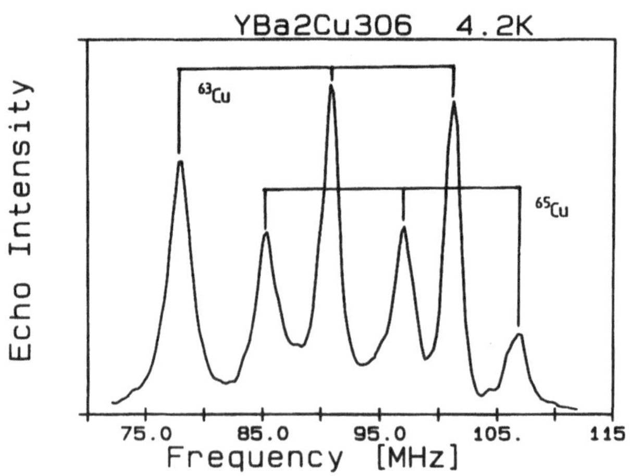

Fig. 2: Antiferromagnetic zero–field spectrum of Cu(2) in $YBa_2Cu_3O_6$ at 4.2 K.

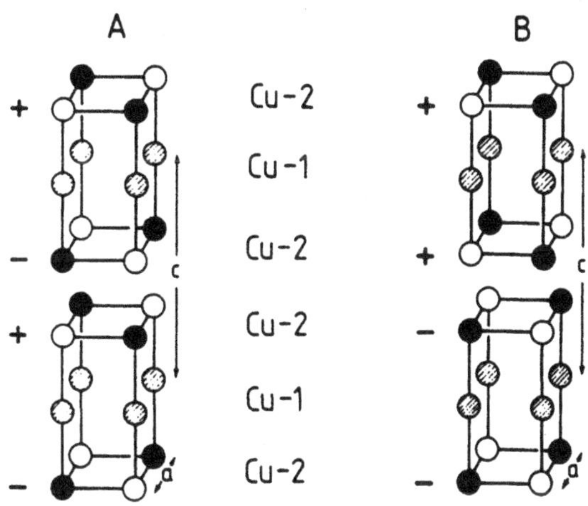

Fig. 3: The two different AF structures of Cu in 123 compounds with different stacking sequence of the magnetic Cu(2) sites.
A: structure of pure samples without a local field at Cu(1).
B: structure in samples containing Fe impurities with a dipolar field at Cu(1).

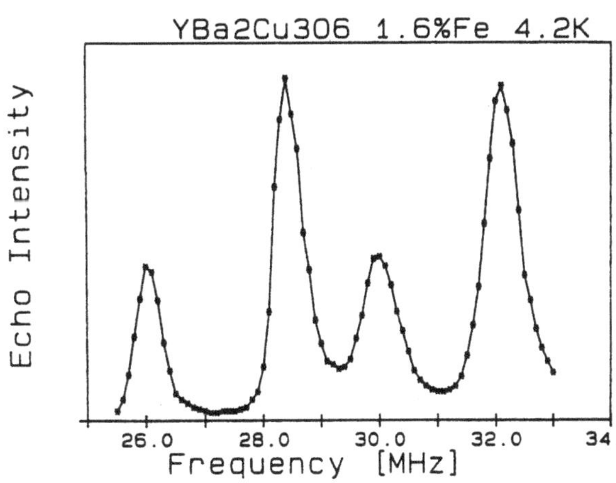

Fig. 4: NQR spectrum of the Cu(1) sites in $YBa_2(Cu_{.984}Fe_{.016})_3O_6$ at 4.2 K showing the dipolar splitting by a field of 0.16 T.

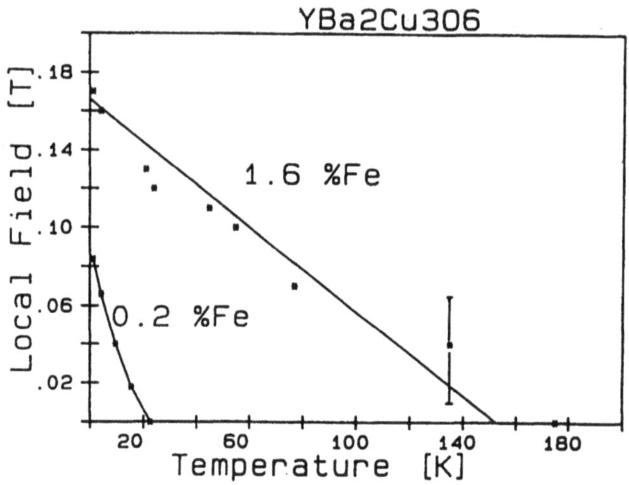

Fig. 5: Temperature dependence of the dipolar splitting in $YBa_2(Cu_{1-x}Fe_x)O_6$ for x = 0.002 and 0.016.

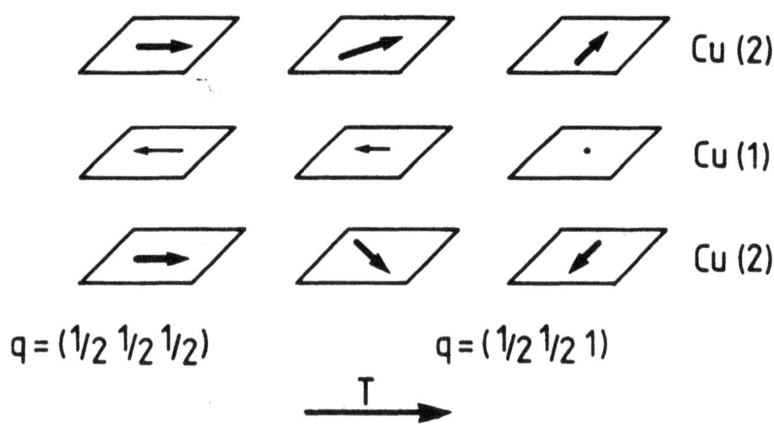

Fig. 6: Model for the continuous transition between the high– and low temperature structures with the intermediate canted spin–structure. The bold arrows indicate the staggered magnetization in the Cu(2) planes, the weak arrows the dipolar field at the Cu(1) sites.

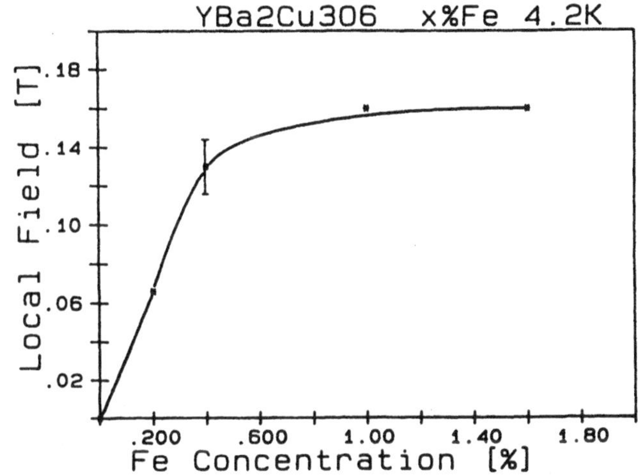

Fig. 7: The dipolar field at the Cu(1) site at low temperature as function of the iron concentration x.

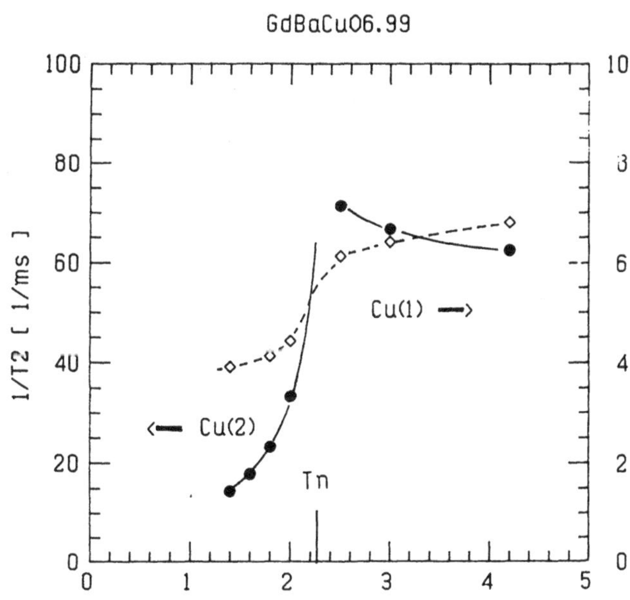

Fig. 8: Temperature dependence of the transverse relaxation rate, $1/T_2$, of Cu(2) sites (solid left scale) and Cu(1) sites (dashed, right scale) in $GdBa_2Cu_3O_7$. The lines are only guides to the eye.

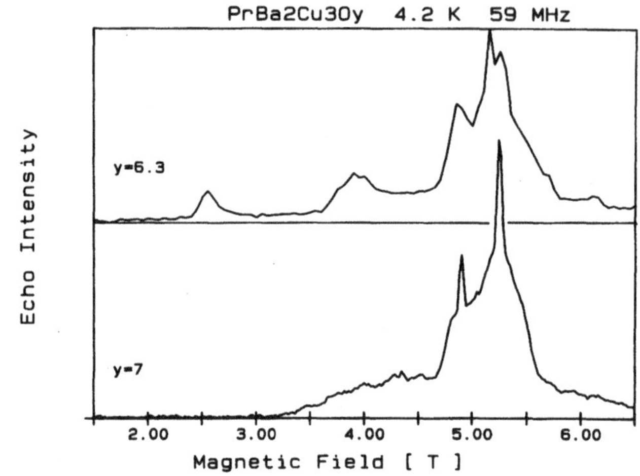

Fig. 9: NMR powder spectra of Cu in $PrBa_2Cu_3O_y$ at 4.2 K and 59 MHz. y = 7.0: Spectrum of both Cu isotopes in a non–symmetric EFG ($\eta = 1$). y = 6.3: A symmetric EFG leads to the 3/2, 1/2 satellite at 3.9 T, the central transition is mixed with the spectrum of a different site with non–symmetric EFG.

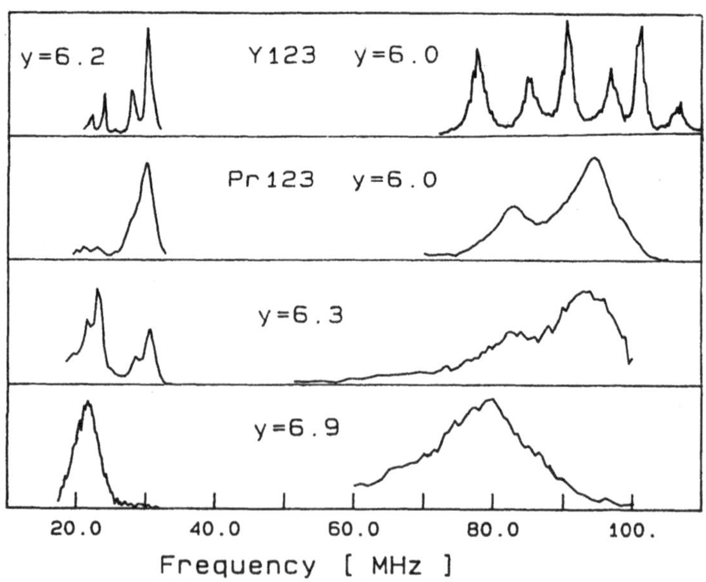

Fig.10: Zero–field spectra of antiferromagnetic 123 compounds at 4.2 K (low frequency part) or 1.2 K (high frequency part).

NMR STUDY OF PHASE TRANSITIONS IN $(NdY)_2Co_{14}B$.

E. Jedryka, M. Wojcik,

Institute of Physics, Polish Acad. Sci., 02 668 Warsaw.

P. Panissod

I.C.P.M.S. G.E.M.M.E. 67070 Strasbourg, France

A.T. Pedziwiatr

Jagiellonian University, 30 059 Krakow, Poland.

M. Slepowronski

Institute of Atomic Energy, 05 400 Otwock-Swierk, Poland.

ABSTRACT

^{59}Co NMR has been studied at 4.2 K and as a function of temperature in the mixed $(Nd_{1-x}Y_x)_2Co_{14}B$ compounds throughout the full substitution range. It was found that the anisotropy of the Nd sublattice determines the canted magnetic structure of the system at 4.2 K up to the threshold Y content of 92%. Above 92% Y the system switches rapidly to a planar structure which is enforced by the Co sublattice. Below the threshold Y content, the system undergoes two spin reorientations with increasing temperature : a second order phase transition from canted to uniaxial state and a first order phase transition from uniaxial to planar structure.

Due to anisotropic properties of hyperfine interaction tensor, NMR spectrum in magnetic materials depends critically on the crystallographic direction in which the magnetization vector is oriented. Our previous ^{59}Co NMR study of the temperature induced spin reorientation in tetragonal $Nd_2Co_{14}B$ compound has shown that among six crystallographic Co sites, three lines ($16k_2$, $16k_1$ and 4c) develop a doublet structure when magnetization is undergoing a rotation from the tetragonal c-axis. This effect is due to the strong anisotropy in the c-plane for those sites. At the same time two other lines (8j and 4e) reveal frequency shifts which are a result of anisotropy between the c-plane and c-axis [1]. These features will now be used as a probe to study spin arrangements in mixed $(Nd_{1-x}Y_x)_2Co_{14}B$ compounds, where non magnetic yttrium is introduced into the 4f sublattice. $Y_2Co_{14}B$ has a planar anisotropy which is enforced by the Co sublattice. In $Nd_2Co_{14}B$ anisotropy of the 4f sublattice causes at low temperatures a 12° tilt of magnetization direction away from the c-axis while at T_{SR1} = 32 K a spin reorientation towards c-axis takes place. At T_{SR2} = 546 K anisotropy of the 3d sublattice prevails over the 4f sublattice and the material assumes a planar magnetic structure. Upon dilution with non magnetic yttrium T_{SR1} increases from 32 K to 72 K and T_{SR2} decreases from 546 K to 120 K for x = 0.9, as determined from the bulk magnetization measurements [2]. To obtain microscopic information on these phase transitions ^{59}Co NMR has been studied for x = 0, 0.01, 0.02, 0.05, 0.1, 0.25, 0.5, 0.75, 0.8, 0.85, 0.9, 0.92, 0.95, 0.97, 0.99 and 1, at temperatures between 4.2 K and 360 K reaching different magnetic configurations. The samples were in the form of powder and spin echo spectra have been recorded between 40 and 240 MHz using a broad band phase sensitive spectrometer.

The spectra recorded at 4.2 K across the entire composition range are presented in Fig.1. The most evident effect of yttrium substitution is the development of a rich satellite structure which is especially pronounced in case of the 4e (55 MHz) line, $16k_1$ (131 MHz) and $16k_2$ (168 MHz) lines. Such satellites are readily explained as typical local environment effects due to the distribution of a number of Y ions over Co neighbor sites, which modify the on site magnetic moment and thus the hyperfine field. Up to x = 0.25, satellite structures remain the only modification of a spectrum. The absence of a change in doublet separation and in the line position indicates that the canting angle does not change significantly up to this composition. For higher Y content the alloying effects make it difficult to trace particular NMR lines. The doublet lines overlap with their satellites which become stronger than the main lines. However, there is a clear continuation of a spectrum shape evolution up to x = 0.9. At x = 0.92 a sudden change takes place yielding a different spectrum which is identical for all compositions between 95% Y and a pure $Y_2Co_{14}B$ compound. This means that in spite of

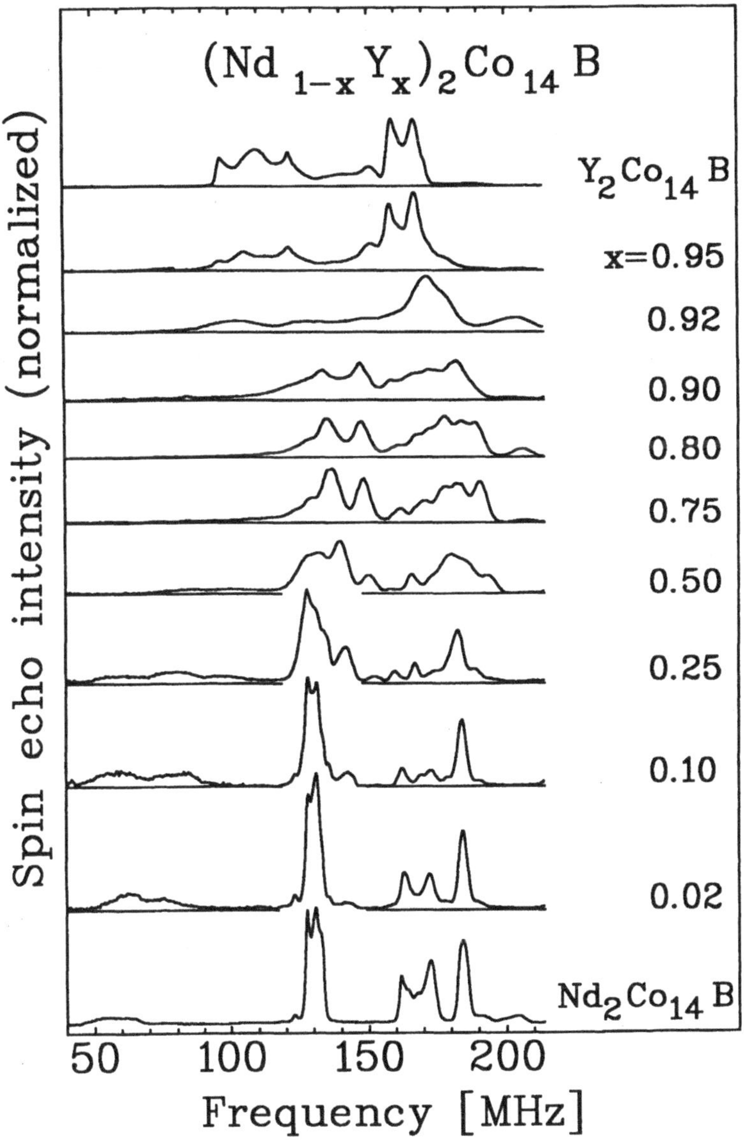

Fig. 1. ^{59}Co NMR spectra in $(Nd_{1-x}Y_x)_2Co_{14}B$ at 4.2 K as a function of yttrium concentration.

significant dilution with non magnetic yttrium, the low temperature magnetic anisotropy of the system is determined by the remaining fraction of Nd. This experimental fact confirms the theoretical results which predict that the anisotropy of this system is of a single ion origin.

Fig.2 illustrates different magnetic configurations encountered in the chosen compositions as a function of temperature. For x > 0.92, planar tendency of Co prevails at all temperatures and our NMR spectra provide a direct evidence of this, as shown in Fig. 2c. A small frequency shift between 77 K and 300 K is due to the obvious decrease of magnetization over this temperature range. For compositions with lower Y content where the temperature interval between T_{SR1} and T_{SR2} is large enough, we can clearly observe all 3

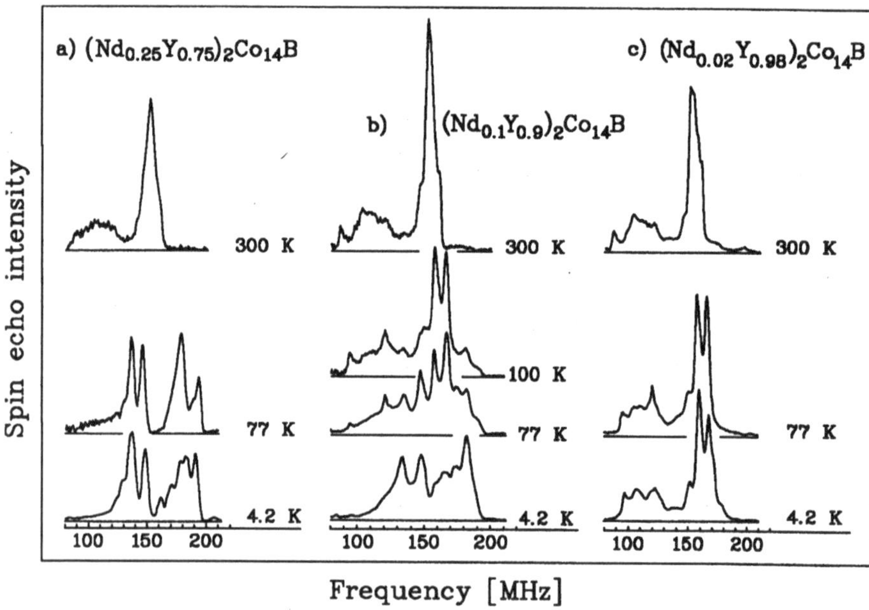

Fig.2. ^{59}Co NMR spectra as a function of temperature in $(Nd_{1-x}Y_x)_2Co_{14}B$

a) x=0.75 : canted, uniaxial and planar configuration,

b) x=0.9 : canted, mixed and planar state,

c) x=0.98: planar structure at all temperatures.

configurations as shown in Fig. 2a for x = 0.75. The transition between the canted state at 4.2 K and uniaxial state at 77 K removes a multiplet structure due to the canting effects. A planar spectrum at 300 K is very similar to that observed for x = 0.98 (Fig. 2c) with some broadening due to the alloying. A very interesting situation is encountered around T_{SR2}, as illustrated in Fig. 2b for x = 0.9. The spectra at 77 K and at 100 K both bear the traces of the 4.2 K spectrum as well as of the planar spectrum recorded at 300 K indicating the presence of a mixture of the two phases over a broad temperature range. This is evidence that the reorientation between the uniaxial and planar state is a first order phase transition, i.e., a certain energy is required to rapidly change the state of the sample. In our powdered samples, the transition temperature is probably scattered over different grains and thus the coexistence of two states is observed in a certain temperature interval. This situation is very different from the reorientation between the canted and uniaxial state as observed in $Nd_2Co_{14}B$ (ref.1), where a smooth and homogeneous change of the canting angle suggested a second order phase transition.

In conclusion, our study demonstrates that at 4.2 K the anisotropy of the mixed $(Nd_{1-x}Y_x)_2Co_{14}B$ system is determined by Nd sublattice up to a threshold Y concentration of 92%. Above the threshold value, the exchange interactions cause the 4f sublattice to follow the 3d anisotropy and planar structures in those compositions have been microscopically confirmed between 4.2 K and the room temperature. Below threshold concentration, planar structures are reached by increasing the temperature in a process which was determined to be the first order phase transition.

References.

1. P. Panissod, E. Jedryka, M. Wojcik, J. I. Budnick, Phys. Rev. B. Rapid Comm. 40 (1989) 2606.

2. A. T. Pedziwiatr, H. Y. Chen, W. E. Wallace, S. E. Malik, IEEE MAG 23 (1987) 41.

Selected Applications of Synchrotron Radiation

Wojciech M. Kwiatek

Henryk Niewodniczański Institute of Nuclear Physics
ul. Radzikowskiego 152, 31-342 Kraków, Poland

Introduction

Since several years synchrotron radiation has been found its application in many fields of science, technology and industry. Such applications have been possible thanks to properties and advantages of synchrotron radiation. It seems that synchrotron radiation techniques are very powerful and their applications will obviously enriched our knowledge about the world and the occurring phenomena.

Synchrotron radiation and its properties.

In circular high energy electron accelerators with the strongly bending magnets and electron energies up to several GeV, when accelerated by the magnetic field the circulating electrons emit electromagnetic radiation with the energies of up to tens of keV. In the reference frame of the electrons the radiation emission is dipolar in nature. The relativistic transformation to the "LAB" frame results in a narrow "fan" of radiation. This kind of radiation produced by relativistic electrons stored in synchrotron storage rings is called synchrotron radiation.

The principle of synchrotron operation can be shortly described as follows. Electrons are accelerated in a linear accelerator of up to several hundred MeV and then injected into an electron synchrotron (booster). There, on a fixed circular orbit they are accelerated up to appropriate energy and transferred to the storage rings. A more specific description of synchrotron operation has been given in several articles and books[1,2]. Fig. 1 shows the emission pattern of synchrotron radiation. The opening angle $\theta = m_0c^2/E$ defines the vertical size of the beam.

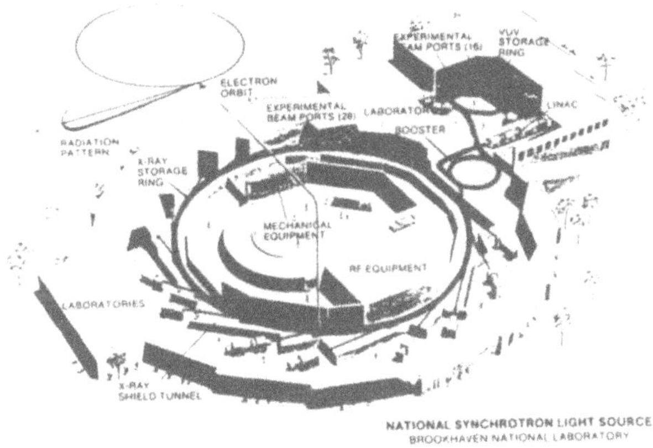

Fig. 1. Diagram of synchrotron radiation

The quantitative description of synchrotron radiation can be given by Schwinger's formula for the total power radiated by a mono energetic relativistic electron in a circular orbit (CGS units).

$$P = \iint I(\lambda,\psi)\, d\lambda\, d\psi = \frac{2}{3}\ \frac{e^2c}{R^2}\ \left(\frac{E}{mc^2}\right)^4 \qquad (1)$$

where:

$$I(\lambda,\psi) = \frac{27}{32\pi^3}\frac{e^2c}{R^3}\left(\frac{\lambda_c}{\lambda}\right)^4 \gamma^8\left[1+(\gamma\psi)^2\right]^2\left[K^2_{2/3}(\phi)+\frac{(\gamma\psi)^2}{1+(\gamma\psi)^2}K^2_{1/3}(\phi)\right]$$

λ_c is the "cut off" wavelength given by

$$\lambda_c = 4/3\ \pi R\gamma^{-3} \quad \text{and} \quad \phi = (\lambda c/2\lambda)[1+(\gamma\psi)^2]^{3/2}$$

In above formulas: $\gamma = E/m_0c^2$, where m_0c^2 is the electron rest mass energy (0.511 MeV) and c is the velocity of light, λ is the wavelength of the emitted radiation, R is the bending radius of the electron orbit, ψ is the elevation angle. Functions $K_{1/3}(\phi)$, $K_{2/3}(\phi)$ are modified Bessel functions of the second kind and they describe the parallel and perpendicular components of the radiation respectively.

The additional basic synchrotron radiation relationships can be summarized as follow[3,4,5].

- energy loss per turn, per electron

$$\delta E(keV) = 88.5\ E^4(GeV)/R(m),$$

- total power radiated

$$P_{tot}(kW) = 26.6E^3(GeV)\ B(T)\ I(A),$$

- critical wavelength

$$\lambda_c(Å) = 5.59R(m)/E^3(GeV) = 18.64/[B(T)\ E^2(GeV)],$$

- characteristic energy

$$\varepsilon_c(eV) = 2218\ E^3(GeV)/R(m) = 2.96\ 10^{-7}\ \gamma^3/R(m),$$

- emission angle

$$\theta_v = 1/\gamma = mc^2/E \quad \text{with} \quad \gamma = E/m_0c^2 = 1957E(GeV),$$

- energy of a photon

$$\varepsilon(eV) = 12.40/\lambda(Å).$$

X-rays from GeV electrons are emitted into emission angles smaller than 0.5 mrad. This opening angle results in an intensity gradient in the vertical direction. The intensity of the radiation as a function of the elevation angle ψ against the orbital plane is shown as an example in fig. 2. The data for fig. 2 are taken from a storage ring with R = 6.875 m and an energy of E = 2.52 GeV (NSLS) for several photon energies.

- advantages and disadvantages

Synchrotron Radiation is a powerful scientific tool. It has several important properties:

- wide spectral range (intense continuum from infrared to the X-ray region)
- excellent directional properties, high degree of collimation (small opening angle)

- spectral brilliance and high brightness
- high degree of polarization in the plane of the electron orbit
- pulsed time structure
- quantitatively known characteristics
- high stability of storage rings ($p<10^{-9}$ Torr)

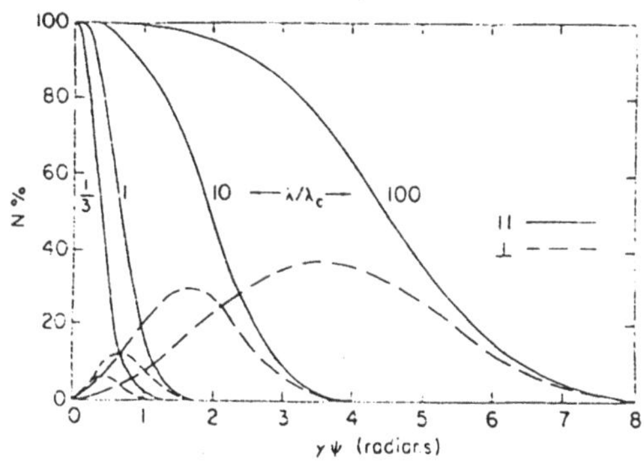

Fig. 2. Vertical intensity of parallel, perpendicular, and total radiation for several energies as a function of the elevation angle ψ against the orbital plane.

Typical spectra of synchrotron radiation (total emitted flux) for several storage rings used as synchrotron radiation sources are shown in fig. 3.

<u>The brilliance</u> of the source is defined as the number of photons emitted per unit source area per second and per unit solid angle and per 0.1% bandwith (BW) in energy.
[photons/(s mm^2 $mrad^2$ 0.1% BW]

<u>The brightness</u> corresponds to the number of photons emitted into a given solid angle.
[photons/(s $mrad^2$ 0.1% BW)]

<u>The flux</u> is the number of photons integrated over the vertical opening angle and beam cross section.
[photons/(s $mrad_{horizontal}$ 0.1 % BW)]

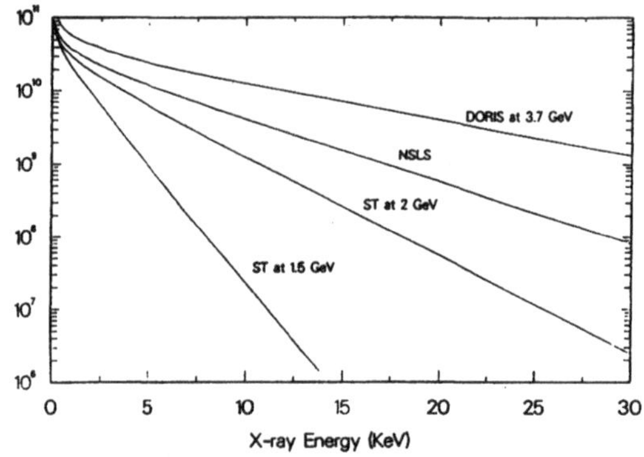

Fig. 3. Spectra of total emitted flux

The degree of polarization is defined as follow:

$$P = (E_{pa} - E_{pe}) / (E_{pa} + E_{pe}) \qquad (2)$$

where:

E_{pa} is a parallel component of the electric field vector and
E_{pe} is a perpendicular component of the electric field vector

It can be shown that the degree of polarization P can be defined as a function of ψ and θ. For $\psi = 0$ (in the orbital plane) the synchrotron radiation emitted by an electron is 100% polarized with the electric vector parallel to the orbital plane. Fig. 2. shows the vertical angular distribution of parallel and perpendicular polarization components for different photon energies.

Circulating electrons in both the synchrotron and the storage rings are grouped in bunches. The lengths of those bunches are determined by the Radio-Frequency (RF) system. Thus, the radiation is pulsed and the typical length of the bunch is 50 ps to 1 ns. The orbital period must be an integral multiple of the RF period. This integer is called the harmonic number of the ring. Very often this number is quite large (greater than 100).

The amount of current stored in a single bunch is limited. A short current pulse excites electrical oscillations in some parts of the vacuum chamber that can result in power dissipation and overheating. Those effects are some of the disadvantages of the system.

Required ultra high vacuum environment for synchrotron radiation production can be both an advantage and disadvantage of the source. It makes the experiments in surface physics which require window less monochromators possible, when however the costs and technical problems are considered the ultra high vacuum requirement is a disadvantage.

Selected applications

Synchrotron radiation produced at NSLS, DESY and other synchrotrons has been applied in physics, chemistry, materials science, metallurgy, biology, and medicine. Various number of different techniques, like: EXAFS (Extended X-ray Absorption Fine Structure), scattering, diffraction, topography, fluorescence (SRIXE), gas phase spectroscopy, lithography, tomography, and microscopy, are used.

The Trace Element Analytical Group in Cracow collaborates with the Division of Atomic and Applied Physics from Brookhaven National Laboratory (USA) and the Synchrotron Radiation Fluorescence Analysis Group from HASYLAB, DESY (Germany), mainly on trace element analysis in different materials. We have applied different types of set-ups in order to improve sensitivity and efficiency of our measurements. Those set-ups include: white-, monochromatic-, focused-, apertured-, and diffracted- beams. Some of the set-ups are shown in fig. 4.

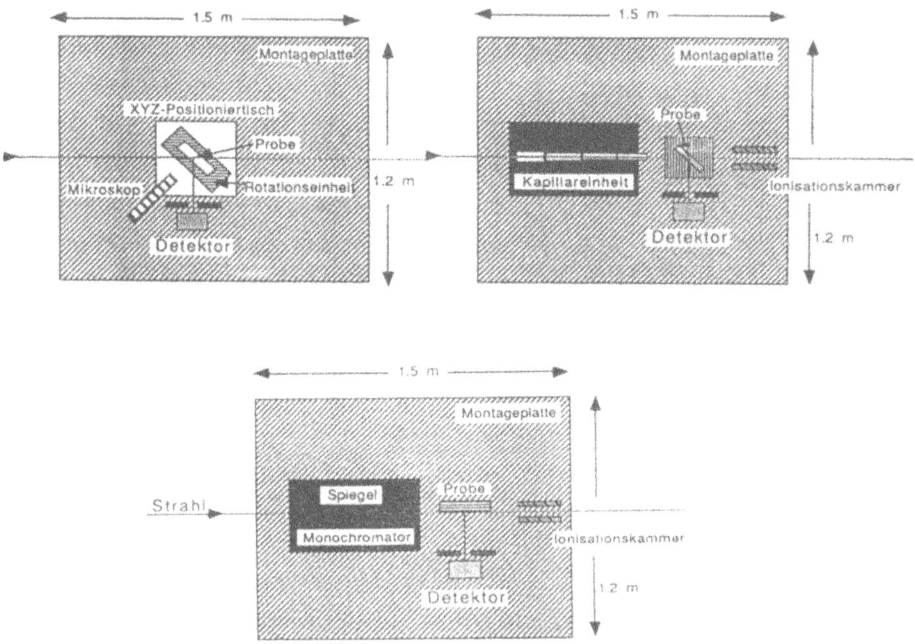

Fig. 4. Different set-ups for synchrotron radiation techniques.

For the SRIXE and micro-SRIXE experiments the samples to be irradiated are placed at 45° to the beam on X, Y, Z, Θ stepper motor driven stage. Such a stage is controlled by a computer. During the measurements the samples are viewed outside the hutch by the TV camera attached to the microscope. For the TRXF analysis thin samples are placed on a crystal that totally reflects incoming X-rays. In this case one can perform bulk analysis.

<u>white apertured beam</u>

The latest application of the SRIXE technique in HASYLAB was the study of trace element content in the cancerous and normal kidney tissues[6]. The kidney cancer is a disease that expands seriously and therefore we have analyzed distribution of TE in thin cancerous kidney sections.

All samples were obtained during surgical operations and then frozen in the liquid nitrogen. The samples were cut into 10 μm thick sections on a cryo-microtome. The sections were placed on 2.5 μm Mylar foil. Adjacent sections were histologically investigated in order to choose the right area for TE analysis and in order to define the cancer type. Majority of the samples were mostly classified as "carcinoma clarocellulare". There was one sample classified as "angiomyolipoma". In each case a boundary between the normal and the cancerous part of a kidney was seen.

Each sample was irradiated for 300 s with a 70 μm x 100 μm photon beam from DORIS storage ring. The beam was filtered with a 125 μm thick aluminum foil. There was no absorber on a 50 mm^2 Si(Li) detector apertured by tantalum collimator with a hole of 3 mm in a diameter. Having such experimental conditions one could achieve a good MDLs (Minimum Detectable Limit) values for multi-elemental standards. An example of MDL curve is shown in fig. 5.

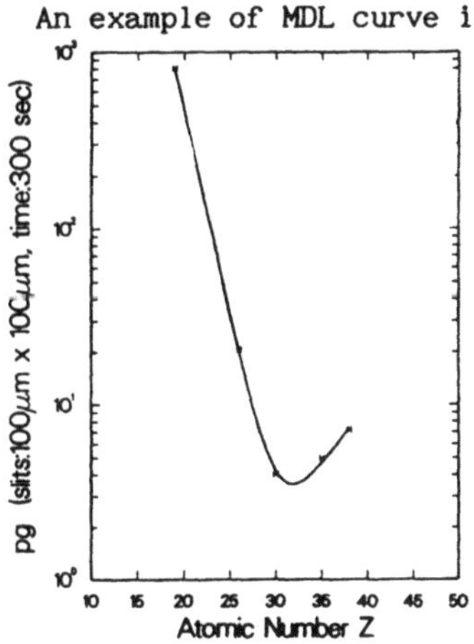

Fig. 5. MDL curve for X-rays.

Making point-to-point analysis on the normal and cancerous sections one can observe significant differences in TE concentration levels between those samples. Basing on this results one can conclude that the SRIXE technique is very powerful, useful and valuable technique for TE analysis in thin tissue sections. Fig. 6 shows three spectra of different kidney sections.

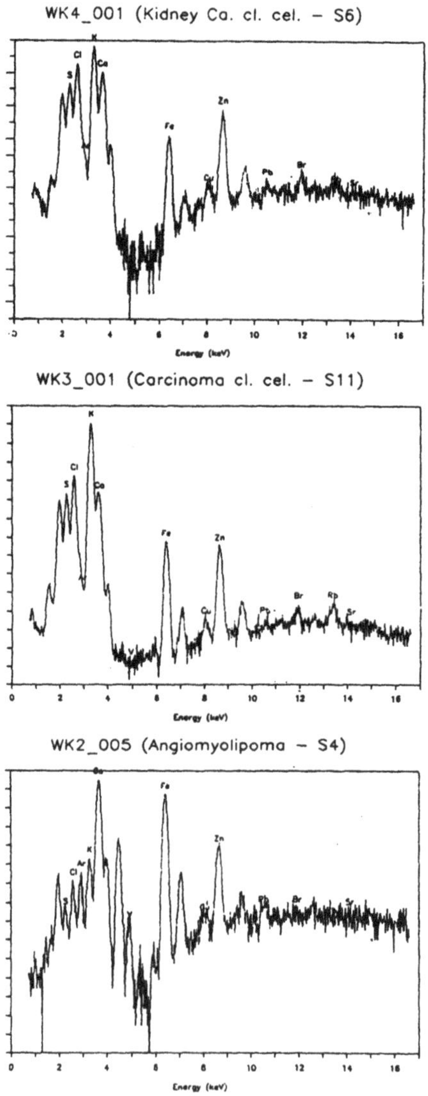

Fig. 6. Different SRIXE spectra of the kidney sections.

-focussed beam with capillary

A swedish group from Göteborg has developed microbeam technique by using the total reflecting properties of X-ray inside glass capillary tubes[7].

According to the Snell's law total reflection of X-ray occurs if the incidence angle is greater than critical angle defined as:

$$\phi_c = \sqrt{2\delta} \qquad (3)$$

where: $\delta = 4.12 \cdot 10^{-4}\ (Z/A)\ \rho\ E^{-2}$

where: Z - atomic number

A - mean atomic mass

ρ - density [g/cm^3]

E - energy of X-rays [keV],

also δ fulfils the equation:

$$n = 1 - \delta \qquad (4)$$

where: n is a refractive index.

Fig. 7 shows the schematic diagram of the total reflection principal in capillary[8].

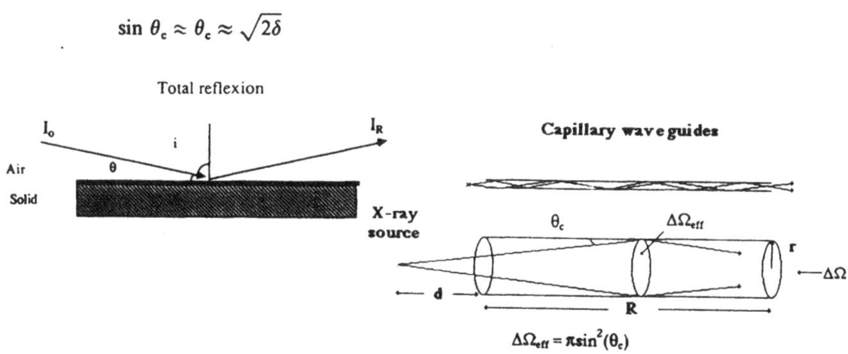

Fig. 7 Schematic diagram of total reflection in capillary.

diffracted beam

White synchrotron radiation from NSLS was applied to obtain monochromatic fluorescent beam. The aim of such an experiment was to get better sensitivity by selecting one specific line from the fluorescent beam and by reducing the number of scattered primary photons from the white beam. The energetic line of 10.54 keV was chosen to be observed. The line selection was not random. This particular line was selected in order to improve the sensitivity of lead measurement. Later, lead toxicity and the importance of the role of trace elements in biological materials will be discussed.

The experimental setup was based on Bragg's diffraction law:

$$\lambda = 2d \sin(\alpha), \qquad (5)$$

were:

d - lattice spacing of a crystal,

λ - Bragg's wavelength,

α - diffraction angle.

Formula (13) can be presented also as

$$\frac{hc}{2dE} = \sin(\alpha), \qquad (6)$$

were:

E - photon energy,

h - Plank constant,

c - light velocity.

Using Be crystal with d = 3.5832 the diffraction angle for Pb L-α line is $9^{\circ}26'55.39''$.

Several apertures were used to cover silicon detector in order to reduce the scattered photons. Beryllium crystal was used as a monochromator and installed on the fluorescent beam. The schematic diagram of the experimental setup is shown in fig. 8.

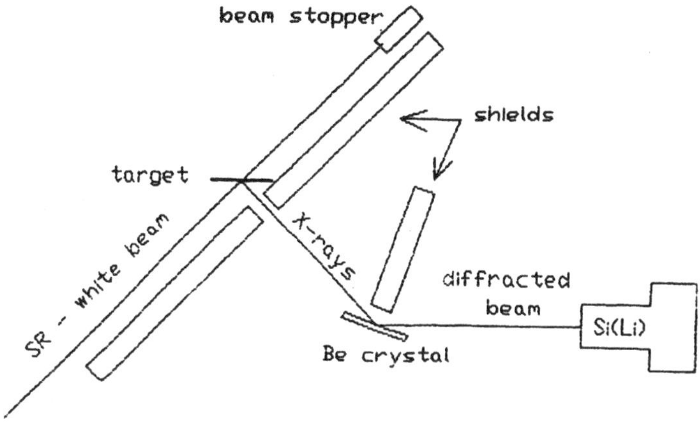

Fig. 8 Schematic diagram of experimental setup for diffracted beam

Fig. 9 shows an overlay of three diffraction spectra obtained with the Be crystal. The intensity of the fluorescent L-α lead peak strongly depends on the crystal rotation angle, while the other components do not.

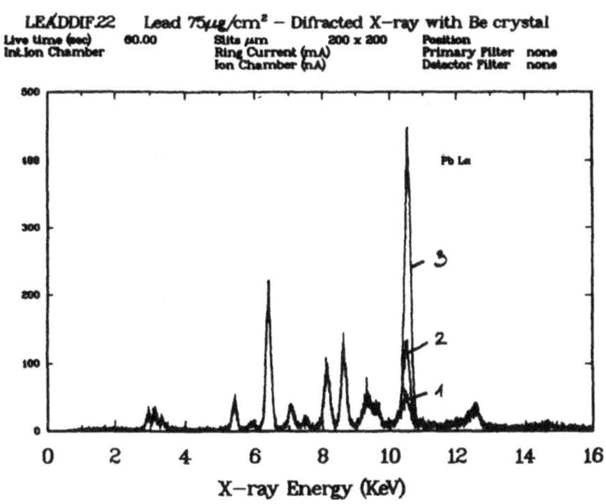

Fig. 9. Overlay of three fluorescent spectra obtained with Be crystal

Since the thin carbon foil coated with 75 $\mu g/cm^2$ Pb was used the above example proves feasibility of such an experiment. The use of Be-crystal as a monochromator of the fluorescent beam obviously increases the sensitivity of a particular element detection. It should be noticed that spectrum labeled "1" was taken without Be-crystal. Then the crystal was employed (spectrum "2") and oriented at Bragg's angle for Pb Lα line (spectrum "3"). The selected peak (Pb Lα) has been increased with a factor of 9 while the background has been not. This made the better MDL and proved the usefulness of such a technique. In order to reduce the background and other peaks in the spectrum the experimental setup has to be improved by better definition of shields. Better alignment would also be required.

Summary

As was shown above it is evident that synchrotron radiation is a very powerful tool in many different field of science. Although the work with synchrotron radiation is not trivial as was previously described[9] such analytical techniques that use synchrotron radiation will improve the further development of synchrotron radiation technique and will find new applications.

The work presented in this paper was partially performed and supported by both Brookhaven National Laboratory, USA and HASYLAB, DESY, Germany.

References

1. Koch Ernst-Eckhard - "Handbook of Synchrotron Radiation" North-Holland Publishing Co. 1983

2. Winnick H., and Doniach S. - "Synchrotron Radiation Research", Plenum Press, New York 1980

3. Schwinger J. Phys. Rev. 70 (1946) 798, Phys. Rev. 75 (1949) 1912, Proc. Nat. Acad. Sci. U.S. 40 (1954) 132

4. Ivanienko D. and J. Pomeranchuk Phys. Rev. 65 (1944) 343

5. Green G.K. -"Spectra and Optics of Synchrotron Radiation", BNL Report (1976) 50522

6. Kwiatek W. M., Cichocki T., Galka M., Paluszkiewicz C. - Microanalysis using synchrotron radiation, To be published, NIM (1991).

7. Buttkewitz A., Grabe S., Gaul G., Knochel A., Lechtenberg F., Engstron P., Larson S., Rindby A. - A submicron synchrotron X-ray Beam generated by capillary Optics, HASYLAB Progress Report (1990) 577.

8. Klockenkamper Reinhold - Total reflection X-ray fluorescence spectrometry: Principals and applications, Spectrometry Intern. Vol.2 No.2 (1990) 26.

9. Kwiatek Wojciech M. - Trace element analysis using synchrotron radiation, Cracow, PhD-thesis 1989.

seminars

Positron Annihilation in a Pyroelectric Medium - Phenanthrene

W. Górniak, T. Goworek and C. Rybka[1)]
Institute of Physics, UMCS Lublin (Poland)
[1)]Lublin Polytechnic, Lublin (Poland)

The pyroelectric effect consists in the appearance of sample polarization P as a result of the change of its temperature

$$P = \int_{T_1}^{T_2} p(T)dT$$

where p(T) is called the pyroelectric coefficient. This effect was observed in various media: crystalline and amorphous ones, and even in the biological systems.

The polarization currents in phenanthrene were observed by Arndt and Damask [1] in the samples biased by external field, but in 1988 Kroupa et al. [2] have shown that similar polarization vs. temperature dependence appears also without bias, thus proving the appearance of pure pyroelectric effect. The polarization current (proportional to the pyroelectric coeffecient when the temperature change rate is constant) is peaked just below the solid-solid phase transition point 73°C and closes to zero at ~ 30°C. Essentially there are no pyroelectric effects above 73°C, in the disordered phase.

It is known that the electric field influences the parameters describing the positron annihilation in solid, thus we performed a series of experiments with positrons to gain some informations about the pyroelectricity of phenanthrene.

Positron entering the insulator can annihilate as a free particle, in collisions, or it can form the bound state with one of the electrons in the medium - the positronium atom (Ps). Positronium can exist in two states: singlet, short-lived one (p-Ps, mean lifetime ~ 120 ps), and triplet state (o-Ps) which annihilates mainly via so called pick-off process, its lifetime in a molecular solid is approximately 1 ns. Thus, the positron lifetime spectrum consists of three exponential components, the longest-lived of them is the hallmark of triplet positronium.

It was shown [3] that in paraffins and in some polymers the electric field reduces the intensity of long lived component I_3 almost without change of its lifetime. At high fields the intensity I_3 approaches the saturation value which is approx. 50% of that in zero field. The sensitivity of phenanthrene to the external electric field is very similar to that of paraffins. The measurements of this kind (Fig.1) we performed at room temperature, thus outside the range where the pyroelectric effects appear. A sum of exponential and a constant was fitted to the experimental data. The value of constant corresponding to the saturation value of I_3 was (11.3 ± 0.8)%.

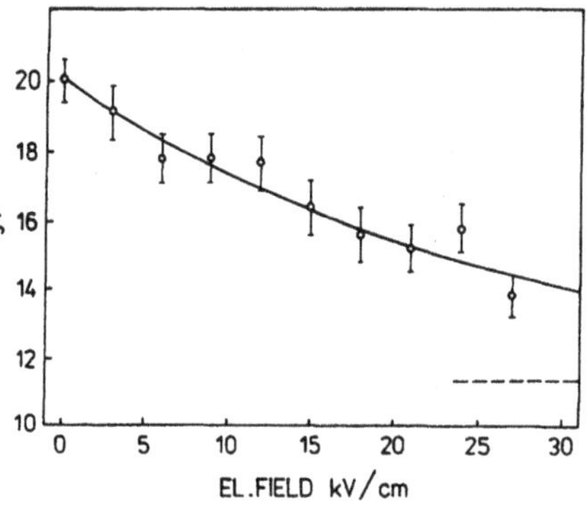

Fig.1 The intensity of long lived component I_3 at room temperature as a function of applied external electric field.

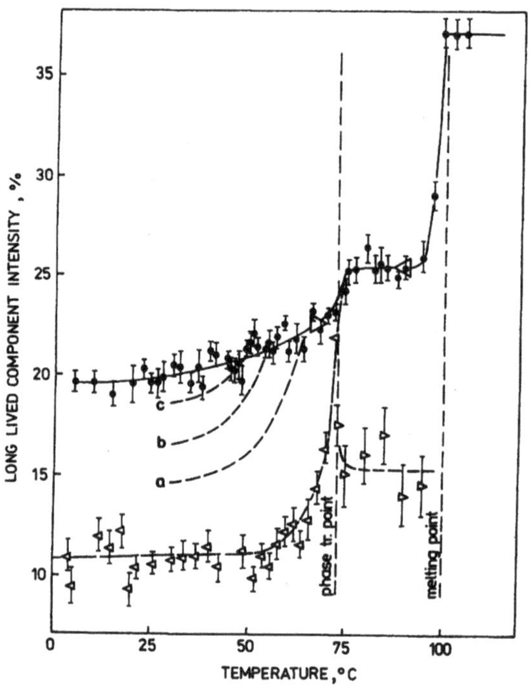

Fig.2 The intensity of long lived component I_3 in the time spectrum of positrons in phenanthrene. Full circles represent the stable values, open ones - the values observed after the change of temperature from the start value indicated by big triangle. Broken lines a, b, c represent the intensities observed for start temperatures 65, 55 and 50°C respectively (experimental points omitted for clarity).

The pyroelectric polarization was generated in the sample as follows: the sample was kept several hours in 90°C, i.e. above the range of pyroelectricity, then the temperature was changed with the rate 0.2 K/s to a preselected value below solid-solid phase transition point and the lifetime spectrum of annihilating positrons was registered in the time interval as short as possible. This time limitation is necessary, because the reduced I_3 observed after the thermal jump is not stable and rises slowly to the value corresponding to the zero field. The "stable" intensity and the intensity reduced after the skip from 90°C are shown as a function of temperature in Fig.2 (triangles pointed to the left). The relation $-\Delta I_3/I_3$ vs. temperature is very similar to that of polarization vs. T [2], but it is more flattened at low temperatures. It is due to the saturation of o-Ps inhibition at high fields. The value of I_3 in polarized sample at low temperatures is very close to saturation value determined in the experiments with the external field.

The measurements in an external electric field were also performed at room temperature with the samples directly after the change of temperature (low I_3). No field dependence was observed, as it was expected - addition of small external field to the internal spontaneous polarization did not change measurably I_3 in the saturation region (Fig.3).

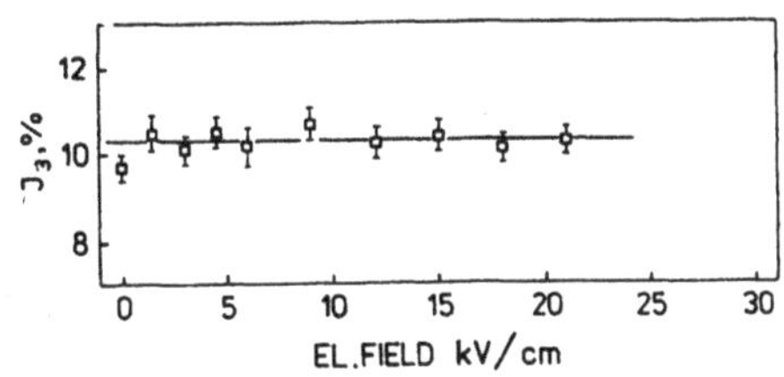

Fig.3 The intensity I_3 in polarized sample (at room temperature) as a function of applied external electric field.

When the polarized sample is stored at a fixed temperature, the intensity I_3 rises as a result of depolarization by internal currents. This rise can be described (at least in high temperatures) by the exponential curve (Fig.4), the time constant t_d is strongly temperature dependent. If the reciprocal of t_d is the measure of detrapping rate of the pyroelectric charges its variation with temperature enables to estimate the depth of trap level. The values of t_d, particularly in the region of low temperature, are not fully reproducible, only the rough estimate of the trap depth can be given. This value, $\sim$ 0.5 eV, is close to the maximum of trap distribution determined in the luminescence experiments [4].

The positron before annihilation is already thermalized and diffuses in the sample, thus the orientation of the field is not important from the viewpoint of Ps inhibition observations. The change of temperature from a start point below the phase transition to the temperature above that transition, generating the polarization in reverse direction, gives also the reduction of positronium formation yield, as can be seen in Fig.2 (triangles pointed to the right).

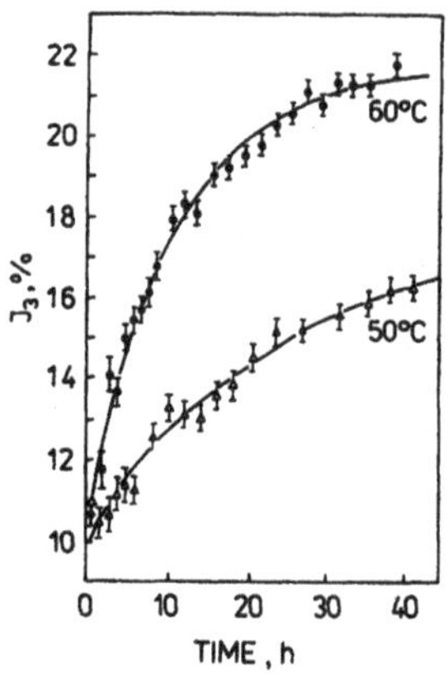

Fig.4 Long lived component intensity I_3 after the change of sample temperature 90° C - 25° C as a function of storage time for two selected temperatures.

The pyroelectric coefficient $p(T) = \partial P/\partial T$ is peaked slightly below the phase transition point but it has the form of a broad distribution covering the range of about 40°C, thus when the start point for the change of temperature lies in that range, the polarization of sample should be observed. The lower the start temperature is, the smaller is the effect. A series of measurements with initial temperatures 65, 55 and 50°C was performed. The intensities I_3 are shown in Fig.2 by the curves a, b, c respectively. The jump 50°C - 30°C gives a very small effect.

The closed cycle of temperatures should result in the lack of polarization, if there is no hysteresis in p(T) temperature up - temperature down dependence. The positron lifetime spectrum was measured at 78°C, then the temperature was lowered for 0.5 h to 68°C and then increased again to 78°C. The spectra before and after the cycle were identical in the limits of error, indicating the lack of detectable hysteresis.

The examples of measurements described above indicate that the positron annihilation can be an useful tool in the investigations of pyroelectricity. An additional advantage of the method is the possibility to use the policrystalline samples, a disadvantage - long duration of the experiment.

References

[1] R.A. Arndt and A.C. Damask, J.Chem.Phys. 45, 4627 (1966).

[2] J. Kroupa, J. Fousek, N.R. Ivanov, B. Březina and V. Lhotska, Ferroelectrics 79, 189 (1988).

[3] W. Brandt and J. Wilkenfeld, Phys.Rev. B12, 2579 (1975).

[4] Z. Dreger, J. Kalinowski, I. Davoli, S. Stizza and S. Feliziani, phys.stat.sol. (b) 149, 363 (1988).

HELIUM ASSISTED TRAP MUTATION IN TUNGSTEN OBSERVED BY MEANS OF PAC

M. S. Abd El Keriem[a)], D. P. van der Werf and F. Pleiter

Laboratorium voor Algemene Natuurkunde, Materials Science Centre
University of Groningen, The Netherlands

ABSTRACT. We have seen with PAC that monovacancies in tungsten decorated with He can mutate to form He decorated divacancies while pushing out self-interstitials.

Thermal Helium Desorption Spectrometry (THDS) measurements on He filled monovacancies in bcc metals have shown that, up to a certain number of He atoms per vacancy, the dissociation energy decreases with increasing occupancy. When more He is added, it seems that the dissociation energy increases[1)]. An explanation of this phenomenon is that the decorated monovacancy tranforms into a decorated divacancy plus an interstitial:

$$VHe_n \rightarrow V_2He_nI \rightarrow V_2He_n + I \qquad (1)$$

This is called trapmutation. Unfortunately, VHe_n and V_2He_n have approximately the same dissociation energies[2)] and, therefore, can not be seen separately with THDS. With PAC it is possible to observe this phenomenon directly because the two complexes will give rise to different frequencies and symmetries in the PAC spectra. Since an In atom is part of the defect structure, reaction equation (1) should be rewritten as :

$$InV_2He_n \rightarrow InV_3He_nI \rightarrow InV_3He_n + I \qquad (2)$$

To investigate this process, we implanted two W foils with 50 keV ^{111}In ions to a dose of 7×10^{12} cm^{-2}, and subsequently annealed one of them at 620 K, and the other at 750 K. The PAC spectra show components with hyperfine parameters ω_0 = 133 Mrad/s and η = 0, which is attributed to the InV_2 complex, and ω_0 = 300 Mrad/s and

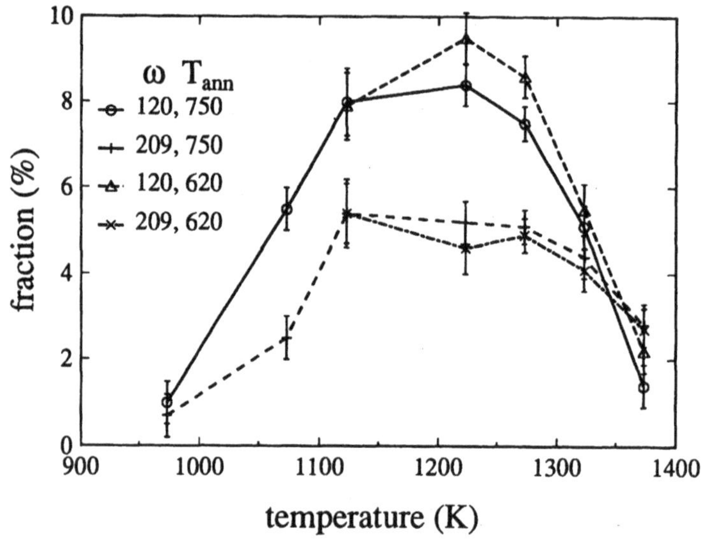

Figure 1 *Fractions of the 122 and the 109 Mrad/s components as function of the annealing temperature.*

$\eta = 1$, which is attributed to the InV_3 complex[3,4]. The fractions of the 133 Mrad/s component were 25% in the former case and 22% in the latter case, while for the 300 Mrad/s component these values were 3% and 9%, respectively. We then implanted the samples with 300 eV He to a dose of 3×10^{15} cm^{-2} and observed that in both samples the fractions of InV_2 and InV_3 decreased by approximately 50 %. Hereafter, we measured the annealing behaviour (see fig. 1), beginning at about 1000 K which is far above the dissociation temperatures of the InV_2 and InV_3 complexes. We saw two distinct components: one with a frequency of 122 Mrad/s and $\eta = 0$, and one with a frequency of 209 Mrad/s and an asymmetry parameter that varied from $\eta = 0.9$ at 1100 K to $\eta = 1.0$ at 1300 K. A tungsten foil which received the same treatment except that the annealing after ^{111}In implantation was carried out at a temperature of 820 K, did not show the presence of InV_2 and InV_3 clusters. After He implantation and annealing above 1000 K no clear frequencies were observed.

We ascribe the 120 Mrad/s and the 209 Mrad/s components to He filled monovacancies and He filled divacancies, respectively. The asymmetry parameter of 0 and 1 is what is to be expected for these defect configurations. The change of η from 0.9 to 1 in the second case is attributed to "cooking off" one by one the He atoms. We see further that both components disappear simultaneously, which is to be expected according to THDS measurements[5]. We conclude from our observations that the 120 Mrad/s and 209 Mrad/s components are generically related to the InV_2 and InV_3 clusters. From the fact that the fraction of InV_3 after preparative annealing of 750 K was significantly smaller then the fraction of He decorated InV_3 (see fig 1), we conclude that trap mutation has occurred indeed.

This work forms part of the research programme of the "Stichting voor Fundamenteel Onderzoek der Materie" (FOM) with the financial support from the "Nederlandse Organisatie voor Wetenschappelijk Onderzoek" (NWO)

[a] Permanent address : Ain Shams University, Cairo, Egypt.

References

[1] Veen, A. van, Materials Science Forum 15-18, 3 (1987)

[2] Veen, A. van, Evans, J. H., Buters, W. Th. M., and Caspers, L. M., Radiation Effects 78, 53 (1983)

[3] Pütz, U., Hoffmann, A., Rudolph, H.J., and Vianden, R,. Z. Phys. B46, 107 (1982)

[4] Post, K., Pleiter, F., Kolk, G. J. van der, Veen, A. van, Caspers, L. M., and Hosson, J. Th. M. de, Hyperfine interactions 15/16, 421 (1983)

[5] Kolk, G. J. van der, Post, K., Veen, A. van, Pleiter, F., and Hosson, J. Th. M. de, Radiation Effects 84, 131 (1985)

PAC STUDY OF INTERNAL NITRIDING OF HAFNIUM IMPLANTED IN IRON

J.M.G.J. DE BAKKER and F. PLEITER
Laboratorium voor Algemene Natuurkunde, MSC,
Westersingel 34, 9718CM Groningen, The Netherlands

ABSTRACT

Using the PAC-technique, we studied the substitutional fraction of ^{181}Hf implanted in Fe single crystals as a function of temperature and the trapping of low-energy post-implanted nitrogen as a function of dose.

Two Fe single crystals were implanted at 80 K and 295 K, respectively, with 110 keV ^{181}Hf to doses lower than $1x10^{14}$ at.cm^{-2}. The first sample was annealed for 20 minutes at successively higher temperatures in the range 80 K - 295 K. The second sample was post-implanted at 450 K with increasing doses of 500 eV N_2. After each annealing treatment and after each post-implantation, the perturbed angular correlation of the 133 - 482 keV γ-ray cascade of ^{181}Ta was measured with a conventional four-detector setup equipped with BaF_2 crystals[1)]. The samples were magnetized in an external field oriented in the plane of the detectors at an angle of 45° with the detector axes[2)]. One measurement was done with the field oriented perpendicular to the detector plane.

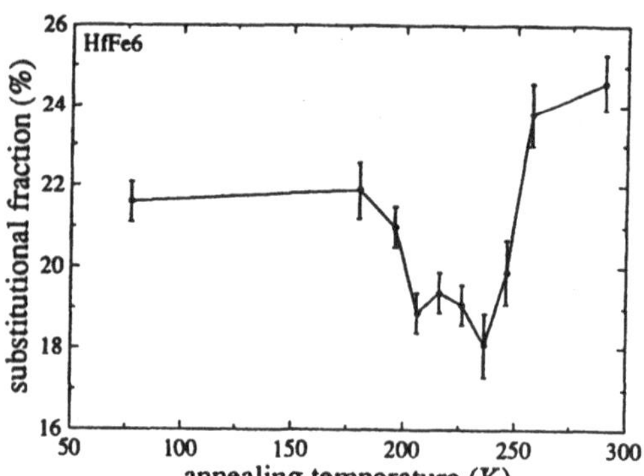

Fig. 1. Temperature dependence of the substitutional fraction of hafnium implanted in iron as measured with PAC.

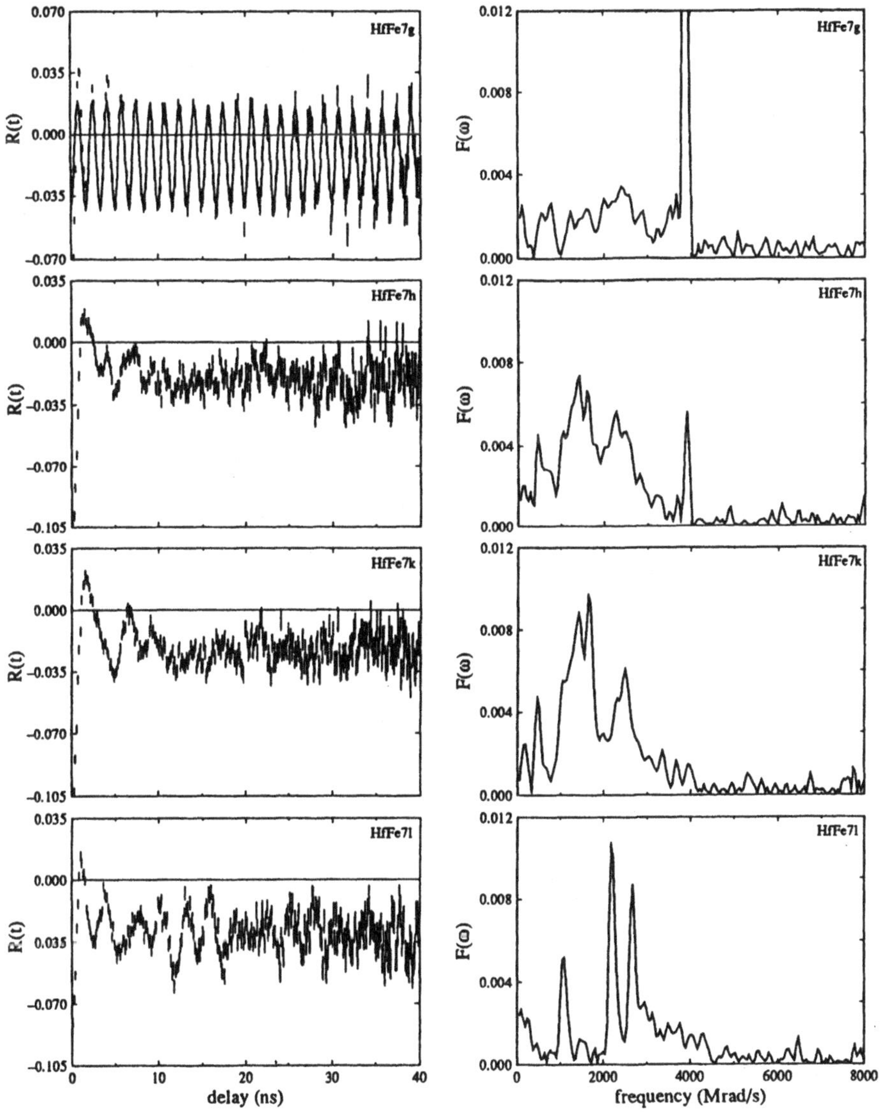

Fig. 2. Ratio R(t) and its Fourier transform for ^{181}*HfFe post-implanted with nitrogen to doses of 1.5x10*^{16} *cm*$^{-2}$ *(HfFe7g), 2.5x10*^{16} *cm*$^{-2}$ *(HfFe7h) and 4x10*^{16} *cm*$^{-2}$ *(HfFe7k and HfFe7l). The measurements were conducted in the Raghavan geometry, except for HfFe7l where the external magnetic field was oriented perpendicular to the detector plane.*

The Fourier transform of the perturbation function measured on ^{181}Hf<u>Fe</u> as implanted at 80 K shows one well-defined peak with a Larmor frequency of ω_L = 3.788(2) Grad/s. This interaction frequency is assigned to substitutional ^{181}Hf atoms in a damage-free close environment and corresponds to a magnetic hyperfine field of $|H_{hf}(^{181}\mathrm{Hf}\underline{\mathrm{Fe}})|$ = 60.8(5) T. The substitutional fraction f_S as a function of annealing temperature is shown in fig. 1. The decrease and increase of f_S after annealing at 200 K and 250 K, respectively, is attributed to trapping and subsequent detrapping of vacancies by the ^{181}Hf probe atoms. A similar change of f_S has been observed for the system ^{111}In<u>Fe</u> by Pleiter et al.[3]

Plots of the ratio R(t) and its Fourier transform for the sample post-implanted with different doses of nitrogen are shown in fig. 2. Up to a dose of about 1.5×10^{16} cm^{-2} there is no significant change in the close environment of the substitutional ^{181}Hf atoms. At about 2.5×10^{16} cm^{-2}, a series of new frequency components appear at the expense of the substitutional peak. Above 3.5×10^{16} cm^{-2} f_S is zero. The new frequencies do not change significantly up to the highest dose of about 3×10^{17} cm^{-2}.

From a comparison of two measurements with different magnetization orientations (fig. 2, HfFe7k and HfFe7l) we conclude that the interaction leading to the observed frequencies must be of combined electric and magnetic nature. The relative magnitude and orientation of the electric field gradient and the magnetic hyperfine field is not yet known. The interaction is attributed to trapping of nitrogen by substitutional ^{181}Hf atoms.

REFERENCES

1) Arends, A.R., Hohenemser, C., Pleiter, F., De Waard, H., Chow, L. and Suter, R.M., Hyp. Int. <u>8</u>, 191 (1980)

2) Raghavan, R.S. and Raghavan, P. Nucl. Instr. Meth. <u>92</u>, 435 (1971)

3) Pleiter, F., Hohenemser, C. and Arends, A.R., Hyp. Int. <u>10</u>, 691 (1981)

QUADRUPOLE INTERACTION OF ^{111}Cd IN THE SYSTEMS $Sb_{1-x}M_x$ (M=Zn, Ga, Ge)

H. R. Blank, M. Frank, M. Kaltenhäuser, H. Köchner, W. Kreische, N. Müller, T. Wagner, G. Weeske

Physikalisches Institut der Universität Erlangen-Nürnberg, Erwin-Rommel-Str. 1, D-8520 Erlangen, Germany

ABSTRACT

Using the TDPAC method the electric quadrupole interaction and its temperature and concentration dependence were measured for the semimetal Sb with small amounts of admixtures of the metals Zn, Ga, and Ge. A strong dependence on the concentration of the admixture can be observed. These measurements seem to indicate a general trend for $Sb_{1-x}M_x$ systems.

INTRODUCTION

The time differential perturbed angular correlation (TDPAC) technique was applied to our measurements of the electric field gradient (EFG) of ^{111}Cd in the systems $Sb_{1-x}Zn_x$, $Sb_{1-x}Ga_x$, and $Sb_{1-x}Ge_x$. For metals the EFG shows a $T^{3/2}$ dependence as reported by Christiansen et al. [1]. However, for semiconductors and semimetals no general behaviour has been found yet. Therefore the semimetal Sb mixed with small amounts of some elements of the fifth period of the periodic table had been investigated in previous measurements [2]. In this paper the corresponding measurements with elements of the forth period are presented. We used ^{111}Cd as probe nuclei. From the experimental TDPAC spectra the effective amplitude, the quadrupole coupling constant ν_Q, the relative frequency distribution, and the asymmetry parameter η can be extracted.

EXPERIMENTAL PROCEDURE

The samples were prepared by electroplating radioactive ^{111}In onto antimony foils. Afterwards the antimony pieces were melted together with inactive Sb and Zn, Ga, or Ge in argon atmosphere. Subsequently the samples were quenched to room temperature. In all cases an annealing procedure has been performed to heal out structure defects in order to get considerable amplitudes.

We have used the common experimental setup for TDPAC measurements [2]. A typical spectrum is shown in Fig. 1. All samples were investigated in the temperature range 20 K $\leq$ T $\leq$ 700 K. The R(t) spectra were fitted with one frequency and $\eta = 0$. The amplitude $A_{22}(t)$, corrected for finite solid angle, decreases with increasing temperature. Fig. 2 shows the concentration and temperature dependence of the systems $Sb_{1-x}Zn_x$, $Sb_{1-x}Ga_x$, and $Sb_{1-x}Ge_x$. The EFG is plotted against a $T^{3/2}$ scale.

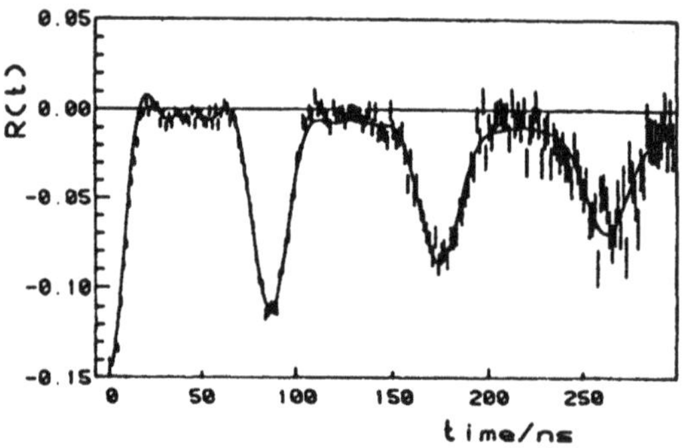

Fig. 1: R(t) spectra of $Sb_{1-x}Zn_x$ (x=0.0023) at room temperature (295 K).

With increasing amount of impurity the temperature dependence approaches the well known $T^{3/2}$ behaviour, which is valid for almost all metals. For $Sb_{1-x}Zn_x$ at 20 K the EFG increases from 70 MHz at x=0.1 wt.% to 91 MHz at x=1.65 wt.%. The EFG in $Sb_{1-x}Ga_x$ at 110 K increases from 65 MHz at x=0.14 wt.% to 112 MHz at x=0.97 wt.%. For $Sb_{1-x}Ge_x$ at 20 K we observed an interaction frequency increasing from 78 MHz at x=0.07 wt.% to 133 MHz at x=1.85wt.%.

RESULTS

The obtained results are in accordance with earlier measurements of antimony containing the impurities Cd, In and Sn [3]. For the system $Sb_{1-x}Cd_x$ it has been shown there that the quantity of the added impurity Cd is the reason for the varying temperature dependence. In this case it is possible to remove the impurity as CdSb by a phase transition. This is done by heating the sample up to more than 720 K. During the subsequent cooling process Cd precipitates as CdSb, while the probe nuclei stay in the regular lattice sites. Afterwards the measured coupling constants are the same as in pure Sb. To ensure that there is no segregation of metal clusters, generating a quadrupole frequency of the respective system $^{111}Cd\underline{M}$, the results were compared with measurements of ^{111}Cd in Zn, Ga [4]), and Ge [5]). The EFG of ^{111}Cd in $Sb_{1-x}Zn_x$, for example, increases from 71 MHz (x=0.1 wt.%) to 86 MHz (x=1.65 wt.%) at room temperature, while the EFG of ^{111}Cd<u>Zn</u> equals 134 MHz at room temperature.

All metal admixtures which have been investigated yet show the same temperature dependences of the EFG (except Ag, which could be an artefact). In the observed systems $Sb_{1-x}M_x$ the EFG highly depends on the concentrations of the admixtures. For vanishing impurities the behaviour approaches that of pure antimony. For impurities approximately greater than 1 % the observed behaviour is equivalent that of pure metals. Further measurements will be performed to verify that general behaviour of $Sb_{1-x}M_x$ systems for other metals as well as to investigate any possible pressure dependence.

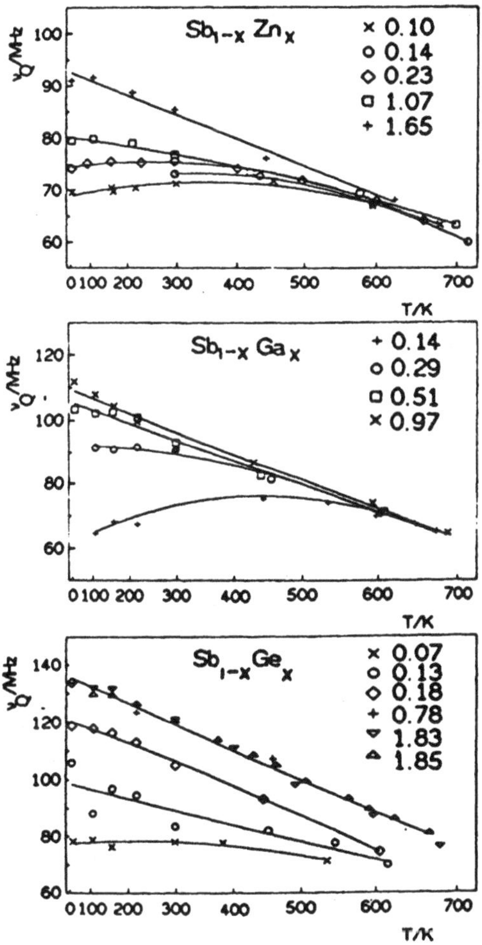

Fig. 2: ν_Q versus $T^{3/2}$ for different concentrations (in wt.%) of Zn, Ge, and Ga.

REFERENCES

/1/ J. Christiansen, P. Heubes, R. Keitel, W. Klinger, W. Loeffler, W. Sandner, W. Witthuhn: Z. Phys. B 24 (1976) 177

/2/ H. Barfuß, G. Böhnlein, F. Gubitz, W. Kreische, B. Röseler: J. Mol. Struc. 111 (1983) 21

/3/ G. Schatz, A. Weidinger: Nukleare Festkörperphysik. B. G. Teubner, Stuttgart (1985)

/4/ R. Vianden: Hyp. Int. 35 (1987) 1079

/5/ D. Forkel: Thesis. Erlangen 1987

MAGNETIC MOMENTS IN ULTRATHIN FILMS DETERMINED BY SPIN-POLARISED NEUTRON REFLECTION

H J Lauter, J A C Bland[1], R D Bateson and A D Johnson[2]

Institut Laue Langevin, Avenue des Martyrs, Grenoble 38042, France
1 Cavendish Laboratory, Madingley Road, Cambridge, CB3 OHE, UK.
2 Department of Physics, University of Leicester, Leicester, LE1 7RH, UK.

ABSTRACT

Polarised neutron reflection (PNR) studies of the absolute value of the in-plane magnetic moment per atom of sandwiched monolayer thickness ferromagnetic single films are discussed. It is demonstrated that the non-magnetic overlayer of the sandwich structure results in an increased spin-dependence of the reflectivity arising from the magnetic layer. PNR is also shown to provide valuable information relating to the interface quality in such ultrathin structures. We discuss recent results for sandwiched Co and Fe films in the 1-10ML thickness range as an illustration of PNR as a magnetometric technique and as a test of recently developed theories of modified magnetic moments in ultrathin films.

INTRODUCTION

Many recent experimental studies of the novel magnetic properties of ultrathin epitaxial films have focussed on the thickness-dependence of the magnetic anisotropy, magnetic order and Curie temperature in the few monolayer thickness range [1]. However, a key question concerns the value of the magnetic moment per atom, the most fundamental of magnetic properties. Calculations of the electronic and magnetic structure of epitaxial films using recently developed spin resolved band structure techniques [2] demonstrate that the reduced atomic coordination, modified volume per atom and the interface-induced electronic states all strongly affect the moment. In this contribution we discuss the application of the recently developed technique of polarised neutron reflection (PNR) to the study of the absolute value of the magnetic moment per atom of sandwiched ultrathin epitaxial films and magnetic layers [3]. We demonstrate that PNR also provides a tool for characterising the structural quality of the interfaces which strongly influence magnetic properties [4,5].

Predictions of strongly modified magnetic moments in epitaxial transition metal monolayers supported by noble metal films [2] have motivated our recent PNR studies of coated Fe [4-8] and Co [9] films in the 1 - 10 ML thickness range supported by Ag(001) and Cu(001) substrates.

SPIN-POLARISED NEUTRON REFLECTION

The refraction of neutrons at a magnetic surface has been extensively discussed [12,13] and we briefly outline the utility of PNR in directly determining the magnetisation of sandwiched epitaxial films [14,15]. The sample is modelled as a multilayer of homogeneous media of neutron refractive index n_i, thickness t_i, magnetic induction $\mathbf{B}_i$, atomic density ρ_i and coherent nuclear scattering length b_i where the subscript refers to the medium i. The neutron beam is incident at a glancing angle θ with respect to the sample surface and q_i, the component of the neutron wavevector normal to the interface in the medium i can be written as [15]:

$$q_i^2 = q^2 - 4\pi\rho_i b_i + (8\pi^2 m_N/h^2)\,(\mu.\mathbf{B}_i) \qquad (1)$$

where μ, $\mathbf{k}$ denote the neutron magnetic moment and wavevector respectively and $q=k\sin\theta$ corresponds to the perpendicular wavevector of the incident beam. It follows from (1) that the discontinuity in the in-plane component of the magnetic induction gives rise to refrac[illegible] [illegible]nterface and so reflectivity measurements determine the in-plane magnetisation M_1 [illegible]. The reflection amplitude r_{ij} at the interface i,j can be written in exact analogy with the Fresnel formula for s wave (TE) polarised light. The sandwiched single magnetic layers can be modelled as a four medium system of vacuum(1)/overlayer(2)/magnetic layer (3)/substrate (4) with amplitude reflectivity r_{1234} [14]. The reflected beam polarisation S(q) is related to the ratio of spin dependent intensity reflectivities F(q) (flipping ratio) and beam polarisation P_o by:

$$S(q) = (F(q) - 1)/(\,P_o(F(q) + 1)) \qquad (2)$$

The peak value of F - 1 is directly proportional to the layer magnetisation B_3t_3 and is enhanced by the overlayer [14]:

$$(F(q) - 1) \;\alpha\; B_3t_3 q\sin(2q_2t_2) \quad \text{for } q_3t_3 << 1 \qquad (3)$$

For 'thin' overlayers the substrate determines the wavevector q_c correponding to critical reflection and the polarisation is therefore zero for $q < q_c$. Equation 3 shows that for all ultrathin sandwiched ferromagnetic films the flipping ratio has the same form as a distorted sine wave with the position in wavevector of the first peak determined by the overlayer thickness. In figure 1 we compare the flipping ratio calculated for a coated (curve A) and an uncoated (curve B) 10 ML fcc Co film. The inset of the figure shows schematically the scattering geometry and the sandwich sample structure. The role of the overlayer in enhancing the magnetic response is similar to that of the antireflection coating layer used in magneto-optic media. For thicker magnetic films the wavector dependence is modified by the form of the magnetisation profile within the film.

The effect of short lengthscale interface roughness of r.m.s. amplitude $\langle z_r^2\rangle^{1/2}$ can be modelled by reducing the sharp interface reflectivity r_{ij} by a multiplicative factor of $\exp(-4q_iq_j\langle z_r^2\rangle)$ [16]. The correlations in the interface fluctuations cause the local refractive index to significantly differ from the average value over an extended region. The coherence length w for such correlations can be estimated from the width in wavevector Δq_t of the diffuse scattering in the vicinity of the specularly reflected beam and the FWHM is proportional in a first approximation to w^{-1} [17]. The surface 'waviness' is modelled as part of a total effective wavevector spread Δq of the incident beam and can be determined from the width in wavevector of the specularly reflected beam.

Experimental arrangements for controlling the incident wavevector of cold neutrons using both sample rotation at fixed wavelength in combination with a position-sensitive multidetector and also time of flight have been previously described in detail [14,15]. In figure 2 we show schematically the rotating sample arrangement we have adopted for the studies presented here based on an adaptation of the D17 diffractometer at ILL, Grenoble. The neutron spin is aligned parallel or antiparallel to the in-plane magnetisation induced along the sample rotation axis as shown in the inset of figure 1.

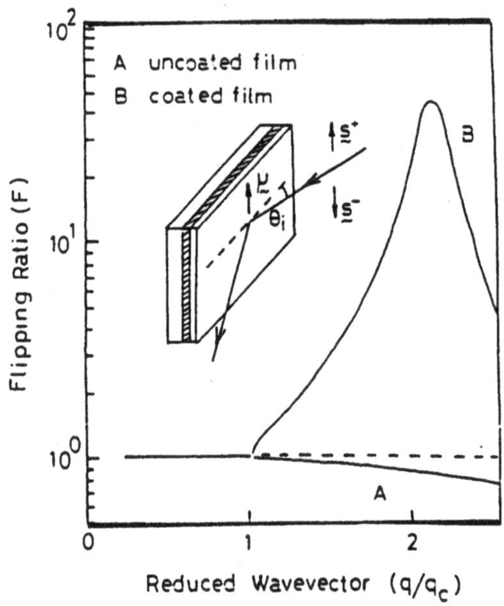

Fig.1: The flippling ratio F(q) calculated for 42 Å Cu/18 Å Co/Cu(001)-curve B, μ_{Co} =1.8 μ_B; for 18 Å Co/Cu(001)-curve A, μ_{Co} =1.8 μ_B and Cu(001) (dashed line).

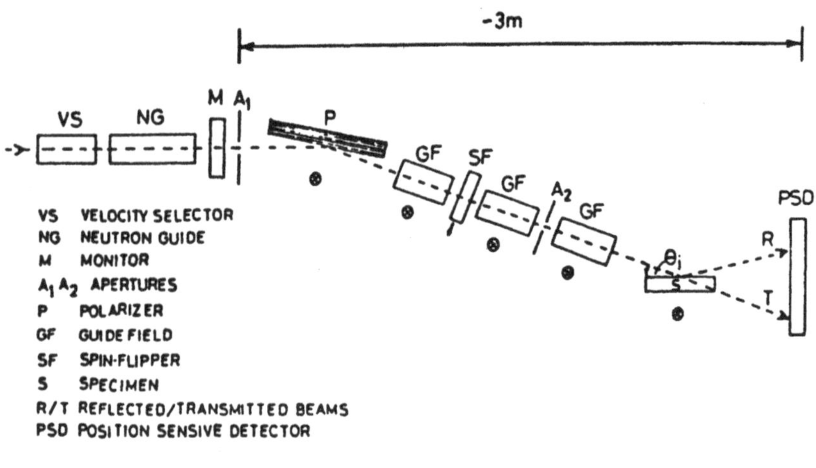

Fig.2: Spin polarized neutron reflectance beam line showing the perpendicular (⊗) and parallel (↓) configuration of the H fields with respect to the scattering plane.

FCC Co(001) FILMS

The close lattice match of the high temperature phase fcc Co lattice parameter (3.544Å) and room temperature fcc Cu lattice parameter (3.61Å) permits the epitaxial growth of fcc Co/Cu(001) at 300K up to large thickness. Surprisingly largely thickness-independent magnetic properties are observed for [14,22-3] epitaxial fcc Co films prepared on Cu(001) substrates in the 1 - 10 ML thickness range, with moments comparable with the bulk phase value of 1.7μ_B and very high Curie temperatures. In figure 3 we show the flipping ratio observed for a 4ML Co film and also for a 10ML Co film. The polarisation of the thicker film is consistent with a moment per atom between 1.5 μ_B and 2.1 μ_B whereas the data for the thinner film is consistent with a moment per atom greater than 2μ_B. The data are of importance in confirming that the enhancement mechanism due to wave interference within the sandwich structure does indeed occur in practice and that the predicted form of equation 3 is valid.

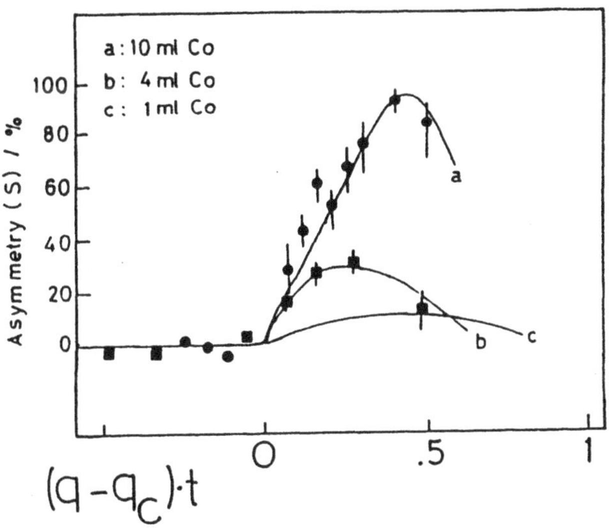

Fig.3: Experimental spin asymetry S in function of (q-qc)t. t is the thickness of the Cu-overcoating. Curve a: d=18 Å and t=42 . Curve b: d=7.2 Å and t=130 . Curve c is the projection over 1 ML film with overcoating thickness 75 Å. The solid curves present the fit to the experimental data.

Recent angle resolved Auger measurements of the growth of Co/Ag(001) have confirmed the stabilisation of the fcc phase [24]. For the present study [25] Co films were prepared by evaporation in UHV at 300K. Three well defined breaks in the Auger signal versus time (As-t) curve are observed upon completion of successive monolayers. Beyond 3ML the LEED pattern fades. The existence of the fcc phase at low coverage is surprising in view of the 15.4% in-plane expansion required for true epitaxial growth and it is possible that the initial growth proceeds with the development of fcc islands with the lattice parameter of the high temperature bulk fcc phase. The angle resolved measurements show a variation in the angular position of the local emission maximum which coincides with 45 degrees at 2ML and moves to lower angles at 5ML. The observed flipping ratio for 2ML sandwiched Co films at 4K with a 0.83 kG field applied fits to μ_{Co} = 2.0μ_B

with the Co film lattice matched to the Ag substrate. However the magnetisation of the 5ML film shows a very much reduced value[25] No increase in the neutron polarisation is observed for the 5ML film upon increasing the applied field in the range range $.83 < H <$ 5kG suggesting that the easy magnetisation axis lies in-plane with a saturation field $H_S <$ .83kG. SMOKE measurements confirm in plane easy axis switching for the 2ML Co film with $H_S <$ .83kG and an extremely high signal to noise ratio is observed, indicative of a large specific Kerr rotation [8]. By contrast very low signal to noise is obtained for longitudinal SMOKE measurements on the 5ML sample. A determination of the exact value of the μ_{Co} for the 2ML film requires knowledge of the Co lattice parameter a_{Co} but by calculating F(q) for a_{Co} = 3.54Å we conclude that the moment significantly exceeds the bulk-like moment of 1.64μ_B predicted by Li et al [9] for a 5ML fcc Co(001) slab. The observed collapse of the 5ML magnetisation is attributed to the film being unable to sustain the high substrate-induced strain energy at this thickness or alternatively to the reduced structural quality of the film weakening the wave interference process.

BCC Fe(001) FILMS

The unusual magnetic properties of ultrathin bcc Fe/Ag films have been extensively studied by several groups [1-2,4-6]. The Fe lattice constant is expanded in-plane by 0.8% with respect to the bulk equilibrium value and pseudomorphic growth can be obtained at room temperature with interdiffusion limited to one atomic layer [1]. Fe films were prepared by evaporation in UHV at 300K as previously described [8]. Layer by layer growth of the bcc phase is observed with island formation ocurring beyond 5ML thickness. A sharp p(1 x 1) LEED pattern is observed at 1ML thickness correponding to a sharp break in the Auger signal versus time (As - t) curve. No reflected beam polarisation

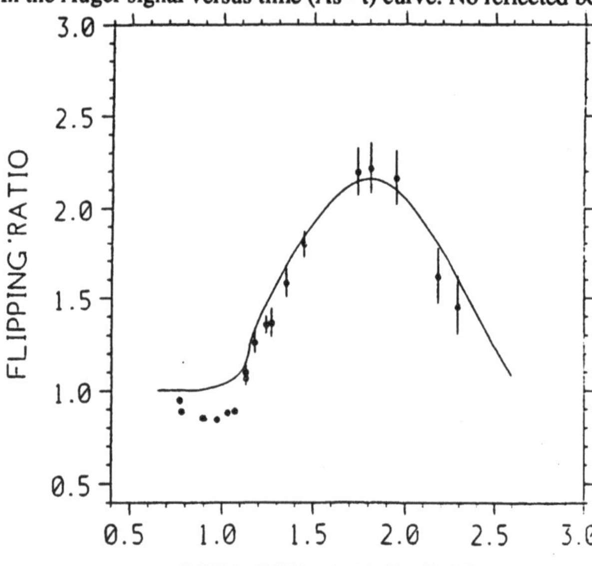

Fig.4: The flipping ratio observed at 4K for a Ag/8ML bcc Fe/Ag(001) film magnetised in-plane by a 0.83kG field. The solid line refers to calculations described in the text for μ_{Fe} = 1.0μ_B.

was detected for PNR measurements on a 150Å Ag/1ML bcc Fe/Ag(001) sandwich at 4K with a field of 0.5kG applied in plane. This behaviour is consistent with the existence of a dominant perpendicular anisotropy at this thickness as previously reported [19]. In fig.4 we show the flipping ratio observed for a 90Å Ag/8ML bcc Fe/Ag(001) film at 4K. The model fit (solid line) accurately yields a layer averaged ground state moment of $\mu_{Fe} = 1.0$ μ_B [8] independently of temperature in the range 4 - 300K, in sharp contrast with the near bulk-like value of $\mu_{Fe}= 2.0$ μ_B obtained for a 20ML/9ML bcc Fe/Ag(001) structure prepared at 300K by MBE [7]. The observed wavevector dependence of F(q) imposes an upper limit of 2ML on the extent of interdiffusion in the 8ML sample. The reduced moment may be indicative of islanding but the bulk-like temperature independence of the magnetisation anticipated for this thickness and the existence of planar anisotropy observed by surface magneto-optic Kerr effect (SMOKE) measurements exclude superparamagnetic behaviour in the 8 ML sample [8]. SMOKE measurements on both samples confirm that the films are magnetically saturated in-plane during the PNR measurements [7,8]. Since theoretical calculations [6] indicate that it is unlikely that the local moments have collapsed, the experimentally observed behaviour suggests that the long range magnetic order is modified or alternatively that the interface structure has modified the interference process.

By carefully fitting the specular reflectivity at large wavevector for the 8ML Fe sample we are able to extract a value for the roughness amplitude of 28 Å which is double that of the 9ML Fe sample. The FWHM of the diffusely reflected beam from the 8 ML Fe sample is also approximately double that obtained for the 9 ML sample indicating that the lateral coherence of the interface is significantly smaller in the 8ML layer. These observations are consistent with a larger substrate roughness and a lower degree of planarity in the 8ML Fe sample. The reflected beam profile obtained for an etched GaAs(001) substrate permits Δq to be estimated which is found to correspond to the instrumental resolution. The differences between the two thick Fe samples can be related to the Ag substrate microstructure prior to growth. The data of fig.4 show that in the critical reflection region, diffuse scattering is sufficiently strong for the 8ML sample to redistribute intensity away from the specular beam. The strong spin dependence of the diffuse scattering in the 8ML sample confirms that the scattering is associated with the Fe/Ag interface. Such effects cause a strong dipping of the flipping ratio in the range $0.6 < q/q_c < 1$ which are not present in the 9ML sample [7]. Experimentally, we find that the extent of the dip increases with substrate roughness and magnetic layer thickness. Such behaviour is consistent with roughness scattering controlled by the spin-dependent illumination of the substrate [14].

We conclude that small changes in interfacial roughness amd growth mode markedly affect the Fe layer magnetisation or the interference process is modified by the interface structure. A layer averaged moment of $2.41\mu_B$ is predicted by Ohnishi et al [6] for a Ag coated 5 ML bcc Fe film and an enhanced value can also be expected in 8 and 9 ML films. Recently we have observed moments close to the predicted values for Fe films prepared on extremely flat substrates [7]. This is depicted in figure 5, where the difference between the fit with $\mu_{Fe} = 2.4$ μ_B and $\mu_{Fe} = 2.2$ μ_B can be clearly seen.

CONCLUSIONS

Our recent PNR measurements provide partial support for the predicted values of the moments per atom of bcc Fe and fcc Co films and demonstrate the utility of PNR as an absolute magnetometric probe of ultrathin films.

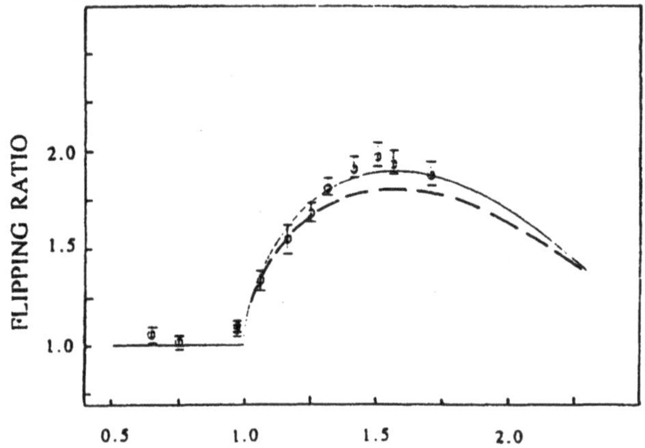

Fig.5: The flipping ratio observed at 4K for a 52ML Au/ 5ML bcc Fe/ Ag(001) film. The solid line is a fit with μ_{Fe} = 2.4 μ_B and the dashed line represents μ_{Fe} = 2.2 μ_B.

REFERENCES

[1] B Heinrich, K B Urquhart, J R Dutcher, S T Purcell, J F Cochran, A S Arrot, D A Stiegerwald and W F Egelhoff, J. Appl. Phys. 63, 3863 (1988).
[2] C L Fu, A J Freeman and T Oguchi, Phys. Rev. Lett 54, 2700 (1985).
[3] J A C Bland, D Pescia and R F Willis, Phys. Rev. Lett. 58, 1244 (1987).
[4] J Araya-Pochet, C A Ballentine and J L Erskine, Phys. Rev. B 38, 7846 (1988).
[5] F A Volkening, B T Jonker, J J Krebs, N C Koon and G A Prinz, J Appl. Phys. 63, 3869 (1988).
[6] S Ohnishi, M Weinert and A J Freeman, Phys. Rev. B 30, 36 (1984).
[7] J A C Bland, R D Bateson, A D Johnson, B Heinrich, Z Celinski and H J Lauter, to be published
[8] J A C Bland, A D Johnson, C Norris and H J Lauter, Proc. of the 34th Conference in Magnetism and Magnetic Materials, Boston 1989, J Appl. Phys. (in press).
[9] C Li, A J Freeman and C L Fu, J. Magn. and Magn. Mat. 75, 53 (1988).
[10] S Blugel, M Weinert and P H Dederichs, Phys. Rev. Lett. 60, 1077 (1988).
[11] M N Baibich, J M Broto, A Fert, F Nguyen Van Dau, F Petroff, P Eitenne, G Creuzet, A Freiderich and J Chazelas, Phys. Rev. Lett. 61, 2472 (1988).
[12] M Lax, Rev. Mod. Phys. 23, 287 (1951).
[13] G P Felcher, Phys. Rev. B 24, 1595 (1981).
[14] J A C Bland, D Pescia and R F Willis, Physica Scripta T19, 413 (1987).
[15] J A C Bland and R F Willis in "Thin Film Growth Techniques for Low Dimensional Structures", NATO ASI series, ed. by R C Farrow, S S P Parkin, P J Dobson, J H Neave and A S Arrot (Plenum, New York 1987) p405.
[16] L Nevot and D Croce, Rev. Phys. Appl. 15, 761 (1980).
[17] A Steyerl, Z. Physik 254, 169 (1972).
[18] B T Jonker, K H Walker, E Kisker, G A Prinz and C Carbone, Phys. Rev. Lett. 57, 142 (1986).

[19] J A C Bland, R D Bateson and G Hird, J. Phys. CM 1, 4403 (1989).
[20] G Bayreuther, J. Magn. and Magn. Mat. 38, 273 (1983).
[21] Chun Li, A J Freeman and C L Fu, J. Magn. and Magn. Mat. 75, 201 (1988).
[22] D Pescia, R F Willis and J A C Bland, Surf. Sci. 189/190 724 (1987).
[23] R F Willis, J A C Bland and W Schwarzacher, J. Appl. Phys. 63, 4051 (1988).
[24] A D Johnson, PhD Thesis, University of Leicester (1989).
[25] J A C Bland, A D Johnson, R D Bateson, H J Lauter, in "VI International School on Neutron Physics",Alushta, Oct.1990, and to be published.
[26] A J Freeman and C L Fu, J. Appl. Phys. 61, 3356 (1987).
[27] D A Newstead, C Norris, C Binns and P C Stephenson, J. Phys. C 20, 6245 (1987).
[28] A D Johnson, J A C Bland, C Norris and H J Lauter, J. Phys. C 21, L899 (1988).
[29] A D Johnson, J A C Bland, C Norris and H J Lauter, to be published.
[30] S Blugel, D Pescia and P H Dederichs, Phys. Rev. B 39, 1392 (1989).
[31] C L Fu and A J Freeman, Phys. Rev. B 33, 1755 (1986).
[32] G Binasch, P Grunberg, F Saurenbach and W Ziunn. Phys. Rev. B 39, 4828 (1989).
[33] J J Krebs, P Lubitz, A Chaiken and G A Prinz, Phys. Rev. Lett. 63, 1645 (1989)
[34] B A Gurney, D R Wilhoit, V S Speriosu and I L Sanders, to be published.
[35] J A C Bland, S Blundell, R D Bateson, V S Speriosu, B A Gurney, H J Lauter, S Shackleton, J Penfold to be published.
[36] J A C Bland, R D Bateson, P C Riedi, R G Graham, H J Lauter, S Shackleton and J Penfold, to be published.
[37] F Xu, J J Joyce, W M Ruckman, H W Chen, F Boscherini, D M Hill, S A Chambers and J H Weaver, Phys. Rev. B 35, 2375 (1987).
[38] P C Riedi, T Dumelow, M Rubinstein, G A Prinz and S B Qadri, Phys. Rev. B 36, 4595 (1987).
[39] B I Min, T Oguchi and A J Freeman, Phys. Rev. B 33, 7852 (1986).

^{4}He FILMS ON GRAPHITE STUDIED BY NEUTRON SCATTERING

H.J. Lauter[1], H. Godfrin[1], V.L.P. Frank[1,2] and P. Leiderer[2]

1 Institute Laue-Langevin, BP 156X, F-38042 Grenoble, France
2 University of Konstanz, D-7750 Konstanz, West Germany

ABSTRACT

A ^{4}He film shows particular properties. For example the superfluid onset temperature is shifted to lower temperatures with decreasing film thickness and a reduced superfluid fraction is detected [1-3]. This may be caused by a different excitation spectrum with respect to the bulk [e.g. 3]. In addition a ^{4}He film has two interfaces the gas-liquid boundary and the liquid-solid boundary. At the gas-liquid boundary quantized capillary waves (ripplons) [4-7] can be excited. Concerning the liquid-solid boundary it was measured that in a film deposited on a graphite substrate, the two layers adjacent to the graphite are solid [8]. The interesting interface is thus the liquid-solid ^{4}He interface. Here the freezing-melting wave could be excited [9]. In addition this interface could play a role in the anomalous small Kapitza-resistance. In this report we will give some new information about excitations in a ^{4}He film and at its boundaries. It shows also that neutron scattering can be used to detect signals from monolayers and interfaces.

EXPERIMENT

In contrast to previous neutron scattering experiments of ^{4}He films [10,11] using graphite powder as substrate, we used this time Papyex [12]. Papyex is exfoliated graphite with about 20 m^2/gr surface area. It has besides a powder contribution oriented graphite crystallites which show a 30° (FWHM) mosaic distribution with respect to the c-axis. A densest first layer (which under consideration of the pressure of the second layer has a density of 0.115 atoms/Å^2 [8]) consisted of 312 cc (STP) ^{4}He. The subsequent layers are taken to have densities of 0.95 atoms/Å^2 [8], 0.078 atoms/Å^2, 0.078.atoms/Å^2...

The inelastic neutron scattering experiments were performed on the time-of-flight spectrometer IN6 [13] at the ILL with a chosen wavelength of 5.12Å. The detectors are located in an angular range of 11.9° - 113° ; this corresponds, for elastically scattered neutrons, to momentum transfers (Q) between 0.254 - 2.046Å^{-1}. The energy resolution is only slightly dependent on Q and is about 0.055 meV for elastically scattered neutrons. The sample was mounted with the c-axis perpendicular to the scattering plane. The data obtained from the sample cell before any ^{4}He was adsorbed were used as background and subtracted from subsequent measurements with adsorbed ^{4}He.

EXCITATIONS AT THE LIQUID-SOLID BOUNDARY

In previous experiments it was shown that excitations can be measured at the liquid-solid ^{4}He interface, which have a constant dispersion ("flat modes") indicating localized excitations [10,11]. The proof that these flat modes are localized at the liquid-solid ^{4}He boundary was given by the fact that they still persist in a sample cell filled completely with ^4He and that its intensity does not depend on the amount of adsorbed ^4He beyond a total coverage of 4 layers [11]. These modes appeared at energies around 0.4, 0.6, 0.8 and 1.0 meV. The main origin of the flat modes was suggested to be an exchange of atoms between the second solid ^4He layer and the liquid ^4He. In order to test this idea we replaced the solid ^4He layers adjacent to the graphite by two layers of H_2 [14]. It is shown in Ref.14 that indeed the intensity of these flat modes is greatly reduced by the suppression of the solid ^4He - liquid ^4He interface, at which the flat modes must consequently be located.

These solid ^4He layers are present in most examples of Kapitza-resistance measurements and should be taken into consideration, which means that the excitation of a localized mode can render possible a transmission of a phonon from a solid through an interface to liquid ^4He.

PHONONS AND ROTONS IN A ^4He-FILM

A long outstanding question is how the dispersion curve of bulk liquid ^4He is changed if the ^4He is confined in a film. A very simple approach was given in Ref.3, where a change in the static structure factor was calculated in going from 3-dimensional ^4He to 2-dimensional ^4He. This change results in a shift towards lower energies of the roton minimum and towards a lower wave vector. Our scans hardly approached the roton minimum in Q and on the other hand this theory was not made for the phonons and maxons. Thus an other theory was considered [5,15]. Here a clear increase of the energy in the maxon region was predicted due to the higher density regions of the liquid layers near the solid. However our fits to the experimental data, exhibited in fig.1, show that a lowering of the dispersion curve takes place in the maxon region if the film thickness is reduced. The lowering of the dispersion curve is continued towards the roton minimum. However the roton minimum was not covered by the scans. Our data from the cell filled completely with ^4He agree with the bulk dispersion curve [16]. But up to now we had no explanation for the softening of the maxon and the roton branch with a negative group velocity for the rotons [14]. In recent studies [17] of ^4He in porous Vycor glass it turned out, that again the maxon energy was lowered but that in addition the roton minimum is slightly increased and shifted to a lower wave vector in qualitative agreement with a reduced pressure in the film. Indeed, if the surface tension is pulling on a monolayerfilm, e.g. due to capillary condensation, a negative pressure of about -10 bar can be calculated. If the pressure dependence of the parameters of the phonon-roton curve is extrapolated to our measured values again this value of -10bar is deduced. This new feature must be incorporated into the data analysis of thermodynamic measurements of thin ^4He films.

The behavior of the intensity of the phonon is plotted in fig.2 against the film thickness for a Q of 1.2 $Å^{-1}$. The intensity extrapolates to zero intensity at a total coverage of 3.5 layers. This coincides with measurements in Ref.18, where a loss of superfluidity is detected at about 3 adsorbed layers

RIPPLON

Quantized capillary waves (ripplons) are the elementary excitations of a free liquid surface. Their existence at the bulk ^4He surface and in films has been predicted by theory and indirectly confirmed by experiment [4,19]. At long wavelengths the ripplon dispersion

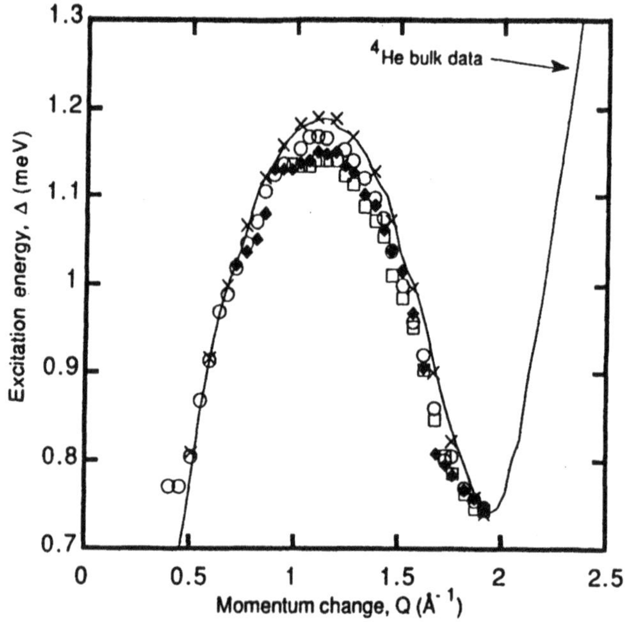

Fig.1: Variation of the phonon-roton dispersion relation as a function of film thickness at a temperature of 0.7 K. The total ^{4}He layer thickness is 5.06 layers (○), 4.16 layers (◆) and 3.85 layers (□). (×) marks data from the cell filled completely with ^{4}He.

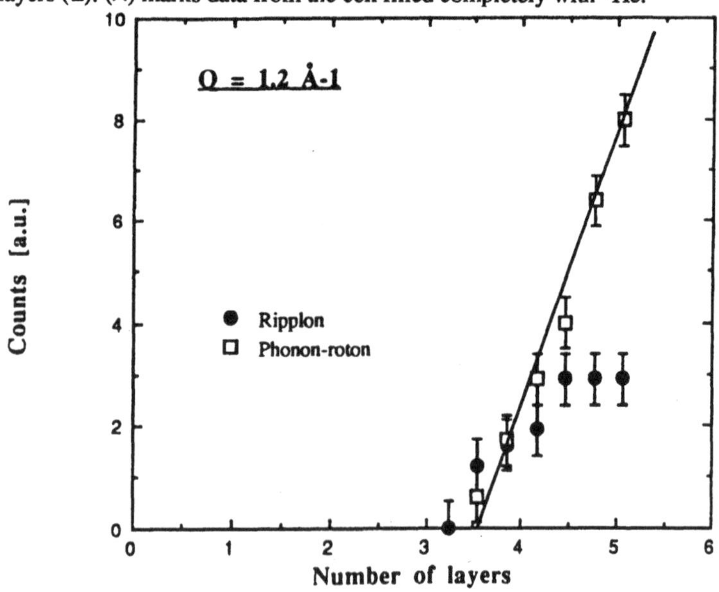

Fig.2: Intensities of the phonon (□) and ripplon (○) as a function of coverage at T=0.7K.

relation is easily evaluated using hydrodynamic relations for an incompressible fluid:

$$\omega^2 = (\alpha_0/\rho_0)\, k^3 \qquad \{1\}$$

where α_0 is the zero temperature surface tension, ρ_0 the ^4He density at zero pressure and k the wavevector. The temperature dependence of the surface tension ($\alpha(T)$) at very low temperatures can be deduced from the ripplon dispersion relation. Detailed measurements [20] of $\alpha(T)$ revealed a much larger temperature dependence than expected from formula {1}. Several modified dispersion curves have been proposed which differ mainly for wavevectors above $0.5Å^{-1}$. The idea of a 'surface roton', with a minimum at ~2K, was introduced by Reut and Fisher [21] improving the agreement with the available thermodynamic data. Edwards et. al. [4,20], taking into account the curvature dependence of α, were able to fit the experimental data on the excess surface entropy. Their model involves two parameters: a length $\delta = d(\ln \alpha_0)/dK$ where $K = (r_1^{-1} + r_2^{-1})$ is the curvature of the surface, and an area $a = d\delta/dK$. Within the precision of the entropy data, several sets of parameters have been used ($a=+1.5Å^2$, $\delta=0$ [20] and $a=+1.0Å^2$, $\delta=-0.336Å$ [4]), the latter giving a better agreement. Such a large variation in the parameters corresponds to very different ripplon dispersion curves at wavevectors $\sim 1Å^{-1}$, with a common trend indicating the presence of a downward curvature. Little direct experimental evidence was available [10], however, on the ripplon dispersion curve at these wavevectors. Such a study requires a microscopic probe like inelastic neutron scattering (INS), but due to the low neutron cross section of ^4He the measurement has to be performed on samples with a large surface to volume ratio.

A first outline of the experiment of which some results are depicted in fig.3 was already given in Ref.22. The top figure (fig.3a) shows the behavior of the intensity as a function of energy and momentum transfer from 5.06 adsorbed layers of ^4He on graphite and the bottom figure (fig.3b) the signal from the cell completely filled with ^4He. In the grey-graduations marks white the lowest intensity and black the highest intensity, e.g. at the roton minimum in fig.3a. The lines represent the bulk phonon-roton curve [6], the ripplon dispersion [4] and the constant-Q trajectories. It is clearly seen in fig.3a that besides the intensity on the phonon-roton curve there is intensity on an energetically lower lying branch. Evidence of the existence of this branch has been given previously on measurements [10,11]. This branch coincides with a calculated dispersion of the ripplon [4.] The agreement is very good. At higher Q than $1.5A^{-1}$ the roton intensity combined with the one of the flat modes becomes too high to distinguish the ripplon signal. This good agreement allows to say that the temperature dependence of the surface tension is really based on an experimentally verified dispersion relation. It still remains to prove the modified parameter set [4] by a theory.

In fig.3b the result of the completely filled sample cell is shown. In the region where the ripplon should show up, the gray-graduations are the same as in fig.3a. Only near the phonon-roton intensity and the flat bar of the multiple scattering [10,11] the attribution of the greys to intensities has been modified. Thus no signal of the ripplon intensity is visible in fig.3b, the filled cell without liquid-gas interface, although below $0.7\ Å^{-1}$ it would have been distinguishable from the overwhelming quasi bulk phonon roton intensity. This disappearance proves that the ripplon signal is really bound to the gas-liquid interface.

The ripplon intensities are plotted as a function of coverage for a Q of $1.2\ Å^{-1}$ in fig.4. The ripplon intensity seems to saturate beyond a coverage of 5 layers. This is in agreement with the picture of a surface wave, whose intensity does not depend on the amount of bulk liquid below the surface. On the other hand the ripplon intensity disappears near 3 layers. This is in agreement with a penetration depth of about an atomic layer [4,6]. However this intensity disappears definitely at a lower coverage than the phonon-ripplon intensity. It

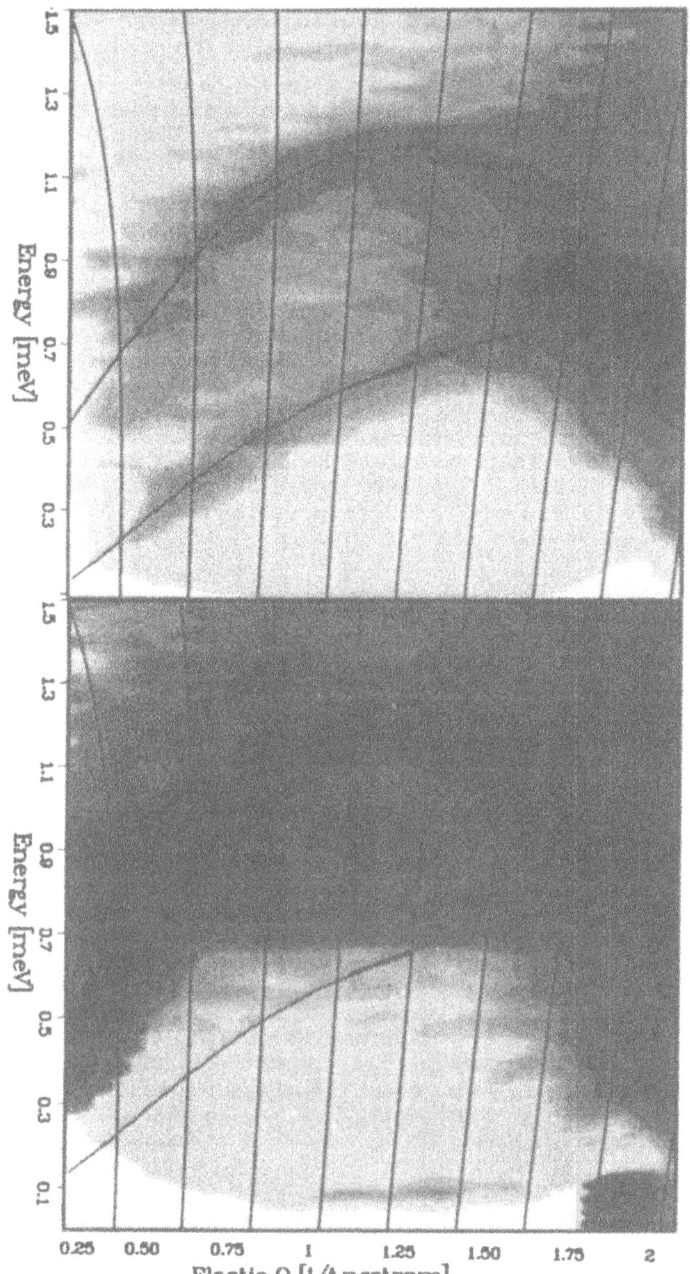

Fig.3: Intensity on the phonon-roton curve and on the ripplon curve in the energy-Q plane.

seems that the excitations on the phonon-roton curve cannot be excited anymore below a thickness of the film of 3.5 layers (which means 1.5 liquid layers). Obviously also a two-dimensional roton cannot be excited, because this excitation would require a movement only in two dimensions in the plane of the film, however the atoms can still move in the direction perpendicular to the film and thus the ripplon is still excited, which is visible down to one liquid layer.

CONCLUSION

In summary it turns out that a liquid helium film exhibits a lot of interesting features in the excitation spectrum. The excitation spectrum of the bulk phonon-roton curve is modified in a helium film as well as the dependence of the intensity on the momentum transfer. There are in addition excitations which have no dispersion (localized modes) at the solid helium-liquid-helium interface. Finally the dispersion curve of the ripplon at the gas-liquid boundary could be measured for the first time to relatively high momentum transfers.

This work has been partially supported by the West Germany Federal Ministry of Research and Technique (BMFT).

[1] M.Chester, Comments Solid State Phys.10, 91 (1981)
[2] Saam W.F. and Cole M.W., Phys.Rev.B11, 1086 (1975)
[3] Götze W. and Lücke M., J.Low Temp.Phys.25, 671 (1976)
[4] Edwards D.O. and Saam W.F., Prog.in Low Temp.Phys., Ed. D.F.Brewer, Vol. VII A, 283 (1978), North Holland
[5] Krotscheck E., Stringari S. and Treiner J., Phys.Rev.B, 35, 4754 (1987)
Epstein J.L. and Krotscheck E., Phys.Rev.B37, 1666 (1988)
[6] Ji G. and Wortis M., Phys.Rev.B34, 7704 (1986)
[7] Vinen W.F., Springer Ser.in Sol.St.Sci., Ed. K.Ohbayashi & M.Watabe, Vol.79, 189 (1989), Springer Verlag.
[8] Lauter H.J., Schildberg H.P., Godfrin H., Wiechert H. and Haensel R., Can.J.Phys., 65, 1435 (1987) and references therein.
[9] Keshishev K.O., Parshin A.Ya. and Babkin A.B., Sov.Phys.JETP53, 362 (1981)
[10] Lauter H.J., Godfrin H and Wiechert H,: "Phonon Physics", Kollar J. et a. edts., (World Scientific, Singapore 1985) p.842
[11] Lauter H.J., Frank V.L.P., Godfrin H. and Leiderer P., Springer Ser.in Sol.St.Sci., Ed. K.Ohbayashi & M.Watabe, Vol.79, 99 (1989), Springer Verlag.
[12]Papyex is a trademark of Carbon Lorraine, France
[13]"Guide to Neutron Research Facilities" from the ILL
[14] Lauter H.J., Frank V.L.P., Godfrin H. and Leiderer P.: NATO-workshop on "Excitations in 2D and 3D Quantum Fluids, Exeter U.K.,1990, in press
[15] Krotscheck E. in Ref.14
[16] Woods A.D.B. and Cowley R.A., Rep.Prog.Phys.36, 1135 (1972)
[17] Lauter H.J., Frank V.L.P., Godfrin H., Leiderer P.and H.Maris, to be published
[18] Chan M.H.W. in Ref.14
[19] Vinen W.F.,Springer Ser.in Sol.St.Sci., Ed. K.Ohbayashi & M.Watabe, Vol.79, 189 (1989), Springer Verlag.
[20] Edwards D.O., Eckardt J.R. and Gasparini F.M., Phys.Rev.A 9, 2070 (1974)
[21] Reut S and Fisher I.Z., Sov.Phys.JETP, 33, 981 (1971)
[22] Godfrin H., Frank V.L.P., Lauter H.J. and Leiderer P., in "Phonons 89",Hunklinger S., Ludwig W. and Weiss G. edts. (World Scientific, Singapore 1989) p.904

INFLUENCE OF AN ABSORBING SUBLAYER ON POLARIZING PROPERTY OF MAGNETIC NEUTRON MIRRORS

Korneev D.A., Pasyuk V.V., Petrenko A.V.

Laboratory of Neutron Physics, Joint Institute for Nuclear Research, Head Post Office P.O. Box 79, Moscow 101000, USSR

H. Jankovski

Institute of Electronics, AMM, Cracow, Poland

ABSTRACT

Measurements of the neutron reflectivity profile from absorbing thin film mirrors deposited onto glass substrates are presented and the results compared with theoretical predictions. The spectral dependence of the scattering length of natural Gd and GdTi alloy has been determined for the first time for the thermal neutron energy range. The unstable behavior of the neutron scattering length of Gd and consequently the principle impossibility of its compensation in a wide interval of neutron wavelengths is the reason of the strong decreasing of the polarizing properties of neutron guides with ferromagnetic mirrors. The possibility is discussed of producing a new absorbing sublayers with an important decrease of reflection in the neutron wavelength range from 1 to 10 Å. The neutron reflectivity was analyzed for absorbing thin BTi, BV and CdV alloys on glass substrates. The calculated reflectivity as a function of neutron wavelength was optimized for concentration and layer thickness.

First experimental data have been performed at the Laboratory of Neutron Physics (Joint Institute for Nuclear Research, Dubna) and are presented.

1. INTRODUCTION

Polarizators in the form of a thin ferromagnetic film sputtered on a polished glass substrate are largely used. Magnetized mirrors have a neutron spin dependent neutron-optical potential and consequently two critical angles for the total reflection of neutrons with different spin polarizations. The principle of the method is that an unpolarized beam is spatially selected into two beams with opposite polarizations, one of which is reflected and the other goes through the film. However, the reflection of the transmitted beam from the boundary of a ferromagnetic film with a substrate prevents from having a high polarizing properties of such mirrors. This difficulty was overcome by using a highly absorbing and weekly reflecting sublayer. As a result one achieves a high polarization of about 95-97% by total reflection from the FeCo film with the GdTi sublayer for a beam averaged over the thermal neutron spectrum [1]. However, the measurements of the spectral dependence of the polarization [2,3] show an essential decrease of the polarization efficiency for larger neutron wavelengths.

An experience acquired in the work with such mirrors as elements of polarizing neutron guides at the Time-of-Flight Spectrometer of Polarized Neutrons (SPN-1) at the IBR-2 reactor demonstrated the necessity of an improvement of the polarization in the large wavelength part of the neutron spectrum ($\lambda > 4$ Å). To elucidate the cause of the polarization decrease the spectral dependence of the reflection coefficients for Gd and GdTi thin films were measured. The results demonstrate that the problem can be solved by using a sublayer with an essentially lower coefficient of neutron reflection.

2. NEUTRON REFLECTIVITY from STRONG ABSORBERS

It is conventional to describe the medium by the neutron-nucleus interaction potential U

$$U = \frac{2\pi\hbar^2}{m} \sum_j N_j b_j \tag{1}$$

where N_j is the atomic number density, b_j is the neutron scattering length for the j-th isotope, m is the neutron mass. In this case the problem of neutron reflection is reduced to a one dimensional quantum mechanical problem of reflection of a plane wave from a potential barrier with the height U. For a semi-infinite medium the coefficient of reflection can be written as

$$R_\alpha(k_{1\perp}) = \frac{|k_{1\perp} - k_{2\perp}|^2}{|k_{1\perp} + k_{2\perp}|^2} \tag{2}$$

where $k_{1\perp}$ and $k_{2\perp}$ are the normal to the surface components of the neutron wave vector in vacuum and the medium respectively, $k_{2\perp}$ is connected with $k_{1\perp}$ and U by the relation:

$$k_{2\perp} = k_{1\perp} \sqrt{1 - 2mU/\hbar^2 k_{1\perp}^2} \qquad (3)$$

For the media containing adsorbing elements, one should take into account the imaginary part of the neutron scattering length, so, for the j-th isotope

$$b_j = b_j' - i\, b_j'' \qquad (4)$$

The formula (2) now becomes:

$$R_\alpha(k_{1\perp}) = \frac{\left|1 - \sqrt{1-4\pi(\sum_j N_j b_j' - i \sum_j N_j b_j'')/k_{1\perp}^2}\right|^2}{\left|1 + \sqrt{1-4\pi(\sum_j N_j b_j' - i \sum_j N_j b_j'')/k_{1\perp}^2}\right|^2} \qquad (5)$$

From equation (5) it follows that requirement $R_\alpha(k_{1\perp})=0$ at any $k_{1\perp}$ is realized only for

$$\sum N_j b_j' = 0 \quad \text{and} \quad \sum N_j b_j'' = 0 \qquad (6') \text{ and } (6'')$$

These conditions imply the absence of an absorber. Therefore while choosing a weekly reflecting sublayer one tries to satisfy only the first condition by mixing the components with the opposite signs of the real parts of the scattering lengths. It should be noticed that with $\sum N_j b_j'' \neq 0$ the condition $R_\alpha(k_{1\perp})=0$ can not be realized at any $k_{1\perp}$. Therefore, the problem of minimization of the function $R(k_{1\perp})$ should be solved for a potential barrier with a finite width and taking into account the effective potential of the substrate.

In the natural Gd the isotopes Gd^{155} and Gd^{157} with an abundance of 14.8% and 15.7% respectively have strong resonances in the absorption cross section at 2.25 Å and 1.7 Å i.e. in the region of the maximum of the intensity of thermal neutron distribution. This leads to a strong spectral dependence of their scattering length up to the change of the sign of its real part at resonance . It is clear that in this case it is impossible to fulfil the condition (6') in the whole thermal neutron energy range by mixing Gd with Ti or any other element with a negative scattering length. The energy dependence of the scattering length in the region of a unique resonance is given by the well known Breit-Wigner formula [4]:

$$b = b_0 + \frac{g\,\Gamma_n\,(E - E_r)}{2k_1((E-E_r)^2 + \Gamma/4)} - i\frac{g\,\Gamma\;\;\Gamma/2}{2k_1((E-E_r)^2 + \Gamma^2/4)} \qquad (7)$$

where $E=\hbar^2k_1^2/2m$ is the neutron energy, E_r is the resonance energy, Γ is the total resonance width, Γ_n is the neutron width at the resonance, g is the spin weighting factor. The quantity b_0 is energy independent and is given as the sum of the contribution into the scattering length of all the far lying resonances and of the potential scattering. The spectral dependences of the real and imaginary parts of the scattering length for a mixture of Gd^{155}(50%) and Gd^{157}(50%) isotopes calculated following formula (7) is shown in figure 1. The resonance parameters were obtained from reference [5]. It is evident that the compensation of the real part of the scattering length in a large interval of neutron wavelength (λ>0.5 Å) is impossible.

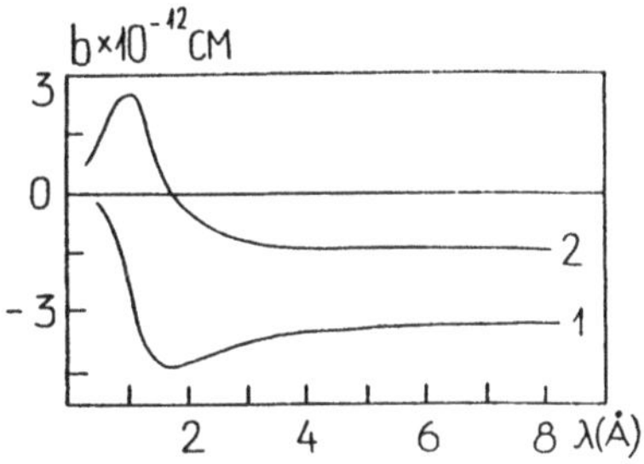

Figure 1. Calculated spectral dependences of the imaginary (1) and real (2) resonance parts of the scattering length for the mixture Gd^{155}(0.50)Gd^{157}(0.50)

The experimental results were interpreted by means of the formula for $R_a(k_{1\perp})$, derived for an absorbing layer with a finite width a evaporated on a semi-infinite substrate:

$$R_a(k_{1\perp}) = \frac{|(k_{1\perp}-k_{2\perp})[(k_{2\perp}+k_{3\perp})\exp(-2ik_{2\perp}a)+(k_{2\perp}-k_{3\perp})]+2k_{2\perp}(k_{2\perp}-k_{3\perp})|^2}{|(k_{1\perp}+k_{2\perp})[(k_{2\perp}+k_{3\perp})\exp(-2ik_{2\perp}a)+(k_{2\perp}-k_{3\perp})]-2k_{2\perp}(k_{2\perp}-k_{3\perp})|^2}, \qquad (8)$$

where $k_{2\perp}$ and $k_{3\perp}$ are the normal to the surface components of the neutron wave vector in the layer and substrate respectively ($k_\perp = k\sin(\varphi)$, where φ is the glancing angle).

3. DISCUSSION OF EXPERIMENTAL RESULTS.

The measurements of the reflectivity spectral dependences were carried out at the Polarized Neutron Spectrometer of the IBR-2 reactor in Dubna [6,7]. By the time-of-flight method were measured the spectra of the incident $N_1(\lambda)$ and reflected $N_2(\lambda)$ beams (λ is the neutron wavelength equal to $2\pi/k$). The experimental spectral dependence of the coefficient of reflection was determined as relation $R_E(\lambda) = N_2(\lambda)/N_1(\lambda)$. The investigated samples were Gd and Gd(0.15)Ti(0.85) thin films sputtered on glass substrates. The reflectivity spectral dependence for the Gd thin film is shown in fig.2. The fitting parameters were the layer thickness a and the energy independent scattering length b_0 (see formula (7)).

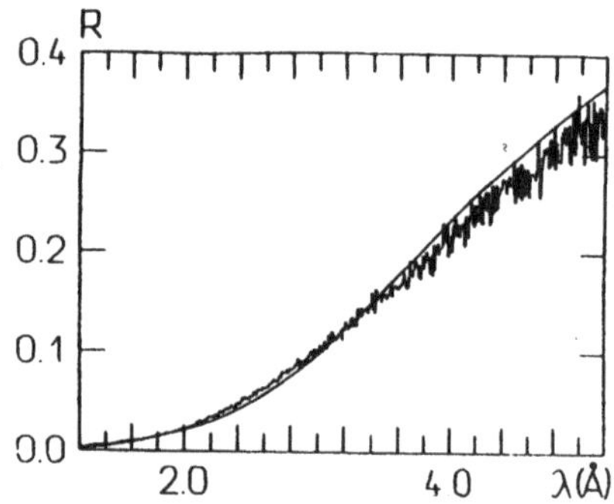

Figure 2. The experimental and theoretical reflectivity profiles for the natural Gd thin film sputtered on a float glass substrate, measured at φ=4.14 mrad. Smooth line is a least-square fit to the data. The constant part of the scattering length b_0=11.853x110^{-12}cm.

The real and imaginary parts of the scattering length of Gd obtained using the fitted parameters are shown in figure 3(a,b).

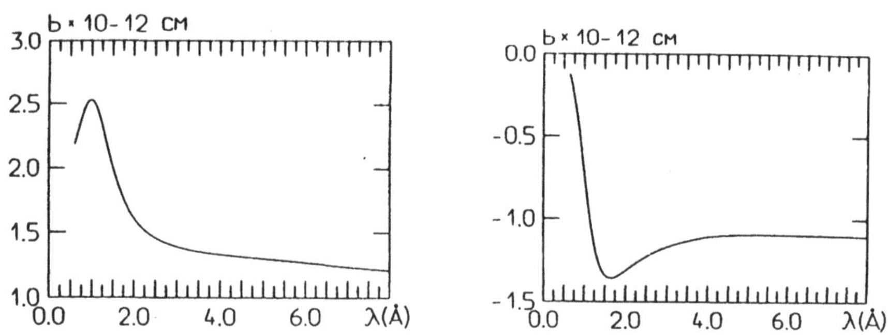

Figure 3. The spectral dependences of the real (3a) and imaginary (3b) components of the scattering length of the naturally occurring Gd, obtained using the parameter b_0 extracted from the fit.

The reflectivity spectral dependence was measured for the Gd(0.15)Ti(0.85) thin film. The theoretical function $R_a(\lambda)$ was fitted to the experimental data using the obtained spectral dependence for the neutron scattering length for natural Gd. The theoretical curve reasonably well describing the experimental data is shown in figure 4. The result of the fitting showed that the negative scattering length of Ti is suppressed by the admixture with the positive scattering length (most probably this is oxygen). In figure 5 are shown the real and imaginary components of the scattering length of the alloy we have got after the fit for this alloy.

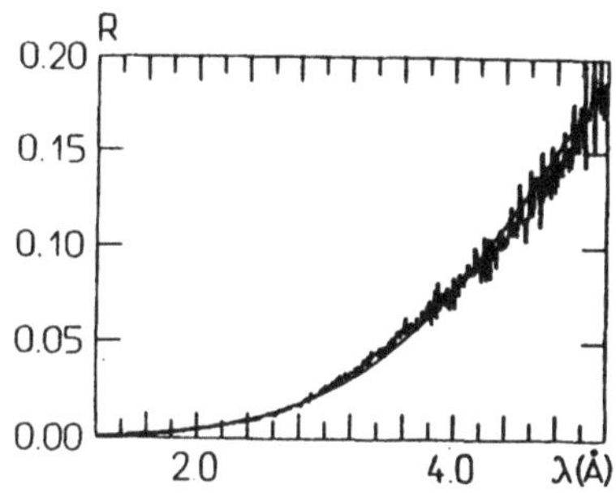

Figure 4. The experimental and theoretical (smooth line) reflectivity spectral dependences for the Gd(0.15)Ti(0.85) thin film on the glass substrate, measured at the glancing angle v=1.64 mrad,

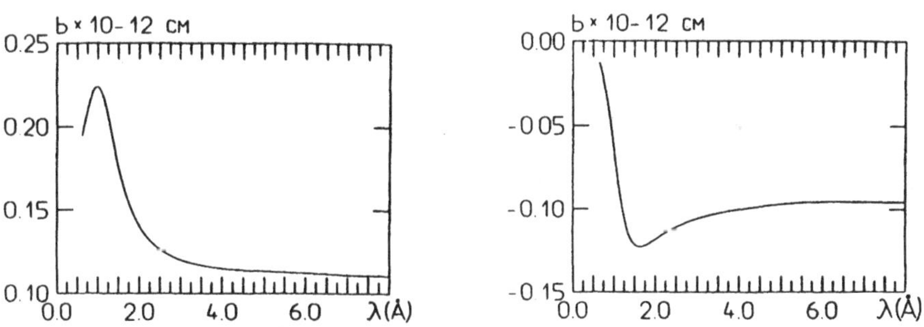

Figure 5. The spectral dependences of the real (a) and imaginary (b) parts of the scattering length of the Gd(0.15)Ti(0.85) alloy, (the results of the fitting following formula (7)).

4. DISCUSSION of the REFLECTING PROPERTIES of Cd ALLOYS.

As it mentioned above, the main obstacle to obtain a weakly reflecting alloys in a wide region of neutron wavelengths on the base of Gd is the presence of resonances in the maximum of thermal neutron spectrum. This inconvenience is absent in Cd which resonance is far from the thermal neutron region, that is for λ=0.36 Å. The scattering length for λ>0.7 Å is practically independent of the neutron wavelength, so the compensation of the real part of the scattering length for Cd seems to be feasible in the full thermal neutron wavelength range.

The coefficient of neutron reflection was analyzed using the formula (8) for a thin layer of CdTi and CdV alloys on glass substrates using the tabular values of the scattering lengths for Cd, Ti, V.

The calculated functions $R_a(\lambda)$ optimized for the concentration and layer thickness for CdV (curve 1) and CdTi (curve 2) are shown in figure 6. The curve 3 corresponds to the experimental values of reflectivity profile for the Gd(0.15)Ti(0.85). Polarizing efficiencies calculated for these alloys are shown in the same figure. It seems that the most perspective absorbing alloy is the Cd(0.20)V(0.80) one with the thickness of about 1.5 μm.

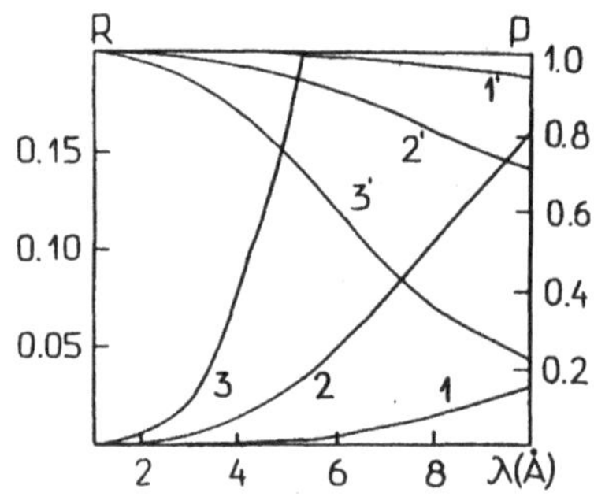

Figure 6. The reflectivity spectral dependences for the optimized alloys: 1 - Cd(0.20)V(0.80), 1.5 mm
2 - Cd(0.68)Ti(0.85), 1mm
3 - Gd(0.15)Ti(0.85), 0.8mm
The curves 1',2',3' are the polarizing efficiencies corresponding to the curves 1,2,3.

5. ABSORBING THIN FILMS on the BASIS of BORON

The other good candidate for the creating of absorbing and weekly reflecting sublayer for neutron polarizing mirrors is boron. Natural isotopic mixture of B contains 20% of isotope B^{10} with the scattering length $b=(0.0 - i1.067)10^{-5}$Å and 80% of isotope B^{11} with $b=6.66\times10^{-5}$Å [5].

Two different isotopic mixture have been investigated:

(i) $B^{10}(0.85)B^{11}(0.15)$

(ii) $B^{10}(0.20)B^{11}(0.80)$

Ti and V were taken as elements with the negative neutron scattering length for the compensation of the real part of the B scattering length. Calculated reflectivity profiles optimized for the minimum of reflection for BTi and BV thin films are shown in figure 7. The very first experimental data we have got for BTi thin film are shown in the insert of figure 7. Now we are on the way of optimization of sputtering BTi and BV thin films.

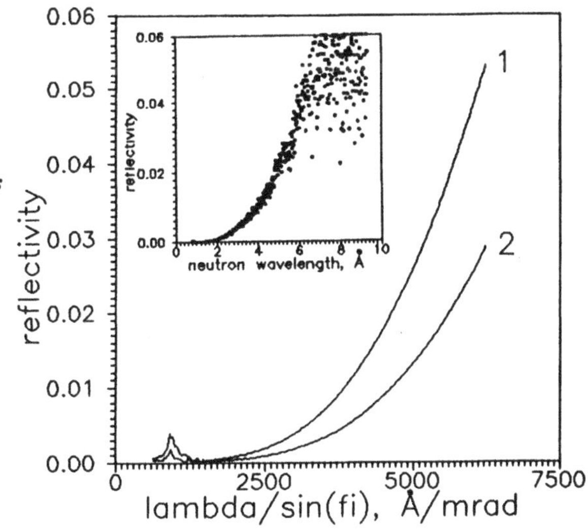

Figure 7. The reflectivity spectral dependences for BTi (curve 1) and BV (curve 2) thin films optimized for the thickness and concentration for the minimum of reflection. The first experimental reflectivity profile for not yet optimized BV thin film, measured at the glancing angle φ= 5.0 mrad is shown in the insert.

6. CONCLUSIONS

The theoretical description of the reflecting ability based on formulae (1,7,9) gives a possibility to describe the spectral dependence of the thermal neutron reflectivity from a strongly absorbing surface.

The experimental data on GdTi alloy demonstrate convincingly that the observed strong neutron reflection in the long wavelength part of the spectrum is a main reason of a bad polarizing ability for mirror with GdTi sublayer.

It has been shown that absorbing alloys on the basis of Cd and B can give good results as absorbing sublayers for polarizing neutron mirrors.

In above described investigations we did not concern a material problem. It can be realized only by really producing mirrors and not only calculating them. It's worth to find a practically realizable alloy. From this point of view the fabrication of the boron-based thin films has some advantages in comparison with Cd-based ones.

REFERENCES

1. G.M.Drabkin, A.I.Okorkov, A.F.Schebetov, N.V.Borovikova, A.G.Gukasov, A.I.Yegorov, V.V.Runov ,JETF 69(1975)1916
2. D.A.Korneev, V.V.Pasyuk, H.Rzany, A.F.Schebetov, JINR Report (1981) P3-81-547
3. G.M.Drabkin, A.I.Okorokov, A.F.Schebetov, N.V.Borovikova, A.G.Gukasov, V.A.Kudriashov, V.V.Runov, D.A.Korneev, Nucl. Inter. and Meth. 133 (1976) 453
4. E.Serge, Experimental Nuclear Physics Vol.II (New York-London 1953)
5. S.F.Mughabhab, Neutron Cross Section, Vol.I, Part B (Academic Press 1981)
6. I.M.Frank, P.Pacher, Physica 120B (1983) 307
7. Yu.M.Ostanevich, JINR Communication, P13-85-310 (Dubna 1985)

PHASE CHARACTERIZATION OF BORON IMPLANTED 100C6 BEARING STEEL

N. Millard, G. Marest, N. Moncoffre and *J. Tousset*
Institut de Physique Nucléaire de Lyon, IN2P3-CNRS et Université Lyon-1
43, Bd du 11 Novembre 1918 - F-69622 Villeurbanne Cedex, France

ABSTRACT

Ion implantation is now an industrial process used to improve wear and corrosion properties of metals and alloys. The alterations of the superficial layers have a great influence on the tribological properties of steels such as microhardness, friction and wear rates. The formation of compounds such as nitrides, carbides and borides near the surface results in the surface layer hardening which will reduce the friction coefficient and improve the wear resistance. Moreover, surface amorphization can also change wear properties and consequently affect friction coefficient. Before performing tribological tests, it is important to well characterize the implanted layers. Such a characterization of boron implanted 100C6 steel is presented in this paper.

EXPERIMENTAL

Flat 100C6 (AISI 52 100) steel samples were implanted with 5×10^{16} and 2×10^{17} $^{11}B^{+}$ ions.cm^{-2} at 20, 35 and 50 keV energies in the isotope separator of the Institut de Physique Nucléaire de Lyon. The nuclear backscattering spectrometry (NBS) using the $^{11}B(\alpha, \alpha')$ nuclear reaction at 6.04 MeV was used to determine the boron distribution profiles.

The physicochemical characterization of implanted layers was studied with Conversion Electron Mössbauer Spectroscopy (CEMS). CEMS spectra were measured in the backscattering geometry, energy integral mode, using a helium flow proportional counter. In addition, to the use of a fitting procedure assuming Lorentzian line shapes which allows to determine hyperfine parameters (Isomer Shifts IS, Quadrupole Splitting QS, Hyperfine Fields HF), a program elaborated by

Le Caer et al.[1] was used to obtain the P(H) hyperfine field distribution of the "amorphous part" in a given spectrum.

RESULTS

1. Boron Distributions

Experimental and theoretical depth profiles obtained using a modified TRIM code are presented in *figure 1*. Sputtering effects have been considered assuming sputtering yields of 0.18, 0.26 and 0.40 for 50, 35 and 20 keV implantation energies respectively, as well as the change in the matrix density during implantation. The corresponding projected Rp (and standard deviation ΔRp) ranges are 84(34), 60(26) and 35(17) nm respectively.

2. Phase Determination

CEMS spectra corresponding to 100C6 steels implanted with 5×10^{16} $^{11}B^{+}$ ions.cm^{-2} at various energies are presented in *figure 2*. The host matrix is rather complex. It is possible to distinguish a doublet with hyperfine parameters (IS = 0.15 mm.s^{-1} and Quadrupole Splitting $\Delta = 0.52$ mm.s^{-1}) characteristic of an austenitic phase and two sextets due to α-Fe (Hyperfine Field HF = 331 kOe) and to a 306 kOe hyperfine field attributed to the presence of chromium alloying element. Implantation induces a new magnetic component (HF $\simeq$ 100 kOe) which amount increases as the implantation energy is lowered. Its hyperfine parameters correspond to the FeB phase.

Implantations with 2×10^{17} ions.cm^{-2} change drastically CEMS spectra (*figures 3a, 3b, 3c*). Thanks to the Le Caer Program [1] matrix contribution can be substracted and the remaining spectra (*figures 3d, 3e, 3f*) can be fitted taking into account a P(H) hyperfine field distribution (*figures 3g, 3h, 3i*). The substracted spectra correspond to spectra obtained with amorphous alloys [2]. The amorphized contribution increases when implantation energy decreases. Some rather well defined peaks in the P(H) distributions are the signature of different boride phases. HF = 262 kOe for Fe_3B [3,4], 236 kOe for Fe_2B [3,4], HF = 100 kOe for FeB. The fact that the last phase increases with implantation energy (i.e when the boron concentration decreases), it means that some boron segregation occurs during the implantation process. The wide HF distributions correspond to Fe_xB_{100-x} phases with a large composition range.

CONCLUSION

These preliminary experiments show that boron implantation can lead to amorphization of the surface together with some phase precipitation. From the tribological point of view, amorphization presents a potential interest concerning friction coefficient diminution whereas precipitation can lead to strengthening of the surface and consequently to a wear resistance improvement. An apparatus for homogeneous implantation of balls has been realized and tribological tests are now in progress.

REFERENCES

[1] G. Le Caer and J.M. Dubois, J. of Phys. E:12 (1979) 1083

[2] J.M. Dubois and G. Le Caer, Nucl. Instr. and Meth. 199 (1982) 307

[3] G. Le Caer, Phys. Stat. Sol.(a), 64 (1981) 275

[4] G. Le Caer, in "Amorphous Metals and Non-Equilibrium Processing", p. 265

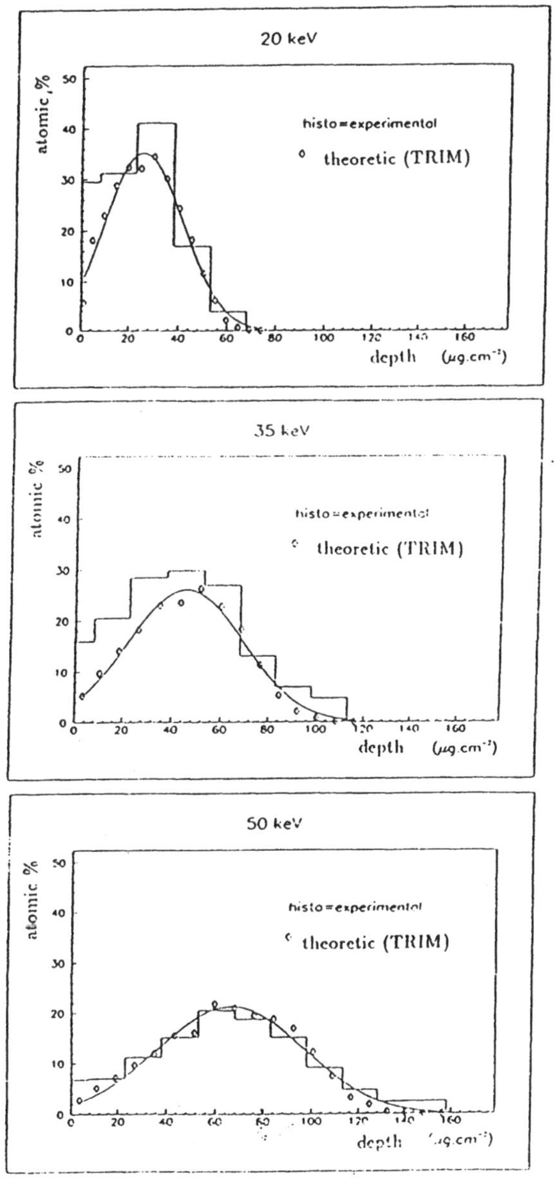

Figure 1 Depth distribution profiles of boron implanted at 2×10^{17} ions.cm^{-2}. Comparison between experimental and theoretical profiles.

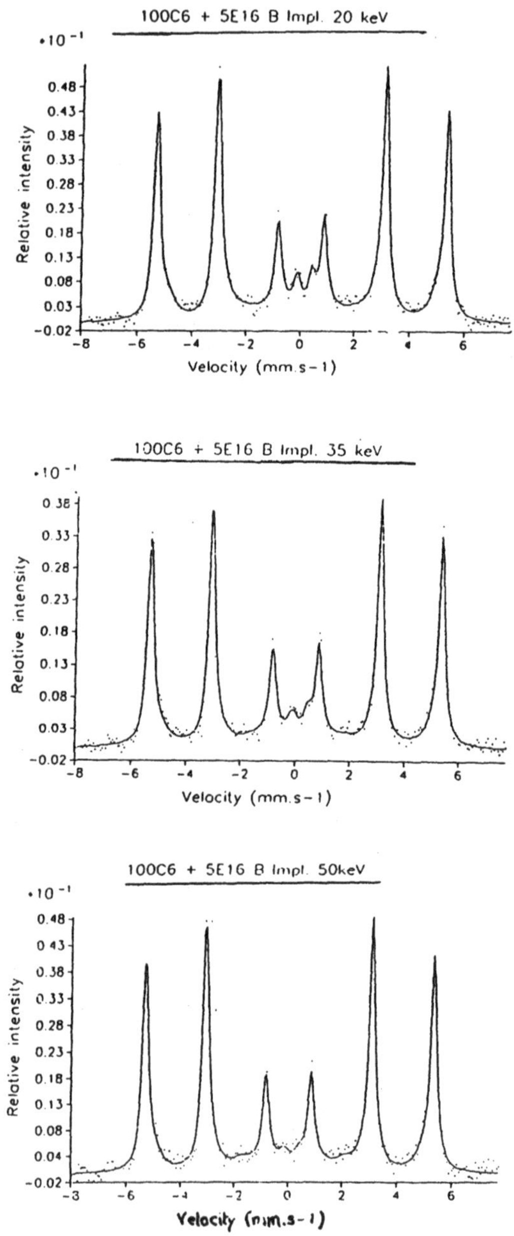

Figure 2 Mössbauer spectra obtained at 5×10^{16} $^{11}B.cm^{-2}$ and with various implantation energies (20, 35 and 50 keV).

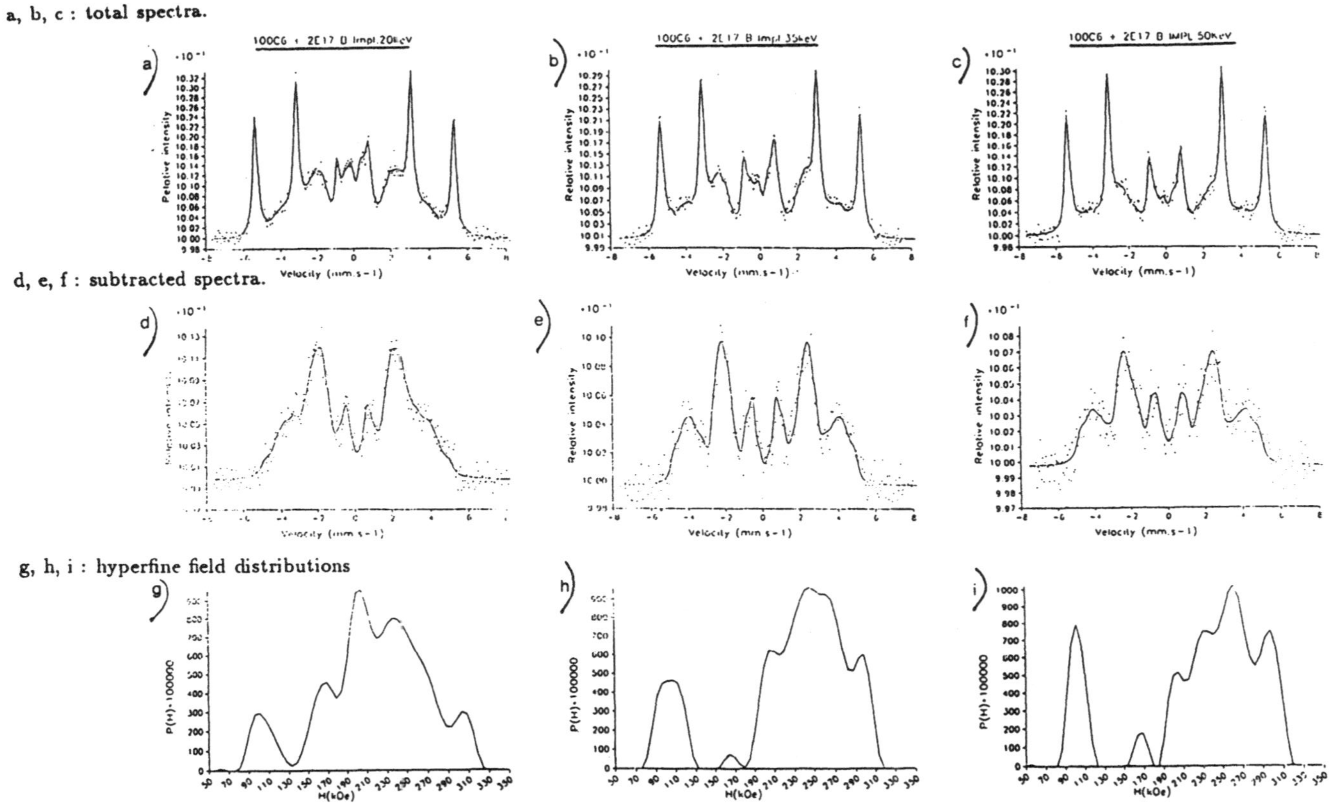

Figure 3 CEMS spectra corresponding to 2×10^{17} $^{11}B.cm^{-2}$ implantation fluence in the 100C6 steel.

HIGH PRESSURE TECHNIQUES IN MÖSSBAUER SPECTROSCOPY

Zhang Li*

Institute of Mineralogy, University of Marburg, Germany

Introduction

Since the first high pressure Mössbauer experiment (HPME) at 0.3 GPa[1] the publications of HPME increase continually[2-11]. This is closely related to the development of high pressure (HP) techniques, especially in the past twenty years. Progresses such as minimization of the size of pressure cell[9,11,12], capability of reaching extreme HP[9,12], and precise in situ pressure (P) determination [13-15] made it possible to carry out experiment both at HP and low temperature (T) [11, 16-18] or HP and high T[19]. In this paper HP techniques used in Mössbauer experiment (ME) is briefly introduced with a example of one of our studies on chain silicates.

High pressure cell

The HP cells used in HPME can be generally divided into two groups according to anvil materials. The anvils of first group (non-diamond anvil cell) are usually made of boron carbide, tungsten carbide, those of the second group are diamonds (DAC). Both groups of pressure cells adopt the opposed anvils arrangement.

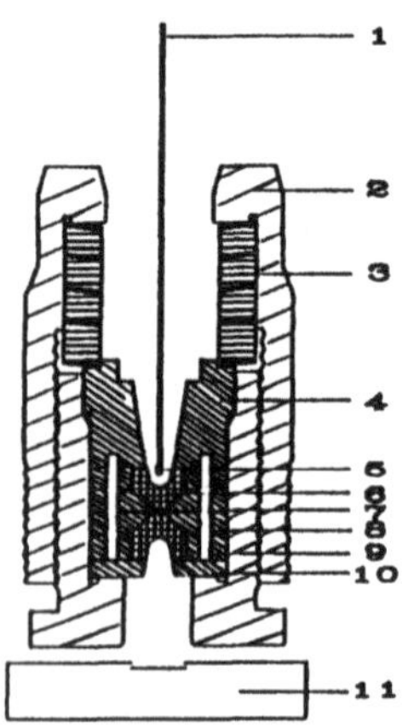

Fig. 1. Clamp technical cell: 1. Driving rod; 2. Clamp; 3. Spring; 4. backing ring; 5. Source; 6. B_4C anvil; 7. Sample; 8. Supporting ring; 9. Binding ring; 10. backing ring; 11. Detector.

Non-diamond anvil cell (NDAC)

There are mainly two types of non-diamond anvil cell used in ME, i.e. the so called Debrunner cell[20,21] and clamp technique cell (CTC)[11]. Because the former is almost out of use, descriptions in the following are about the CTC.

CTC was developed in Munich[11] and is the most versatile NDAC now (Fig. 1). Pressure up to about 15 GPa can be achieved. The pressure cell can be easily put in a metal cryostat, HPME can be performed from 295 K to below 4.2 K. The pressure is applied by tightening the screw on one side of the clamp under a hydraulic press and the force is retained by the springs in the cell. The anvils are made of B_4C which are hollowed to reduce the attenuation of γ rays. The anvil face varies from 8 to 4 mm in diameter so that the absorber can be made as large as 7 and 3 mm in diameter and 0.5 mm thick, respectively. This enables inclusion of internal pressure manometers in the absorber chamber. Pressure transmitting media such as talc, paraffin and 4:1 methanol-ethanol fluid can be used. The P at ambient T is determined by the continual change of electrical resistance of Pb and pressure induced phase transitions of metal

*Permant address: Chinese University of Geosciences, Beijing, P.R. China.

strips, such as Bi[13]. P is calibrated at 4.2 K by the T of metal-superconducting transition of Pb[14]. These internal in situ P calibrants makes determination of P independent of the gasket, absorber preparation as well as material properties as is the case by using Debrunner cell. At the same time pressure gradient can also be easily estimated[14]. Another character of this cell is that the gasket and absorber is situated in a steel belt which also supports the anvils (Fig. 1), this effectively prevents the breaking-up of anvils.

Diamond anvil cell

Different types of DAC have been used in HPME, of which Mao-Bell DAC and Merill-Bassett DAC are the mostly used ones. ME at about 70 GPa and 1000 ^{0}C was carried out[19] with a DAC. The diamond anvil face has usually a diameter of 0.7 - 1 mm, so the absorber chamber has a typical diameter smaller than 0.3 mm and a thickness of 0.2 - 0.04 mm in dependence of P. P can be accurately calibrated by P-induced shift of R1 fluorescence line of ruby chips included in the sample chamber. 4:1 methanol-ethanol fluid is the usual P transmitting media.

HPME of chain silicate hedenbergite

Hedenbergite (Hd) $CaFeSi_2O_6$ is a single chain silicate with space group C2/c and it is characterized by one dimensional chains of corner-linked tetrahedra parallel to c axis and ribbons of edge-sharing connected octahedra (M1, M2). Fe is situated only in M1 site (Fig. 2).

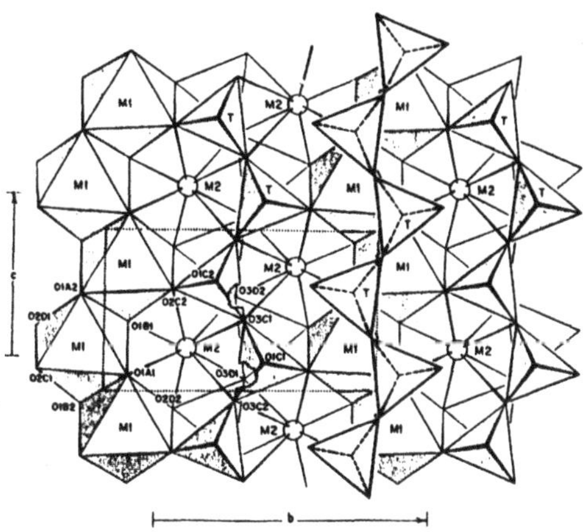

Fig. 2. Crystal structure of C2/c pyroxene (Cameron, 1980).

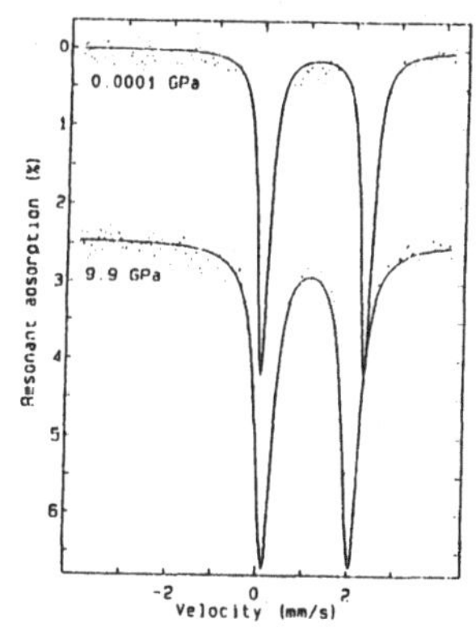

Fig. 3. HP spectra of chain silicate hedenbergite

This sample was studied by using a CTC up to 10 GPa at 295 K and 4.2 K. The absorber thickness was < 5 mg/mm^2. $^{57}Co/Rh$ sources of 10 - 25 mCi with a diameter of 2mm were used. As an example we present some experimental results at 295 K[22].

The ^{57}Fe spectrum of Hd consists of one apparent single doublet (Fig. 3) IS and QS decrease with increasing P. An abrupt change at 4.3 GPa in IS and QS is shown in Figure 4. It is particularly pronounced in QS, where a reduction of 0.1 mm/s (4.9%) occurs. The magnitude of those discontinuities is well beyond the experimental error.

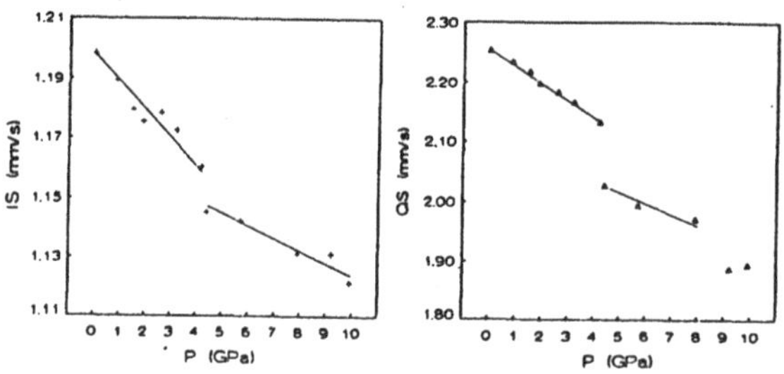

Fig. 4. IS and QS vs P

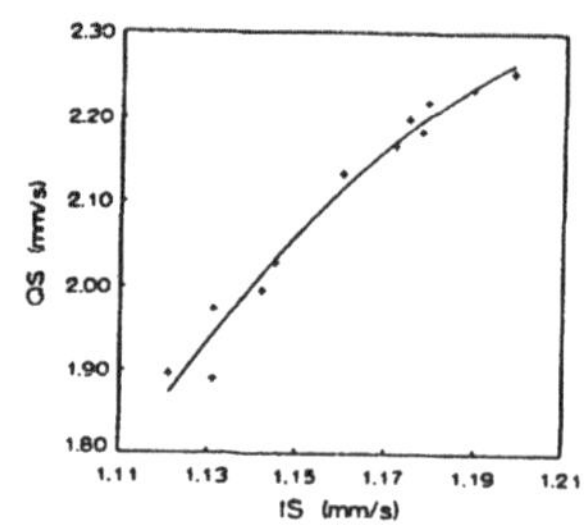

Fig. 5. IS vs QS

Since the magnitude of the reductions of unit cell volume and IS is actually the same (about 3.2%) between ambient P and 4.2 GPa, the decrease of IS in this P range is accounted mainly by the compression of unit cell volume. At higher P $\partial IS/\partial \ln V$ is higher than at lower P. This is interpreted as an increase of overlap between the s orbitals of metal ion and the valence orbitals of oxygen, and polyhedral distortion. The decreasing QS with P between ambient P and 4.2 GPa is attributed to decrease of Vzz^{val} and Vzz^{lat} and at higher P to increasing Vzz^{ov}. A proportional relationship in IS and QS is shown in Fig. 5.

Reference cited

1. Pound, R.V., Benedek, G.B. and Drever, R. Phys. Rev. Letters 7, 405 (1961).
2. Champion, A.R. and Dricakmer, H.G. J. Chem. Phys. 47, 2591-2594 (1967).
3. Champion, A.R., Vaughan, R. W. and Drickamer, H.G. J. Chem. Phys. 47, 2583-2590 (1967).
4. Vaughan, R.W. and Drickamer, H.G. J. Chem. Phys. 47, 1530-1536 (1967).
5. Stanek, J. J. Chem. Phys. 76, 2315-2320 (1982).
6. Chow, L., Deane, P.A., Farrell, J.N., Magill, P.A. and Roberts, L.D. Phys. Rev. B, 33, 3039-3049 (1986).
7. Huggins, F.E., Mao, H.K. and Virgo, D. Carnegie Inst. Washington Year Book 74, 405-410 (1975).
8. Taylor, R.D., Pasternak, M. and Farrell, J.N. Hyp. Int. 351-354 (1988).
9. Taylor, R.D. and Pasternak, M. Hyp. Int. 53, 159-174 (1990).
10. Zou, G.T., Mao, H.K., Bell, P.M. and Virgo, D. Carnegie Inst. Washington Year Book 79, 374-376 (1980).

11. Moser, J., Gal, J., Potzel, W., Wortmann, G., Kalvius, G.M., Dunlap, B.D., Lam, D.J., and Spirlet, J.C. Physica, 102B, 199-205 (1980).
12. Mao, H.K. and Bell, P.M. Carnegie Inst. Washington Year Book 74, 402-405 (1975).
13. Eiling, A. and Schilling, J.S. J. Phys. F, 11, 623-639 (1981).
14. Eichler, A. and Wittig, J.Z. Angew. Phys. 25, 319-327 (1968).
15. Barnett, J.D., Block, S. and Piermarini, G.L. Rev. Scien. Instrum. 44, 1-9 (1973).
16. Bruzzone, C.L. and Ingall,R. Phys. Rev. B, 28, 2430-2440 (1983).
17. Abd-Elmequid, M.M., Sauer, C. and Zinn, W. Phys. Rev. Lett, 55, 2467-2470 (1985).
18. Zaker, A.M., Sauer, C. and Zinn, W. J. Magn. Magn. Mater. 27, 337-42 (1982).
19. Mao, H.K., Virgo, D. and Bell, P.M. Carnegie Inst. Washington Year Book 76, 522-525.
20. Pipkorn, D.N., Edge, C.K., Debrunner, P., Pasquali, G. De., Drickamer, H.G. and Frauenfelder, H. Phys. Rev. 135, A1604-A1612 (1964).
21. Debrunner, P., Vaughan, R.W., Champion, A.R., Cohen, J., Moyzis, J. and Drickamer, H.G. Rev. Sci. Instr. 37, 1310-1315 (1966).
22. Zhang Li and Hafner, S.S. Submitted to Am. Mineral. (....).

Mössbauer Studies of Spin and Charge Density Changes in Ordered Alloys

J.E. Frąckowiak
Institute of Physics and Chemistry of Metals,
Silesian University, Katowice, Poland

Abstract

Using the Mössbauer effect the ordered and disordered Fe-Al and Fe-Co alloys were studied. From both isomer shift and hyperfine field data we deduce that in ordered alloys an additional changes of spin(charge)-density of s-like electrons at Fe nucleus takes place. The measured values of the mean isomer shift for Fe-Al alloys agree very well with the values expected from t he Miedema - Van der Woude model. The tight-binding model with rectangular density of states worked well for Fe-Co alloys and was used to analyze the changes in number of d-electrons.

1. INTRODUCTION

In recent years, much effort both experimental and theoretical has been devoted to studies of properties of atomic ordered alloys. The atomic order phenomenon reflects the strength of interatomic interactions and for this reason investigations on ordered alloys give possibility understanding phase transformations in terms of their underlying electronic structure.

In chemical bonding theories [1] and in tight-binding theories [2], the entalpy of formation of ordered alloys is explained in terms of electron sharing between neighboring atoms in the alloys. According to Miedema model [1], formation of ordered AB alloys can be described in terms of interatomic charge transfer from atoms A to atoms B which is related to the difference in the two electronegativities $\Delta\Phi=\Phi_B-\Phi_A$, for a metallic state of pure constituents. The difference, Δn_{WS}, in electron densities at the Wigner-Seitz cell boundary in pure metals also leads to intraatomic changing of electron configurations. This picture finds support in a recent analysis of experimental data about the Mössbauer isomer shift for nuclei of Au, Fe, Pt, Ir and Ru [3-5].

It has to be noted that in Miedema model the interstitial charge density n_{WS} is determined only by the free-electron-like *s* and *p* bands and not by the *d* band. However, it is well theoretically established that cohesive energies of pure transitions metals can be understood in terms of a simple model for *d*-band bonding [6]. Detailed calculations confirm that the heats of formation of transition metal alloys can be described by considering only the *d*-bands and attributing to them rectangular densities of states [2,7]. The success of these models pointed out the importance of the ratio $(E_A-E_B)/W_{AB}$, where W_{AB} is the alloy bandwidth and E_A and E_B are the centers of gravity of d-bands of the two alloy constituents.

The aim of this investigation is to study of the electron sharing in iron-base alloys with nontransition element (Al) and transition element (Co) using the Mössbauer spectroscopy. The Mössbauer effect is a useful tool for this investigation because the hyperfine field (H) and isomer shift (IS) are sensitive to changes in the s- and d- electron spin and the charge densities, respectively. After the experimental details are given, the results are compared with the models mentioned above.

2. EXPERIMENT AND DATA ANALYSIS

The Fe-Al alloys were made from electrolytic iron and aluminum by vacuum induction melting. Chemical analysis showed that Al concentration in these alloys were 19.8, 22.7, 24.3, 25.2, 27.3 and 29.8 at.%. From each sample was obtained powder for X-ray diffraction and Mössbauer spectroscopy investigations. The powder specimens were annealed at 950°C for 6h and then slowly cooled. Disordered samples were prepared by quenching from 950°C into ice brine.

The Fe-Co alloys with 34, 42, 50, 59 and 70 at.% Co were prepared by induction melting high purity Fe and Co. For the measurements the ingots were filed and crushed to the form of powder of average particle size of 40 μm. The ordered phase was obtained by annealing the powder at 1000°C for 3h then annealing at 650°C for 10 h and then slowly cooling to room temperature. Disordered samples were prepared by quenching from 1000°C.

A standard Mössbauer spectrometer operating in the constant acceleration mode was used. A radiation source of 50 mCi ^{57}Co in a Cr matrix was used. The Mössbauer spectra of all samples were computer fitted by means of a hyperfine field distribution method [8]. In order to take into account the asymmetry of lines, the linear relation $IS=B(H-H_o)$ was included in the fitting procedure, where B and H_o are fitting parameters. This method of calculation gave a good description of the shape of measured spectra.

The X-ray measurements were made at room temperature using a Tur M62 type diffractometer and were described elsewhere [9].

3. RESULTS AND THEIR INTERPRETATION

3.1. Fe_3Al alloys

Fe-Al alloys crystallize in the concentration range below 54 at.% Al in the b.c.c. lattice. Therewith two superlattices are observed: Fe_3Al of DO_3 type structure and FeAl of B2 type. In a study of Fe_3Al alloys [10] it was shown that the mean hyperfine field, $\overline{H}$, was linearly correlated with the average number of Al atoms situated within the first two neighbour shells, N_{Al}, for both ordered and disordered alloys. It was determined that one Al atom being first or second neighbour decreases the $\overline{H}$ by 20.75 kOe. Since the mean hyperfine field $\overline{H}$ can be taken as a measure of the average spin density, this result indicates that one Al atom as either a first or a second

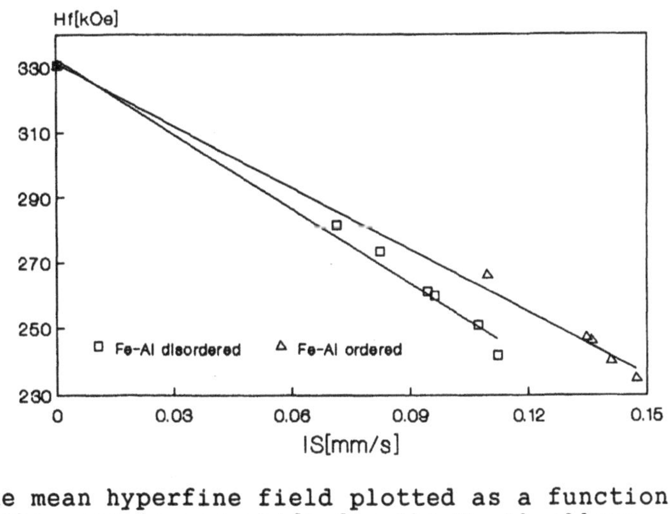

Fig.1. The mean hyperfine field plotted as a function of the mean isomer shift for the Fe_3Al alloys.

neighbour of an Fe nucleus equally reducing effective density of spin-down *s* electrons in both ordered and disordered alloys. Figure 1 shows that $\overline{H}$ and mean value of isomer shift, $\overline{IS}$, are linearly correlated. The correlation between these two quantities can be best described by:

$\overline{H} = 333.2 - 759.2 * \overline{IS}$ for disordered alloys and

$\overline{H} = 333.3 - 647.8 * \overline{IS}$ for ordered alloys.

Using scaling factor [11] for the isomer shift per number of s-like electrons, $dIS/dN_s = -2$ mm/s-electron, we can calculate that the average hyperfine coupling constant $\alpha = dH/dN_s$, is equal, $\alpha_d = 1520$ kOe/s-electron, for disordered alloys and, $\alpha_o = 1296$ kOe/s-el., for ordered alloys. Combination of the slope, $d\overline{H}/dN_{Al} = -20.75$ mm/at. Al, with the above hf coupling constants now yields the change of *s*-like electrons, dN_s/dN_{Al}, per Al atoms equal -0.013 s-el./at. Al for disordered alloys and -0.016 s-el./at. Al for ordered alloys. It means that one Al atom being first or second neighbour reduces much more the charge of *s*-electrons in ordered than in disordered alloy. Finally we can also deduce that in ordered alloys an additional decrease of charge of *s*-electrons both for spin-down and spin-up take place. The simplest explanation of the linear $\overline{H}$-$\overline{IS}$ correlation is to assume that only *s*-electrons from the spin-down subband of Fe are affected by the Al atoms. This is true for disordered alloys however for ordered alloys an additional decrease of charge of *s*-electrons both for spin-down and spin-up takes place.

The mechanism of charge transfer in ordered and disordered Fe_3Al alloys can be understood in the terms of Miedema-Van der Woude model for isomer shift [3]. In this model the strain-free dilute limit IS_{max} of the isomer shift relative to αFe in an alloy A_cFe_{1-c} is given by the expression:

$$IS_{max} = P'(\Phi_A - \Phi_{Fe}) + Q'(n_{WS}^{A} - n_{WS}^{Fe})/n_{WS}^{Fe} \qquad (1)$$

where Φ and n_{WS} represent values for electronegativities and electron densities at the Wigner-Seitz cell boundaries, respectively. P' and Q' are constants characteristic of a given Mössbauer isotope. In principle, the IS of concentrated alloys or any intermetalic compounds can be calculated by using the expression:

$$IS = f_{Fe}^{A}\ IS_{max} \qquad (2)$$

where f_{Fe}^{A} is the degree in which A atoms are surrounded by Fe atoms. For disordered alloys f_{Fe}^{A} equals $c_A^S = cV_A^{2/3}(cV_A^{2/3}+(1-c)V_{Fe}^{2/3})^{-1}$ where V_A and V_{Fe} are atomic volume of A and Fe, respectively. For ordered alloys a formula: $f_{Fe}^{A}=c_A^S(1-8c_A^{S2}(1-c_A^S)^2)$ can be used. It must be stressed that this expression has been derived empirically from a collection of experimental data [1] for heats of formation of ordered phases.

The experimental values of the $\overline{IS}$ as well as calculated values from the Miedema-Van der Woude model ,$\overline{IS}^M$, are shown in Table I. In this calculations the constants P'= 0.75 and Q'=-1.73, very close to those found in [12] for amorphous Fe-base alloys were used. It can be seen from Table I and Fig. 2 that the agreement between the calculated and experimental values of $\overline{IS}$ is excellent especially for disordered alloys. The Miedema - Van der Woude model yields also a good agreement with the experimental data for ordered FeAl alloy (B2 structure) as well as for dilute Al/Fe/ alloys. It is worth notice that this model not only predicts correctly the values of $\overline{IS}$ for disordered and ordered Fe-Al alloys, but it also indicates

Table I. Isomer shift results for Fe_3Al alloys

c_{Al}	$\overline{IS}_d$ [mm/s]	$\overline{IS}_d^M$ [mm/s]	$\overline{IS}_o$ [mm/s]	$\overline{IS}_o^M$ [mm/s]
0.198	0.08	0.08	-	-
0.227	0.086	0.09	0.110	-
0.243	0.094	0.097	0.141	0.130
0.252	0.096	0.100	0.147	0.135
0.273	0.107	0.110	0.136	-
0.298	0.112	0.118	0.134	-
0.500[a]	-	-	0.28	0.280
1[b]	0.315*	0.34		

d-disordered alloys, o-ordered alloys

[a] *G.K. Wertheim and J.H. Wernick, Acta Met. 5 (1967) 297*

[b] *C.Janot and H.Gilbert, J.Phys.F:Metal Phys. 7 (1977) 231*

* *volume correction*

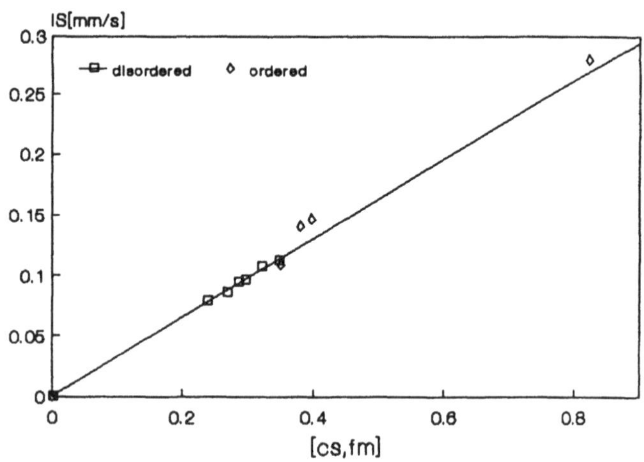

Fig.2. The mean isomer shift for ordered and disordered Fe_3Al alloys plotted vs. (c_{Al}^{s} and f_{Al}^{Fe})

that the s-d electron conversion (the second term in (1)) dominates in the electron sharing. In the ordered alloys the additional s-d electron conversion can be important and this effect leads to decrease of lattice constant in ordered state of Fe-Al alloys.

3.2. Fe-Co alloys

The Fe-Co alloys in the range (0-75) at.% Co have the crystallographic structure of the b.c.c. type. In annealed alloys ranging in composition from about (40-70) at.% Co the ordered structure of B2 type is formed.

Many previous investigations have reported the $\overline{H}$ and $\overline{IS}$ changes during the ordering in Fe-Co alloys [13-15]. In the paper [15] was determined that in range of

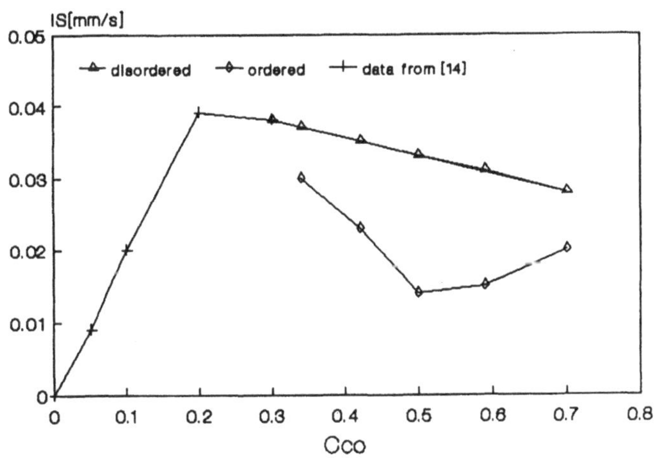

Fig.3. Mean isomer shift as a function of Co concentration for ordered and disordered Fe-Co alloys.

concentration (40-70) at.% Co the Co atom being first or second neighbour of an Fe atom, decreases the mean hyperfine field by 5.5 koe, both for ordered and disordered alloys. The concentrational dependence of the $\overline{IS}$ is nonlinear for ordered alloys (Fig. 3) however the IS is linearly correlated with the difference ,ΔN_{Co},between the values of N_{Co} for ordered and disordered alloys. The dIS/dN_{Co} value is greater for ordered alloys then those for disordered alloys. This result suggests that in ordered Fe-Co alloys, similarly to Fe_3Al alloys, an additional electron charge transfer take place.

Because the Miedema-Van der Woude Model is not appropriate for Fe-Co alloys a calculation of the *d*-electron transfer was performed in the terms of the

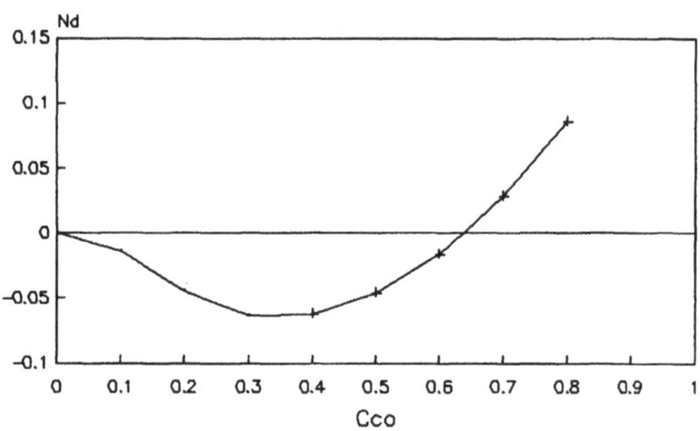

Fig.4. The change in 3d electron count at Fe atom vs. Co concentration calculated from tight-binding model with the rectangular density of states [16].

tight-binding model with the rectangular density of states [16]. The results of this calculations are presented in the Fig. 4. As we can seen the number of *d*-electrons for Fe decreases up to 30 at.% Co then increases and for concentration about 70 at.% Co the *d*-transfer changes sign. This effect can be connected with the transition from b.c.c. lattice to f.c.c. lattice for Fe-Co(75 at.%) alloy.

From the $\overline{H}$ and $\overline{IS}$ data for the whole range of Co concentration we are able to perform a calculation of change of 4s-electrons in Fe-Co alloys. The estimated values of dN_s/dN_{Co} are equal -0.008 s-el./Co atom for (0-30) at.% Co range and 0.003 s-el./Co atom for (30-70) at.% Co range. For ordered alloys this value is about ten times greater. This result suggest that in ordered alloys direct *s*-electron transfer dominates because the IS is much more sensitive to changes in 4s electrons then to changes in 3d electrons. Finally we note that for the range (40-70) at.% Co the substitution of Fe atoms by Co atoms causes an increase of number of spin-up s-like-electrons. The observed decrease of $\overline{H}$ and $\overline{IS}$ in ordered alloys is connected with the simultaneous increase in spin-down and spin-up of *s*- electrons. This process is in simple relation to observed positive volume contraction in ordered Fe-Co alloys.

REFERENCES

1. A.R. Miedema, P.F. de Chatel and F.R. de Boer, Physica 100B (1980) 1
2. D,G. Pettifor, Phys.Rev.Lett. 42 (1979) 846
3. A.R. Miedema and F. van der Woude, Physica 100B (1980) 145
4. F. van der Woude and A.R. Miedema, Solid State Commun. 39 (1981) 1097
5. A.M. van der Kraan and K.H.J. Buschow, Physica 138B (1986) 55

6. J. Friedel, in Physics of Metals I Electrons, edited by J.M. Zeeman, Cambridge Univ. Press, Cambridge 1969, pp. 361-364
7. C.M. Varma, Solid State Commun. 31 (1979) 295
8. J.E. Frackowiak and J. Kansy, Proc. Int. Conf. on Applications on Mössbauer Spectroscopy, Budapest 1989,
9. J.E. Frackowiak, Hyp. Int. 28 (1986) 1067
10. J.E. Frackowiak, Hyp. Int. 54 (1990) 793
11. L.R. Walker, G.H.K. Wertheim and V. Jaccarino, Phys.Rev.Lett. 6 (1961) 98
12. A.M. van der Kraan and K.H.J. Buschow, Phys.Rev. B27 (1983) 2693
13. B. DeMayo, D.W. Forester and J. Spooner, J.Appl. Phys. 41 (1970) 1319
14. H.H. Hamdeh, B. Fultz and D.H. Pearson, Phys.Rev. A39 (1989) 11233
15. J.E. Frackowiak, Hyp. Int. 60 (1990) 757
16. J.E. Frackowiak, to be published

KRYPTON INCLUSIONS IN ALUMINUM AND SILICON

M.J.W. GREUTER and L. NIESEN[1]
F.J.M. BUTERS, A. VAN VEEN and J. DE ROODE[2]
F. v.d. HORST and P.M. BRONSVELD[3]

ABSTRACT

Mössbauer spectroscopy is used to investigate the hyperfine inter-action and lattice dynamics of Kr bubbles in Al. The incorporation of Kr in sputtered Si layers is studied by several techniques.

INTRODUCTION

^{83}Kr was implanted at room temperature in 1.3 μm thick Al foils at an energy of 110 keV to a total dose of $1.7 \cdot 10^{16}$ ^{83}Kr/cm^2, corresponding to a maximum atomic concentration of 4 at%. A total area of 70 cm^2 was implanted at both sides from which an absorber of $1.5 \cdot 10^{18}$ ^{83}Kr/cm^2 was made. Ionic sources were made of ^{83}RbX (X=HF$_2$,Cl,I) produced via the ^{85}Rb(p,3n)^{83}Sr reaction at an energy of 45 MeV. Si layers containing Kr on a Si<100> wafer were produced by plasma-sputtering at the Interfaculty Reactor Institute in Delft. The sputtered layers are characterized by Rutherford backscattering (RBS), glancing angle X-ray diffraction, scanning and transmission electron microscopy (SEM,TEM), thermal desorption spectrometry (TDS) and positron annihilation (PA).

[1] Laboratorium voor Algemene Natuurkunde, MSC,
Westersingel 34, 9718 CM Groningen, The Netherlands

[2] Interfakultair Reactor Instituut, Technical University Delft,
Mekelweg 15, 2629 JB Delft, The Netherlands

[3] Vakgroep Technische Fysica, MSC,
Nijenborgh 18, 9747 AG Groningen, The Netherlands

MEASUREMENTS AND CONCLUSIONS

Kr in Al

Previous Mössbauer measurements on ^{83}Kr inclusions in Al were performed with a $^{83}RbCl$ source [1]. The obtained spectra were fitted with two components: a single single line, associated with atoms in the bulk of solid precipitates, and a quadrupole component originating from atoms at the Kr-Al interface. However, the spectrum of the $^{83}RbCl$, when measured against solid Kr, shows a relatively broad single line (Γ=0.38 mm/s) and a quadrupole doublet (eQV_{zz}=+2.1 mm/s) with about equal intensities. After trying several alternatives, we found that dehydrated ^{83}RbI is a better source material. The spectrum shows a dominant single line (Γ=0.28 mm/s), but still a quadrupole component (eQV_{zz}=+1.0 mm/s) with a relative intensity of 30% (fig. 1). New measurements on a ^{83}KrAl sample are presently in progress. They show again s single line and a quadrupole component.

Kr in Si

We produced Si samples containing 2-8 at% Kr by a sputtering proces, in which low-energy ions from a Kr plasma and neutral atoms from a Si sputtering target are deposited on a Si<100> substrate. Characterization of various sputtered layers is reported in an earlier article [2]. Solid Kr precipitates of average size 50 nm were observed in material containing 8 at% Kr, deposited at 600 °C (fig. 2). The sputtered Si layer is polycrystalline with a highly textured ensemble of large grains. Provided we can deposit relatively thick layers (5-10 μm) containing about 5 at% ^{83}Kr, we will study these samples using Mössbauer spectroscopy.

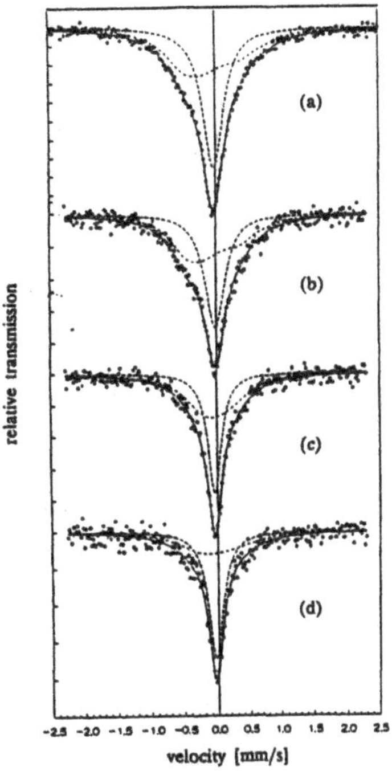

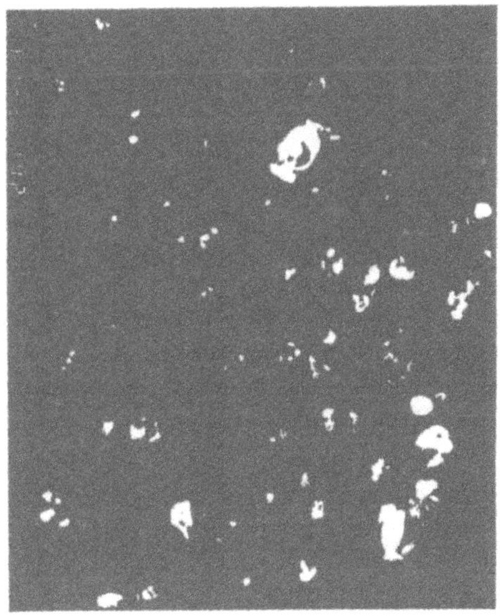

Fig. 1 Comparison of Mössbauer sources
(a) RbCl versus Kr(s)
(b) RbCl versus Kr<u>Al</u>
(c) RbI versus Kr<u>Al</u>
(d) dehydrated RbI versus Kr<u>Al</u>

Fig. 2 Kr inclusions in Si

[1] Zhang, Z.L. and Niesen, L., J. Phys. Cond. Matter 1, 1145 (1989).
[2] Greuter, M.J.W, Zhang, Z.L., Niesen, L., Buters, F.J.M. and v. Veen, A., "^{83}Kr Mössbauer Spectroscopy and inert gas inclusions in aluminum and silicon", Bonas Workshop, september 1990.

MÖSSBAUER AND CHANNELING MEASUREMENTS ON BURIED LAYERS OF $CoSi_2$ IN Si

W.G.SPANNINGA, P.J.M.SMULDERS and L.NIESEN

Laboratorium voor Algemene Natuurkunde, MSC,

Westersingel 34, 9718 CM GRONINGEN, The Netherlands

ABSTRACT

Single crystalline multilayered structures of $Si/CoSi_2/Si$ were made by high dose implantation of Co in a Si wafer which was subsequently annealed. These structures were then researched with both Mössbauer spectroscopy and channeling measurements.

INTRODUCTION

The multilayered structures of $Si/CoSi_2/Si$ were produced at Philips Research Laboratories in Eindhoven. This was done by implanting a high dose ($2\cdot10^{17}/cm^{-2}$) of Co^+ in a Si wafer (at 170 keV), which was subsequently annealed at 1000°C for 30 minutes. The resulting structure consists then of a 70 nm buried $CoSi_2$ layer covered with a 45 nm single crystalline Si toplayer. The buried $CoSi_2$ layers are supposed to be single crystalline having little or no point defects.

MEASUREMENTS

Mössbauer Spectroscopy

Two samples (having <111> and <100> surface orientation) were prepared for Mössbauer spectroscopy by postimplantation of ^{57}Co and another anneal treatment. The 14.4 keV transition of ^{57}Fe was used for the Mössbauer spectroscopy. $CoSi_2$ has the cubic CaF_2 lattice structure. In spite of this, the Mössbauer spectra of the sample having <111> surface orientation showed two distinct components having different isomer shifts at temperatures below 150 K. One component is a single line as would be expected for Co atoms in regular lattice sites. The second component however showed a quadrupole splitting, indicating a non-cubic surrounding of the ^{57}Co atoms at these sites. By varying the direction of gamma ray detection with respect to the <111> surface normal, it was found that the electric field gradient (EFG) associated with this splitting is (predominantly) oriented perpendicular to the surface. The magnitude of this EFG changed with temperature in a reversible way (fig. 1). This effect was not observed in the sample having <100> surface orientation.

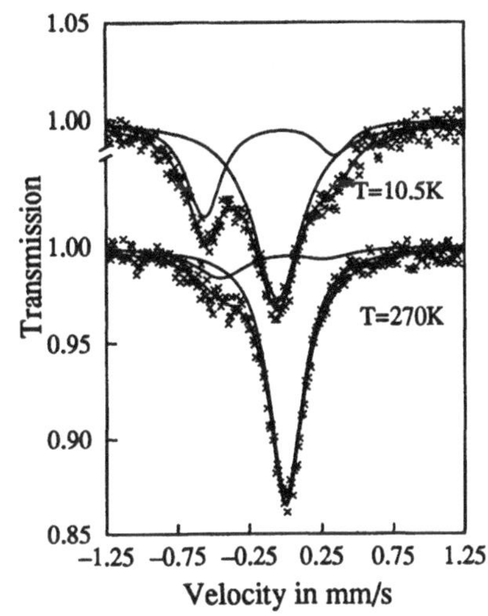

Figure 1 *Mössbauer spectra as a function of temperature.*

Channeling Measurements

On both types of samples, with and without ^{57}Co postimplantation, channeling measurements were carried out, at room temperature and 150 K. Angular scans were taken of the <100>, <110>, <111> and <411> strings. When measured at room temperature all samples showed a low minimum yield in the $CoSi_2$ region, for He^+ ions backscattered from Co as well as from Si atoms. This is in agreement with the expected good quality of the $CoSi_2$ layer. At 150 K however the sample having <111> surface orientation showed a significant increase in minimum yield in the $CoSi_2$ region as well as in the Si substrate. No flux peaking was observed. This indicates a structural change in the $CoSi_2$ layer, in agreement with the results of the Mössbauer measurements. Surprisingly, the high minimum yield decreased as a function of beam dose, down to a value comparable to that at room temperature (fig. 2). This could not be

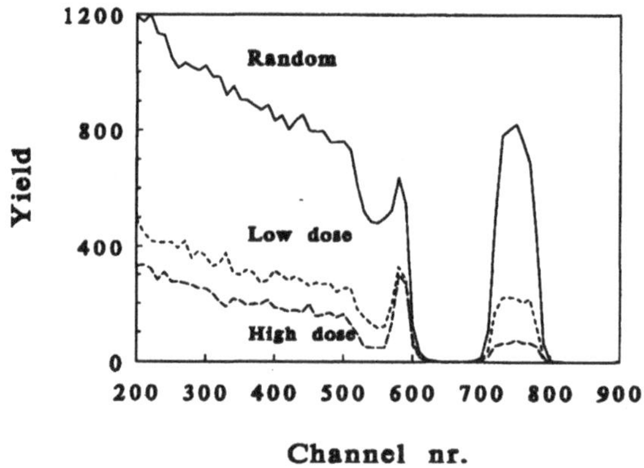

Figure 2 *Yield as a function of beam dose*

attributed to local heating by the beam. The decrease in minimum yield was only observed in the region where the beam hit the crystal. Heating up the sample to room temperature for 30 minutes did not lead to a change back to the original high-yield situation.

CONCLUSIONS

A structural change occurs in buried $CoSi_2$ layers in samples having <111> surface orientation. This was observed independently with Mössbauer spectroscopy and channeling measurements. The nature of this change is still unknown.

STATIC AND DYNAMIC INTERACTIONS OF FE LOCAL MOMENTS IN AUFE-ALLOYS

A. Metz, K.D. Gross, W.D. Brewer and *D. Riegel

*Freie Universität, *Hahn-Meitner-Institut, D-1000 Berlin, Germany*

ABSTRACT

By means of the time-differential perturbed γ-ray-distribution method (TDPAD) following heavy-ion reactions we have studied the spin-rotation of recoil implanted ^{54}Fe nuclei in *Au*Fe alloys with Fe concentrations ranging from 1 ppm to 5 at.%, in the temperature range from 20 K to 500 K. The damping of the spin-rotation-pattern gives quantitative information about the strength of the magnetic interaction between Fe local moments.

INTRODUCTION

The system *Au*Fe is a well known spin glass. Our TDPAD experiments were motivated by the search for dynamic correlation effects above the spin glass temperature T_g, as have been found in e.g. neutron scattering experiments on *Cu*Mn [1].

The TDPAD method is very similar to PAC, as introduced at this school by F. Pleiter [2]. The spin rotation function $R(t)$ is usually approximated by: $R(t) = A_{22}\, e^{-\Gamma t} \cos 2(\omega_L t - \phi)$. Here the amplitude A_{22} is a measure of the number of atoms in "good" sites, the parameter Γ describes the damping of the spin-rotation pattern, and the shift in the Larmor frequency ω_L is a measure of the local susceptibility. In systems, that exhibit stable local moment behavior, the susceptibility follows a Curie-law, and is inversely proportional to the temperature T. Spin-flip scattering of the local moment by conduction electrons, described by a relaxation rate τ_N^{-1}, follows a $1/T$ dependence: $\Gamma = \tau_N^{-1} \propto 1/T$. In the experiments that we will present here, another strong damping of static origin is present, which is proportional to $1/T^2$. This additional damping is so strong, that our measurements are restricted to temperatures far above the freezing temperature T_g of the *Au*Fe spin-glasses. As we will argue below, this damping reflects the width of the distribution of local fields.

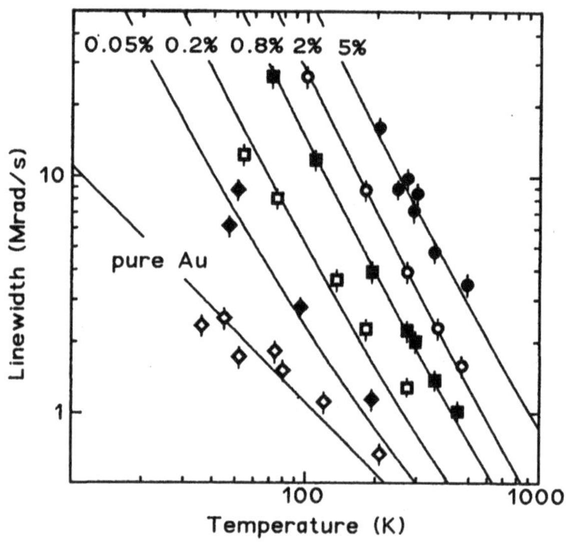

Figure 1. Log-log plot of the damping parameters Γ vs. temperature. The curves are fits to the function $\Gamma = \tau_N^{-1} + 2\Delta\omega_L$ (see text).

RESULTS

The amplitudes A_{22} where all in the range from 0.14 to 0.16, independent of temperature and concentration. This implies that approximately 80 % of the implanted atoms were found in substitutional sites. The temperature dependence of the local susceptibility was found to follow a Curie-law for all concentrations up to 2 at.%. The temperature dependence of the damping parameters Γ is shown in fig.1. It follows a $1/T$ dependence in the case of Fe implanted into pure Au only. For the more concentrated alloys the damping is dominated by a second process, which varies with $1/T^2$.

The field dependence of the damping parameter Γ showed a roughly linear dependence on the external magnetic field. We suggest that the zero field damping reflects the truly dynamic local moment relaxation. The extrapolated zero field values of Γ, as listed in table 1, are close to the values for Fe in pure Au. This implies, that there is no evidence for dynamic correlations between the Fe local moments, which is understandable in view of the fact that the measurements where done at temperatures of at least 8 times T_g.

c_M (at.%)	0.0001	0.05	0.2	0.8	2.0	5.0
τ_N (μs)	2.7(3)	2.8(4)	2.2(7)	2.5(5)		1.0(1)
$2\Delta\omega_L$ (Mrad/s)		0.13(3)	0.44(6)	1.48(4)	2.94(7)	6.22(39)

Table 1. Summary of the experimental results (values at 300 K)

DISCUSSION

The static broadening of the Larmor frequency is the result of the interactions between the magnetic impurities, which is mediated by the Ruderman-Kittel (RKKY) function. We will denote the interaction energy by: $E_{12}(R) = A_{RKKY} \cos(2k_FR)\mathbf{S}_1 \cdot \mathbf{S}_2/(2k_FR)^3$. The impurities, polarized by the external magnetic field B_{ext} cause a local field that is proportional to B_{ext} and to the impurity-susceptibility χ_{imp}. The response of a single impurity to this field is proportional to the strength of the RKKY interaction A_{RKKY}, and to the *square* of χ_{imp}, as has been explained in detail by Narath [3]. The random composition of the AuFe alloys leads to a Lorentzian distribution of Larmor frequencies with a width $\Delta\omega_L$, that increases with the impurity concentration c. Such a distribution of frequencies also causes an exponential damping of the spin-rotation spectra.

Assuming that $\Gamma = \tau_N^{-1} + 2\Delta\omega_L$ we have obtained good fits of the Γ vs. T curves, as shown in fig.1. From the values $2\Delta\omega_L$ we have obtained: $A_{RKKY} = 44$ eV.

The concentration dependence of $2\Delta\omega_L$ at 300 K, as given in table 1, can be described by $2\Delta\omega_L \propto c^{0.83}$. This dependence is remarkably similar to the concentration dependence of the spin glass temperature T_g, which is given by: $k_BT_g = A_{RKKY}c^{0.87}/18\pi$ [4]. The values for T_g, that we find when using our measured value of A_{RKKY}, are roughly 10 times larger than the experimentally observed values. This discrepancy is not yet understood.

REFERENCES

1. A.P. Murani, Phys. Rev. Lett. 41, 1406 (1978)
2. F. Pleiter, these proceedings.
3. A. Narath, C.R.C. Crit. Rev. Solid State Sci. 3, 1 (1972)
4. U. Larsen, Phys.Rev. B18, 5014 (1978)

Muon Spin Rotation Studies of Hydrogenated Polycrystalline $YBa_2Cu_3O_7$ High-Temperature Superconductors

H. Glückler, Ch. Niedermayer, G. Nowitzke, A. Golnik *), E. Recknagel
Fakultät für Physik, Universität Konstanz, W-7750 Konstanz, FRG
A. Weidinger
Hahn-Meitner-Institut Berlin, Glienickerstraße 100, W-1000 Berlin 39, FRG
J.I. Budnick
Physics Department, University of Connecticut, Storrs, CT06268, USA

Polycrystalline superconducting $YBa_2Cu_3O_7$ samples were doped with hydrogen and investigated by the muon spin rotaion technique (μ^+SR). For these samples the magnetic penetration depth which is related to the density of superconducting charge carriers was measured in transverse magnetic field μSR experiments. We find a strong reduction of the carrier concentration by hydrogen doping in $YBa_2Cu_3O_7$. Our data are consistent with the assumption that hydrogen acts as an electron donor passivating the hole states in $YBa_2Cu_3O_7$.

Introduction

A convenient way to change the carrier concentration in $YBa_2Cu_3O_7$ is to remove oxygen from the system [1] or to replace Y by an element with a different valency, e.g. Pr [2]. Here we report our results on a possible third way to change the carrier concentration of $YBa_2Cu_3O_7$, namely by doping the system with hydrogen. The effect of hydrogen doping on the superconducting properties of $YBa_2Cu_3O_7$ was studies in detail by different groups [3-5]. In a previous experiment we have shown that hydrogen doping with hydrogen concentrations above [H] > 0.5 per formula unit $YBa_2Cu_3O_7$ creates an hydrogen induced antiferromagnetic phase in this system [6].

Experimental

Well characterized polycrystalline sintered $YBa_2Cu_3O_7$ pellets (diameter 10mm, thickness 2-5mm) with an oxygen content close to O_7 were charged with hydrogen from the gase phase at temperatures around 170°C. The hydrogen content in the samples was determined from the pressure difference in the charging chamber before and after the hydrogen uptake, and by weight measurements. An observed increase of the weight which is consistent with the hydrogen uptake showed that no oxygen was removed in appreciable amounts from the sample.

The μSR-experiments were performed at the low momentum beam of the Paul-Scherrer-Institut in Switzerland with an usual μSR setup [7]. In these measurements completely spin polarized muons with an energy of 4MeV are implanted in the samples. The implanted muons come to rest at a distance of around 100μm below the surface without loss of their initial spin polarization. The muon spin starts precessing in the local magnetic field at the stopping site. The muon precessing frequency is given by the internal magnetic field at the muon site and the gyromagnetic ratio of the muons: $\upsilon \approx 135.5$ MHz/T. Due to the spin 1/2 property of the muon only magnetic interactions are possible.

μSR results

In a μSR experiment the density of superconducting carriers can be measured via the magnetic field distribution in the superconducting state. The priciple is as follows: After cooling the samples in an external magnetic field (B_{ext} = 0.3T) below the superconducting transition temperature T_c a lattice of flux vortices is build up which generate inhomogeneities of the local field distribution around the muon sites. These inhomogeneities leads to an increase of the depolarization rate σ of the muon signal in the superconducting state.

For the data analysis we assumed a gaussian behavior for the damping $P(t) = \exp(\sigma^2 t^2/2)$, where the depolarization rate σ describes the decay of the initial muon spin asymmetry due to inhomogeneities in the field distribution. Following Uemura and his coworkers [8], we relate σ to the magnetic penetration depth λ_L and the density of superconducting electrons n_s divided by their effective mass m^* by:

$$\sigma \sim \lambda_L^{-2} \sim n_s/m^* \quad (1)$$

Fig. 1 shows that σ increases as the sample enters the superconducting state. It can be seen in Fig. 1 that σ at T = 0K decreases with increasing hydrogen content [H] indicating a reduction of the carrier concentration.

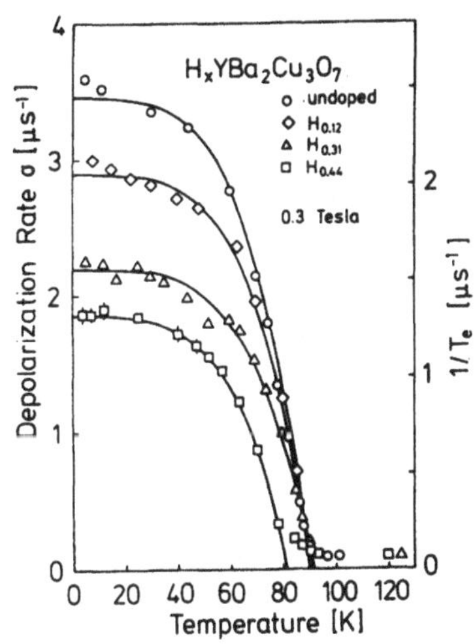

Fig. 1: Depolarization rate σ for hydrogen doped $YBa_2Cu_3O_7H_x$ in an external magnetic field (B_{ext}=0.3T). The solid line is a fit to the data with $\sigma=\sigma_0 \cdot (1-(T/T_c)^4)$

T_c remains constant for hydrogen concentrations below $x = 0.3$. For $x = 0.44$ a small reduction of T_c is observed.

Conclusions

Our results on the compensation of the charge carrier concentration by hydrogen doping of $YBa_2Cu_3O_7$ are summarized in Fig. 2.

It can be seen, that σ_0 - and assuming a constant effektive mass m^* for the charge carriers in equation (1) - the carrier concentration n_S decreases linearly with the hydrogen concentration. The extrapolation to $\sigma_0 = 0$ in Fig. 2 indicates that approximately 0.8 H-atoms per formula unit $YBa_2Cu_3O_7$ are needed to compensate all carriers [10]. Considering the behavior of T_c in Fig. 1 and of σ_0 in Fig. 2, the data indicate that the superconducting carrier concentration can be reduced by approx. 30-40% without changing T_c. This behavior is consistent with the T_c versus σ_0 relation reported by Uemura et al[8].

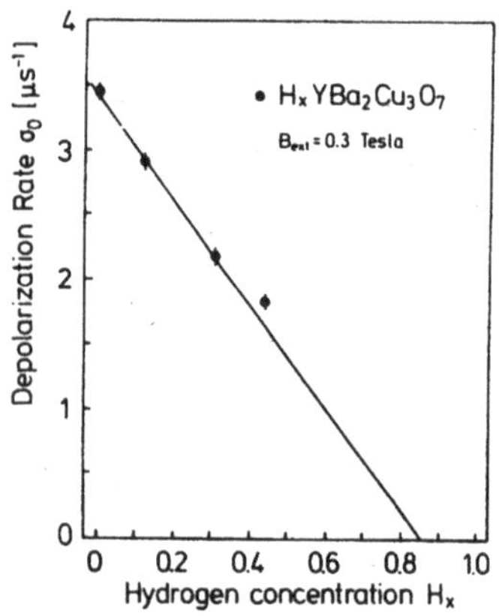

Fig. 2: Relationship between the depolarization rate σ_0 and the hydrogen concentration x in $YBa_2Cu_3O_7H_x$

*) Permanent address: Institute of Experimental Physics, University of Warsaw, 69 Hoza, PL-00681 Warsaw, Poland

References

[1] see for example J.D. Jorgensen et al., Phys. Rev. B36, 3608 (1987)
[2] see for example Y. Tokura et al., Phys. Rev. B38, 7156 (1988)
[3] J.J. Reilly et al., Phys. Rev. B36, 5694 (1987)
[4] H. Fujii et al., Jpn. J. Appl. Phys. 27, L525 (1988)
[5] T. Takabatake et al., Physica C 162-164, 65 (1989)
[6] Ch. Niedermayer et al., Phys. Rev. B40, 11386 (1989)
[7] A. Schenck, Muon Spin Rotation Spectroscopy (A. Hilger Ltd, London 1985)
[8] Y.J. Uemura et al., Phys. Rev. Lett. 62, 2317 (1989)
[9] H. Glückler et al., accepted for publication in Europhys. Lett.

HYPERFINE INTERACTIONS IN $Zr(Fe_{1-x}Al_x)_2$ MEASURED FOR $x \leq 0.20$ BY MOESSBAUER SPECTROSCOPY

J.Sarzyński, M.Budzyński, R.Wasiewicz and M.Subotowicz

Institute of Physics, M.Curie-Sklodowska University, Pl.M.Curie-Sklodowskiej 1, 20-031 Lublin, Poland

We have measured a set of samples $Zr(Fe_{1-x}Al_x)_2$ where the Fe atoms were substituted by nonmagnetic Al. We have observed a strong influence of this replacement on hyperfine interaction parameters.

The pure $ZrFe_2$ compound crystallize in the cubic structure of $MgCu_2$ - type. In the compounds $Zr(Fe_{1-x}Al_x)_2$ both cubic and hexagonal structures are formed depending on the x value. We have limited our study to $x \leq 0.20$ where the cubic structure is retained. This fact was confirmed by the X-ray powder patterns observed for all our samples. The samples were prepared by arc melting of stoichiometric quantities of high purity elements in an argon atmosphere. To recover their crystalline structure, they were annealed in a vacuum $\simeq 10^{-4}$ Pa at temperature 850 - 900 K for 3 days.

Mössbauer spectra were recorded at RT with a standard constant acceleration spectrometer with ^{57}Co in the Cr matrix as a source.

Mean weighted values of hyperfine magnetic field (B), isomer shift (IS) and quadrupole splitting (QS), derived from Mössbauer spectra, are shown in fig.1. It is visible that the field we are interested in can be derived into two part with different slopes of curves. At $x \simeq 0.06$ the inflection point is visible. Such a behaviour, although only for QS has been observed for $Y(Fe_{1-x}Al_x)_2$ at $x \simeq 0.60$

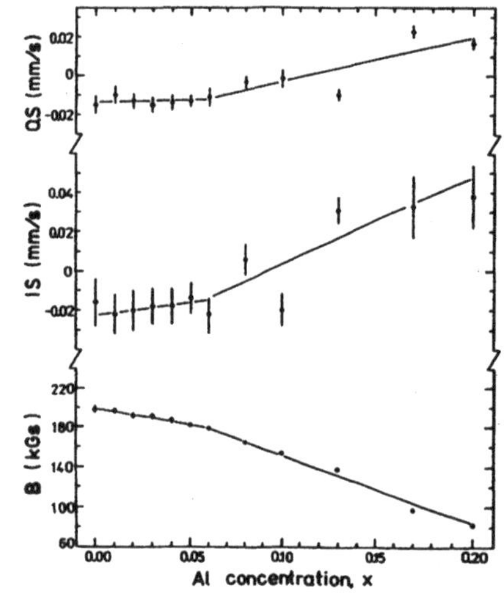

Fig. 1. Mean values of QS, IS and B vs Al concentration x.

by Reissner at al.[1].

It is known, from the measurements of magnetization [2], that the mean magnetic moment of the Fe atom strongly depends on the Al concentration. Such a dependence was also observed by Besnus et al. [3] and was explained by applying the model proposed by Jaccarino and Walker [4].

In this model the nearest neighbour environment is responsible for establishing the value of a magnetic moment. It forms discontinously and is equal to zero or has its maximum value, depending on the number of magnetic neighbours. Our attempt to estimate this critical configuration was made by the extrapolation of B vs x dependence. The limit Al concentration for ferromagnetism in our compounds was estimated as $x_c = 0.32$. It results from the binomial distribution data that the most probable configuration for x = 0.32 is the following: the Fe atom with 4 Al atoms as its nearest and the next-nearest neighbours. But it is a rough estimation, because this value of concentration belongs to the hexagonal structure range and for x = 0.26, where the change of structure occurs, one should suspect a jump in the B vs x dependence.

Since the model of Jaccarino and Walker assumes a discontinuous formation of the Fe magnetic moment, Mossbauer spectra should consist of two sextuplets, one of them showing nearly zero quadrupole splitting reflecting the situation for pure $ZrFe_2$ compound. One should expect also the existence of pure doublett component as an evidence for "nonmagnetic" Fe atoms which lost their magnetic moment.

In our Mössbauer spectra neither the number of hyperfine magnetic fields required by the Jaccarino - Walker model nor the presence of paramagnetic doublett were observed. Almost all our spectra were fitted by more than two sextuplets. The values of hyperfine magnetic fields for all components lie on well - defined curves, fig.2. The members of each pair of curves (a and b, c and d, e and f) differ from each other only by the configuration of the next - nearest neighbours.

It is evident, from our experiment, that the concentration dependence of the hyperfine magnetic field cannot be explained by applying the model of Jaccarino and Walker. One should admit that the decrease of the Fe magnetic moment is rather stepwise. Such an explaination of the Fe magnetic moment behaviour in the $Y(Fe_{1-x}Al_x)_2$ compounds was done by Van der Kraan et al. [5].

Curve	Configuration of nn	nnn	
a	6Fe	12Fe,	11Fe + 1Al
b	6Fe	10Fe + 2Al,	9Fe + 3Al
c	5Fe	12Fe,	11Fe + 1Al
d	5Fe	10Fe + 2Al,	9Fe + 3Al
e	4Fe	12Fe,	11Fe + 1Al
f	4Fe	10Fe + 2Al,	9Fe + 3Al

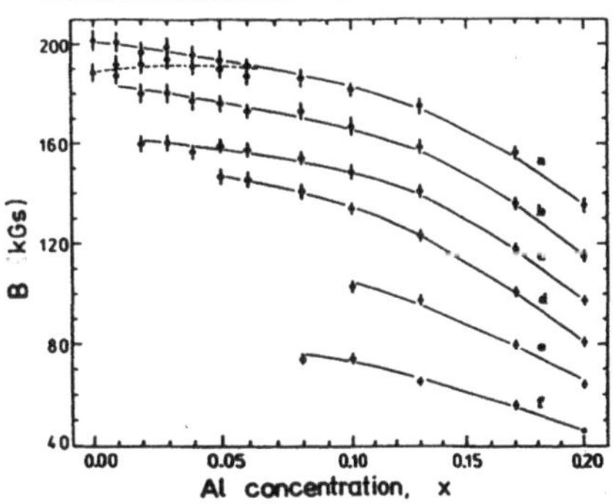

Fig. 2. Fe hyperfine magnetic fields in $Zr(Fe_{1-x}Al_x)_2$ and proposed configurations of nn and nnn atomic shells.

We have estimated from our Mössbauer spectra that the lack of one Fe atom as the nearest neighbour causes the decrease of magnetic field by about 34 kGs. The substitution of one Fe atom as the next - nearest neighbour causes the decrease by about 8 kGs.

REFERENCES

1. M.Reissner and W.Steiner, Hyp.Int. 28(1986)1017
2. M.J.Besnus, J.M.Bouton, R.Cload and A.Herr, Physica B, 85(1977)86
3. M.J.Besnus, P.Bauer and J.M.Genin, J.Phys.F: Metal Phys. 8(1978)191
4. V.Jaccarino and L.R.Walker, Phys.Rev.Lett. 15(1965)258
5. A.M.Van der Kraan, P.C.M.Gubbens and K.H.J.Buschow Proc. Int. Conf. Mössbauer Effect, Bucharest 1977, 121

MÖSSBAUER STUDY OF THE ELECTRODEPOSITED Fe_xNi_{1-x} ALLOYS

J. K. Żurawicz
Technical University of Lublin, ul. Nadbystrzycka 38a,
20-618 Lublin, POLAND

M. Budzyński, E. Jartych
Institute of Physics, University of Maria Curie-Sklodowska,
Pl. M.C.-Sklodowskiej 1, 20-031 Lublin, POLAND

I INTRODUCTION

Recently very intensive investigations of electrodeposited alloys are realized [1,2,3]. The aim of our work was to study changes of hyperfine interaction parameters in the dependence on the relative iron and nickel concentrations.

II PREPARATION AND MEASUREMENTS

Samples were prepared by electrodeposition from the bath, which was consisted of : $FeCl_2 x 4H_2O$, $NiCl_2 x 6H_2O$, HCl (1%) and H_2O. The bath didn not contain brightness additions, which were applied in the industrial aims.

Alloys were deposited on the thin graphite electrode at the steady current density 0.4-1.4 A/dm^2. The concentration of hydrogen ions in the electrolyte was kept on the steady level pH = 2.6. Electrodeposition lasted from 1 to 4 h. It was conducted at the constant cathode potential U = - 2.5 V at room temperature.

The mean thickness of the electrochemically prepared layers was 1.7- 11.3 μm.

The composition of alloys was determined by the chemical analysis (AAS- Atomic Absorption Spectroscopy).

After preparation of samples, Mössbauer measurements and X - ray diffraction studies were done. Then samples were annealed at 600 K in the argon atmosphere during 2 h and after that the same measurements were realized again.

III RESULTS OF X-RAY DIFFRACTION STUDIES

In the Laue's camera the graphite rests and the electro-deposited Fe-Ni alloys were checked, if they had not texture. The continuous concentric circles on the patterns, which were obtained from this camera, testified the absence of the texture.

The lattice constants were calculated on the base roentgen diffractograms obtained by Bragga-Brentano method. The results we can see in Tab.1 and on the other hand they are presented by Fig.1.

	The sample's composition [at %]		The lattice constant before heating [Å]		The mean nanocrystal's length [Å]		The lattice constant after heating [Å]	
	Fe	Ni	fcc	bcc	before heating	after heating	fcc	bcc
1	2.00	98.00	3.521	—	194.7	—	—	—
2	7.26	92.74	3.531	—	124.0	156.8	3.525	—
3	21.98	78.02	3.547	—	76.9	83.5	3.552	—
4	31.75	68.25	3.550	—	67.6	85.4	3.558	—
5	38.87	61.13	3.564	—	65.9	87.4	3.570	—
6	44.60	55.40	3.574	—	69.3	74.9	3.565	—
7	53.51	46.49	3.574	—	58.8	71.6	3.573	—
8	57.62	42.38	3.617	—	50.0	64.2	3.608	—
9	69.51	30.49	—	2.852	105.8	99.7	—	2.857
10	75.25	24.75	—	2.863	118.5	143.2	—	2.860
11	78.90	21.10	—	2.859	93.5	113.0	—	2.856
12	84.81	15.19	—	2.861	100.8	108.2	—	2.857
13	87.40	12.60	—	2.866	112.9	112.2	—	2.869
14	89.21	10.79	—	2.853	84.4	113.0	—	2.853
15	90.73	9.27	—	2.865	118.1	132.4	—	2.868
16	96.30	3.70	—	2.857	109.8	130.3	—	2.858
17	100.00	0.00	—	2.860	173.4	—	—	—

Tab.1 The basic data of crystalline structure of the electrodeposited Fe-Ni alloys

It could be noticed that for iron contents x = 57-69 [%] the crystalline structure changed from the face centre cubic to the body centre cubic (Fig.1). So, in the samples

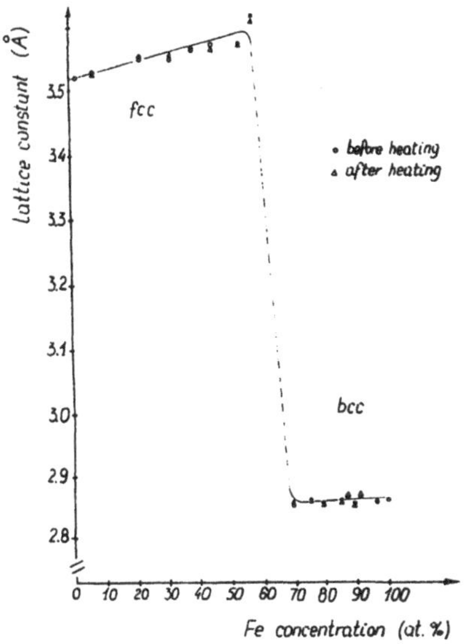

Fig.1 The lattice constant dependence on Fe concentration

of the electrodeposited Fe-Ni alloys the structure change takes place in the same region as in alloys obtained by classical methods [4]. Dimensions of nanocrystals depend on crystalline structure and alloy's composition. The lengthest nanocrystals appear in the fcc structure at minimum Fe contents (Tab.1). Then the mean length decreases and it is the shortest for the region of change crystalline structure. In the region of the bcc structure the mean lengths of nanocrystals are nearly constant.

After heating the mean nanocrystal's length increases, especially in the region of the fcc structure.

IV RESULTS OF MÖSSBAUER STUDIES

Spectra for fcc samples are presented by Fig.2(a) and - for bcc samples by Fig.2(b). These are typical absorption spectra for ferromagnetically ordered samples, where a few sextets appear.

But for x = 0.53 and x = 0.57 , (in the region on the structure change) one isomer line appears in the spectra. It is observed after annealing, too. Isomer line rather is not connected with internal stresses or absorption of hydrogen. Because in the region of the structure change the shortest nanocrystals appear (Tab.1), the isomer line may be the result of superparamagnetism.

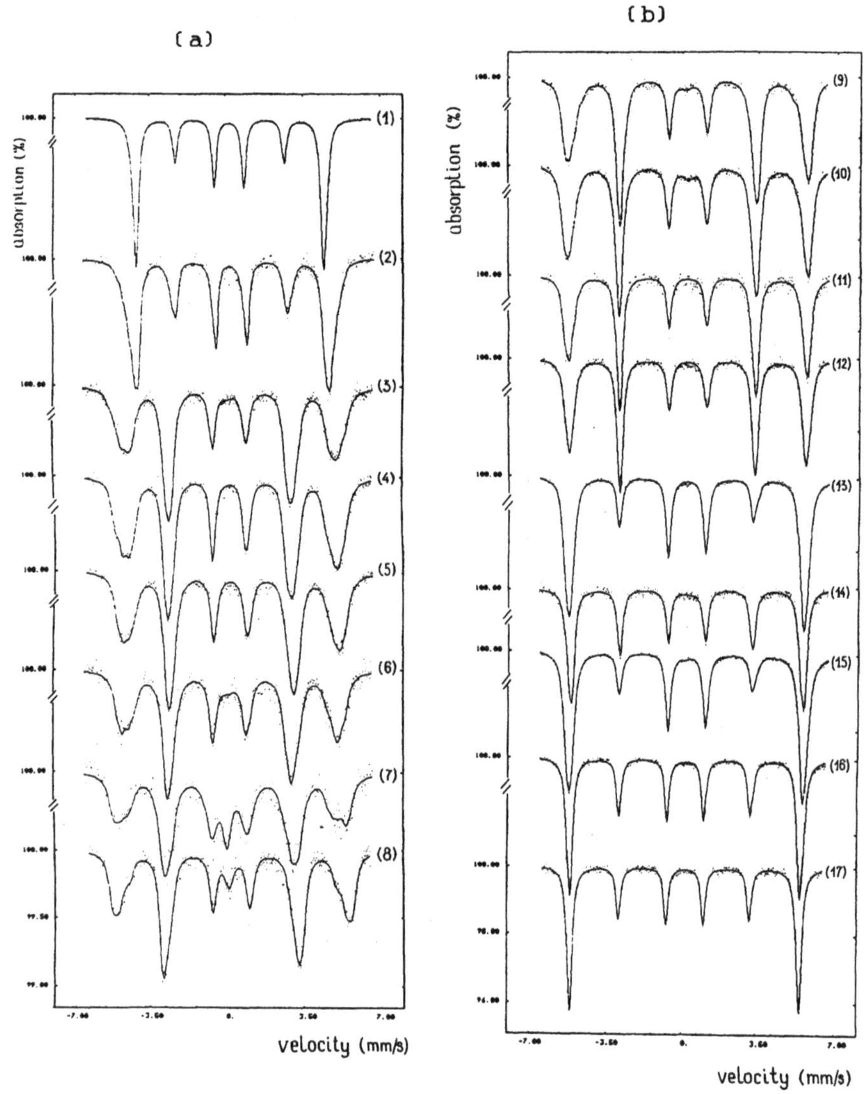

Fig. 2 Room temperature Mössbauer spectra for the respective Fe and Ni contents (see Tab. 1) in the (a) fcc region and (b) in the bcc region .

The mean direction of domains arrangement was determined on

the ground of the ratio of lines' intensities in the sextet. For x= 0.22-0.85 domains of magnetisation take up the position mostly in the sample's plain, for x=0.00-0.21 and x=0.87-1.00 this direction is perpendicular to the plain.

The discription spectra, for fcc and bcc region respectively, were made by a few sextets. The components of hyperfine magnetic field, which are presented in Tab. 2, for the fcc and bcc structures have nearly the same values and they are noticed by the same letters : a, b, c, etc. The field for pure iron is noticed by A. After heating we did not observe essential differences. The correlation between the value of magnetic field and the value of isomer shift for separate sextets is presented on Fig. 3.

The label	HF components values [kGs]	
	fcc	bcc
A	—	329.2(2)
a	324.6(6)	326.2(6)
b	336.0(9)	338.1(4)
c	345.5(1.3)	348.7(5)
d	308.9(9)	308.9(9)
e	293.3(1.1)	300.0(1.3)
f	279.3(4)	—
g	268.4(5)	—
h	264.0(6)	—

Tab. 2 The HF components in the electrodeposited Fe-Ni alloys

(a) (b)

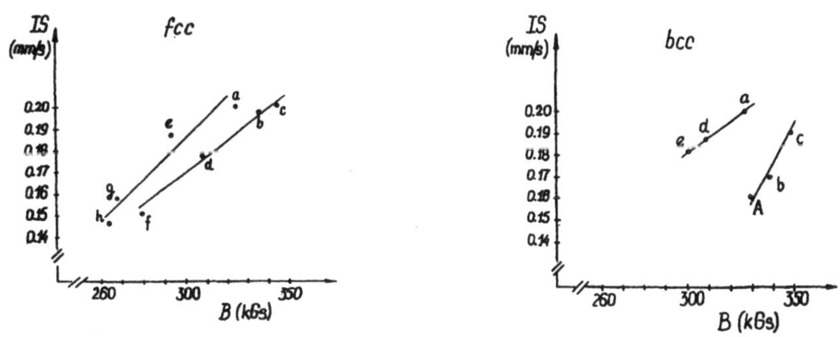

Fig. 3 The correlation between the value of magnetic field and the value of isomer shift for separate sextets for (a) fcc and (b) bcc samples

This dependence and intensities of magnetic fields (Tab.2) will be helpful for determination of nearest neighbours configuration of nuclear probes.

REFERENCES

[1] E.Kuzmann, I.Czako-Nagy, A.Vertes, C.U.Chisholm, A,Watson, M.K.El-Sharif,J.Kerti,G.Konczos, Hyperfine Interactions 45(1989), 397-402

[2] E.Kuzmann, A.Vertes, C.U.Chisholm, A.Watson, M.R.El-Sharif, A.M.H.Anderson, Hyperfine Interactions 00(1989),HFI0324B

[3] Z.Yu, Y.Zhang, Hyperfine Interactions 40(1988), 445-448

[4] Q.Hong, T.Lin, J.Phys.:Condens.Matter 1 (1989),1867-71

MÖSSBAUER STUDY OF STRUCTURAL RELAXATION IN Fe-Ni-B(P) METALLIC GLASSES

R.Męczyński and T.J.Panek

Institute of Physics & Chemistry of Metals, Silesian University, 40-007 Katowice, Bankowa 12, Poland.

Abstract.

The structural relaxations thermally activated in both $Fe_{40}Ni_{40}P_{20}$ and $Fe_{40}Ni_{40}B_{20}$ metallic glasses have been investigated. The Mossbauer temperature scanning method to measure Curie temperature in these metallic glasses was used. The reversible structural relaxation below 500 K for the fastest heating rate of $Fe_{40}Ni_{40}P_{20}$ (above 0.5 K/s) is observed.

The structural relaxation processes in the as-quenched metallic glasses are difficult to observe directly by test techniques. On the other hand the structural changes can influence on the physical properties which can be examined in a simple way [1]. Correlation between two-stage enthalpy relaxation and Curie temperature in $(Fe_{0.5}Ni_{0.5})_{87}B_{13}$ and $(Fe_{0.5}Ni_{0.5})_{87}P_{13}$ was discussed in [2]. The nature of defects, structural relaxation and various properties of amorphous $(Fe_{0.5}Ni_{0.5})_{80}B_{20}$ and $(Fe_{0.5}Ni_{0.5})_{80}P_{20}$ after neutron irradiation were carefully studied in [3,4,5]. More detailed studies of the relaxation behavior in vitrovac 0040 $(Fe_{40}Ni_{40}B_{20})$ reveal several distinct stages, indicating that only a few processes are thermally activated rather that many [6]. In our present studies the method of Mössbauer temperature scanning (MTS) was used for determination of low temperature structural relaxation in $(Fe_{0.5}Ni_{0.5})_{80}B_{20}$ and $(Fe_{0.5}Ni_{0.5})_{80}P_{20}$ metallic glasses. The state of these materials after solidification was analyzed in detail by X-ray diffraction methods and Mössbauer transmission spectroscopy. Measurements of Curie temperature (T_c) and crystallisation temperature (T_k) were conducted using the MTS method. The measured number of gamma quanta $N_o(T)$ recorded after penetrating the absorbent, with constant source velocity equal to zero as a function of the temperature (T) is shown on Fig.1 on the example of curves obtained for the investigated alloys.

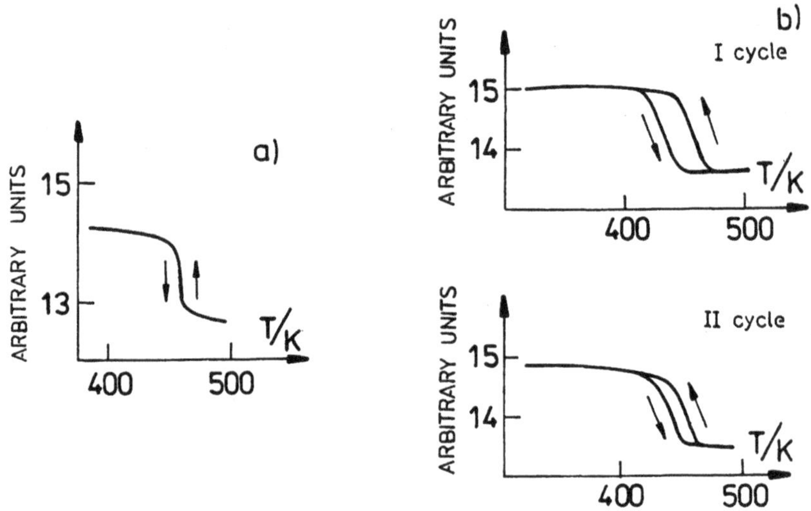

Fig.1 The count rate (N_o) of gamma quanta detected after the absorbent as a function of the heating temperature (solid line is drawing through the experimental points)
a) The heating rate $\eta=(dT/dt)=0.07$ K/sec
b) The heating rate $\eta=0.54$ K/sec (two cycles)
The heating and cooling of the Fe-Ni-P samples are marked by → ← arrows respectively.

A sharp drop on the $N_o(T)$ curve at the Curie point is observed, while above this point there is a plateau (Fig.1a,-1b). The lowering in T_c of Fe-Ni-P and Fe-Ni-B amorphous alloy with heating rate (Fig.2) as well as constant value of T_c during cooling for Fe-Ni-P (Fig.1) have been observed. Second thermal cycle show lowering in T_c at heating in comparison to this value for cooling (Fig.1b) which can be corresponds to the reversible structural relaxation.

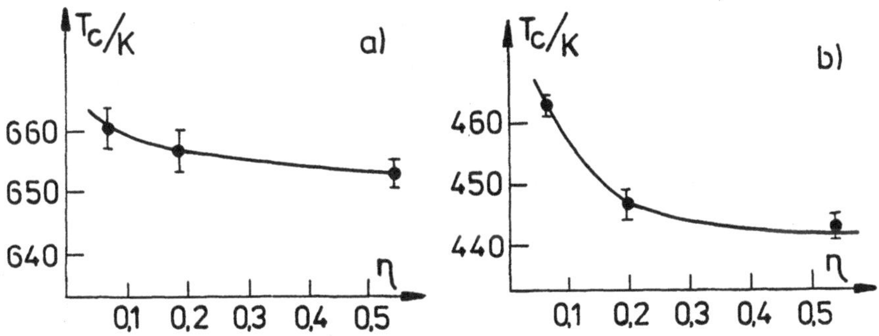

Fig.2 The Curie temperature T_c versus heating rate of a) $(Fe_{0.5}Ni_{0.5})_{80}B_{20}$ b) $(Fe_{0.5}Ni_{0.5})_{80}P_{20}$ The heating rate $\eta = (dT/dt)$ [K/s]

Based on the fact that T_c of Fe-Ni-P amorphous alloys rises with decreasing phosphorus content and with the development of the short range ordering of iron and nickel atoms [2] the reversible change of T_c seems to originate from the cooperative rearrangement among metal and metalloid atoms leading to a decrease in average atomic distance between metal and metalloid elements an increase in average number of metalloid atoms at the nearest neighbor site, resulting in a reduction in T_c.

We are grateful to the Prof. R.Wagner (GKSS Geesthacht) for supplying the amorphous samples.
This work is partial supported by the National Committee of Research.

REFERENCES

[1] H.S.Chen, in *Amorphous Metallic Alloys,* Ed. F.E.Luborsky, Butterworths, Boston (1983) p.169

[2] H.S.Chen, A.Inoue, T.Masumoto, J.of Materials Science 20 (1985) 2417

[3] R.Gerling, F.P.Schimansky and R.Wagner, Int.Journal of Rapid Solidification 5 (1990) 137

[4] F.P.Schimansky, R.Gerling and R.Wagner, Materials Science and Engineering, 97 (1988) 173

[5] R.Gerling, F.P.Schimansky and R.Wagner, in *Rapidly Quenched Metals,* Eds. S.Steeb, H.Warlimont, Elsevier Science Publishers B.V., 1985 p.1377

[6] G.Hunger, B.L.Mordike, in *Rapidly Quenched Metals,* Eds. S.Steeb, H.Warlimont, Elsevier Science Publishers B.V., 1985 p.1397 and private information.

LIST OF LECTURES GIVEN AT THE SCHOOL BUT NOT PUBLISHED IN THE PROCEEDINGS

1. The Investigation of Metal–Hydrogen System by Mössbauer Spectroscopy
   F. E. Wagner (München)
2. Mössbauer Studies of Permanent Magnet Materials
   F. Grandjean (Liege)
3. Magnetic Properties of $REMn_2$: Mössbauer Studies
   K. Krop (Cracow)
4. Mössbauer Studies of Hydrogen Induced Changes of Magnetic Order in RE-Mn Compounds
   J. Zukrowski (Cracow)

CLOSING REMARKS BY GÜNTER SCHATZ:

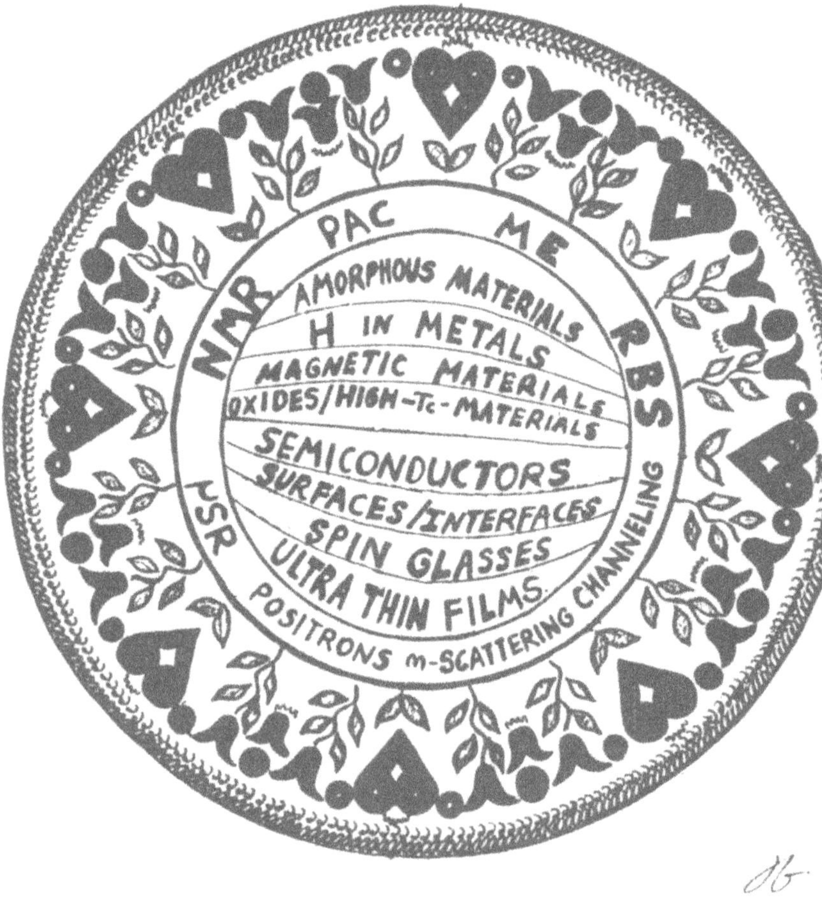

drawing by Joao G.M. Correia, Lisboa

LIST OF PARTICIPANTS

1.	Johan G. M. J. de Bakker	Groningen
2.	Józef Bara	Kraków
3.	Halina Bińczycka	Kraków
4.	Bogdan Bogacz	Kraków
5.	Mieczysław Budzyński	Lublin
6.	Kwietosława Burda	Kraków
7.	Jakub Cieślak	Kraków
8.	Joao G. M. Correia	Lisboa
9.	Michael Frank	Erlangen
10.	Janusz Frąckowiak	Katowice
11.	John A. Gardner	Corvallis
12.	Caroline Gardner	Corvallis
13.	Herald Glücker	Konstanz
14.	Edward Görlich	Kraków
15.	Tomasz Goworek	Lublin
16.	Waldemar Górniak	Lublin
17.	Fernande Grandjean	Liège
18.	Jean Marc Greneche	Le Mans
19.	Jerzy Grębosz	Kraków
20.	Marcel Greuter	Groningen
21.	Roman Hajduk	Kraków
22.	Andrzej Hrynkiewicz	Kraków
23.	Ewa Jackiewicz	Warszawa
24.	Ewa Japa	Kraków
25.	Elżbieta Jartych	Lublin
26.	Ewa Jędryka	Warszawa
27.	Jan Kaczorowski	Warszawa
28.	Józef Kajfosz	Kraków
29.	Wojciech Karaś	Kraków
30.	Józef Komraus	Katowice
31.	Barbara Kopcewicz	Warszawa
32.	Michał Kopcewicz	Warszawa
33.	Józef Korecki	Kraków
34.	Beata Kozłowska	Warszawa
35.	Janusz Kraczka	Kraków
36.	Karol Krop	Kraków
37.	Krzysztof Królas	Kraków
38.	Wiesław Kwaśny	Katowice
39.	Anna Kwiatek	Kraków
40.	Wojciech Kwiatek	Kraków
41.	Jadwiga Kwiatkowska	Kraków
42.	Guido Langouche	Leuven
43.	Hans J. Lauter	Grenoble
44.	Zhang Li	Marburg
45.	Klaus-Peter Lieb	Göttingen
46.	Lubos Lestak	Praha
47.	Hans Lütgemeier	Jülich
48.	Mirosław Łukasiak	Warszawa
49.	Kazimierz Łątka	Kraków
50.	Elżbieta Marczewska	Kraków
51.	Marta Marszałek	Kraków
52.	Jacek Marzec	Kraków
53.	Arnold Metz	Berlin

54.	Barbara Miczko	Kraków
55.	Karen Milants	Leuven
56.	Nathalie Millard	Lyon
57.	Wacław Musiał	Kraków
58.	Denes L. Nagy	Budapest
59.	Małgorzata Niewiara	Kraków
60.	Lech Nowicki	Warszawa
61.	Janusz Oleniacz	Warszawa
62.	Wojciech Osoba	Katowice
63.	Tadeusz Panek	Katowice
64.	Valeria Pasyuk	Dubna
65.	Fritz Pleiter	Groningen
66.	Eustachy Popiel	Katowice
67.	Janusz Przewoźnik	Kraków
68.	Antoni Pędziwiatr	Kraków
69.	Marek Przybylski	Kraków
70.	Bogusław Rajchel	Kraków
71.	Jan Sarzyński	Lublin
72.	Dariusz Satuła	Białystok
73.	Günter Schatz	Konstanz
74.	Jose Carvalho Soares	Lisboa
75.	Izabela Sosnowska	Warszawa
76.	Wigher G. Spaninga	Groningen
77.	Hartmut Spiering	Mainz
78.	Nika Spiridis	Kraków
79.	Jan Stanek	Kraków
80.	Małgorzata Sternik	Kraków
81.	Grzegorz Stoch	Kraków
82.	Mieczysław Subotowicz	Lublin
83.	Marek Sudnik	Kraków
84.	Jan Suwalski	Warszawa
85.	Krzysztof Szymański	Białystok
86.	Stanisław Szymczyk	Kraków
87.	Marek Tuszyński	Katowice
88.	Jagoda Urban	Kraków
89.	Dirk Peter van der Werf	Groningen
90.	Jo De Wachter	Leuven
91.	Fritz E. Wagner	München
92.	Renata Wasiewicz	Lublin
93.	Paweł Wodniecki	Kraków
94.	Wiesława Zarek	Katowice
95.	Ulrich Zimmermann	Villingen
96.	Jan Żukrowski	Kraków
97.	Jan Żurawicz	Lublin

Ode to Zakopane

We hail Zakopane,
Sweet town of inquiry.
Of thee we sing.
Long may your streets be bright
With learning's holy light
And may your fame be bright.
Let knowledge ring.

We come from far-off lands
To share our research plans,
Our knowledge bring.
NMR, PAC,
NQR, PAD,
Mössbauer spectroscopy;
Thy praises sing.

We are not purists all.
Our voices also call
Topics less pure.
μSR, positron,
Polarized neutron,
Bright lighting from a synchrotron,
We can endure.

All is not work and toil.
Our minds we should not spoil.
We also play.
Good food and drink partake,
Good humor ne'er forsake,
And mountain hikes we all must tak.
Hail the HYPERFINE way.

by Carolyn and John Gardner

www.ingramcontent.com/pod-product-compliance
Lightning Source LLC
LaVergne TN
LVHW060543180926
848329LV00007B/39
* 9 7 8 9 8 1 0 2 0 6 8 1 9 *